第五届结构工程新进展论坛暨第七届海峡两岸
及香港钢结构技术交流会文集

钢结构研究和应用的新进展（Ⅱ）

Advances in Research and Practice of Steel Structures(Ⅱ)

李国强　蔡克铨　陈绍礼　刘玉姝　主编

Editors in Chief：Guoqiang Li, K. C. Tsai, S. L. Chan, Yushu Liu

中国建筑工业出版社
China Architecture & Building Press

图书在版编目（CIP）数据

钢结构研究和应用的新进展（Ⅱ）/李国强等主编．
北京：中国建筑工业出版社，2012.11
ISBN 978-7-112-14818-9

Ⅰ.①钢…　Ⅱ.①李…　Ⅲ.①钢结构-学术会议-文集
Ⅳ.①TU391-53

中国版本图书馆 CIP 数据核字（2012）第 252706 号

责任编辑：赵梦梅　刘婷婷　何亚楣
责任设计：赵明霞
责任校对：王誉欣　党　蕾

第五届结构工程新进展论坛暨第七届海峡两岸及香港钢结构技术交流会文集
钢结构研究和应用的新进展（Ⅱ）
Advances in Research and Practice of Steel Structures（Ⅱ）
李国强　蔡克铨　陈绍礼　刘玉姝　主编
Editors in Chief: Guoqiang Li, K. C. Tsai, S. L. Chan, Yushu Liu
*
中国建筑工业出版社出版、发行（北京西郊百万庄）
各地新华书店、建筑书店经销
北京红光制版公司制版
北京市密东印刷有限公司印刷
*
开本：787×1092 毫米　1/16　印张：20¾　字数：500 千字
2012 年 11 月第一版　2012 年 11 月第一次印刷
定价：58.00 元
ISBN 978-7-112-14818-9
　　（22873）

版权所有　翻印必究
如有印装质量问题，可寄本社退换
（邮政编码　100037）

前　言　Preface

"结构工程新进展论坛"由中国建筑工业出版社、《建筑钢结构进展》编辑部和《Advances in Structural Engineering》编委会主办，旨在促进我国结构工程界对学术成果和工程经验的总结及交流，汇集国内外结构工程各方面的最新科研信息，提高专业学术水平，推动我国建筑行业科技发展。论坛每次有一个主题，前四届论坛分别在清华大学、大连理工大学、同济大学和东南大学举办，主题分别为：新型结构材料与体系；结构防灾、监测与控制；钢结构；混凝土结构。

"海峡两岸及香港钢结构技术交流会"由同济大学、台湾大学和香港理工大学合作发起，旨在促进海峡两岸及香港钢结构工程界对学术研究和工程经验的交流和探讨，交流会每隔两年一次，轮流在大陆、台湾及香港举行。继2008年在台湾和2010年在香港分别举行了第五届和第六届交流会之后，2012年的第七届交流会又在大陆地区召开。

以上两个会议都是系列会议，每次的举办都非常成功。这次会议将两个系列会议合二为一，以钢结构研究与应用的新进展为主题，还得到了住建部执业资格注册中心的大力支持，会议除为科研和工程技术人员搭建了交流平台外，还为注册结构工程师提供了很好的继续教育内容，这也使得会议的交流更深入、影响更深远。

感谢中国钢结构协会、中国建筑金属结构协会、上海市金属结构行业协会对本次会议的支持，感谢中国建筑工业出版社、住房和城乡建设部执业资格注册中心、建筑钢结构教育部工程研究中心、香港理工大学《Advances in Structural Engineering》编委会、同济大学《建筑钢结构进展》编辑部为本次会议所做的组织工作。此外，还要感谢中建钢构有限公司对本次会议的鼎力支持。

目 录 Contents

台湾钢结构产业现况与展望/林伟凯　何长庆　陈纯森 …………………………… 1

COLD-FORMED STEEL-AN ALTERNATIVE FOR BUILDING DESIGN
　AND CONSTRUCTION/Ir Prof Paul PANG　Ir Dr CHAN Wai Tai …………… 11

国产钢结构钢材质量现状及质量控制措施/侯兆新　何文汇 …………………………… 30

冷弯成型钢管压弯构件抗震性能研究/沈祖炎　温东辉　李元齐 …………………… 41

高温下冷弯不锈钢螺栓连接结构的试验研究/蔡炎城　杨立伟 ……………………… 51

应用反应谱分析法之阻尼器最佳化配置/黄婉婷　吕良正 …………………………… 62

采用高性能材料和可变形剪切连接件的大跨组合梁有限元分析/钟国辉　陈松基 …… 82

高雄海洋文化及流行音乐中心结构设计概述/苏晴茂　陈陆民　陈焕炜　王胜辉 …… 103

福州市海峡奥体中心体育馆屋盖结构设计概述/傅学怡　周　颖 …………………… 117

大偏心单柱双层高架桥之设计与施工——以台湾国道1号五股杨梅段拓宽工程
　泰山至林口段为例/王泓文　蔡益成　陈光辉　林曜沧　王照烈　张荻薇 ……… 134

空间钢结构无线传感监测技术研究与实践/罗尧治 …………………………………… 152

高层建筑伸臂桁架系统的发展/何伟明 ………………………………………………… 162

CONSTRUCTION OF STEEL STRUCTURE IN HYBRID
　STRUCTURE FOR HIGH-RISE BUILDING/H. Wang ……………………………… 171

型钢高强钢筋混凝土结构柱在地下室逆作施工方案的
　应用与设计/刘志健，李志城 ………………………………………………………… 179

日本 E-DEFENSE 五层楼实尺寸含制震斜撑钢构架振动台
　试验反应预测/蔡克铨　游宜哲　李昭贤　翁元滔　蔡青宜 …………………… 191

脚手架和看台的二阶直接分析/刘耀鹏　陈绍礼 ……………………………………… 208

带竖向加劲肋钢板剪力墙设计研究/范　重　黄彦军　刘学林　肖　坚　王义华 … 221

面内挫屈斜撑之耐震行为与设计/陈诚直，汤伟干 …………………………………… 245

钢造双核心预力自复位斜撑发展与验证：耐震实验
　与有限元素分析/周中哲　陈映全 …………………………………………………… 256

波纹腹板 H 型钢的研究/李国强　张　哲 ……………………………………………… 267

BIM 在钢结构制造中的应用/贺明玄　沈　峰 ………………………………………… 308

台湾钢结构产业现况与展望

林伟凯[1]，何长庆[2]，陈纯森[3]

(1. 金属工业研究中心，高雄 台湾 800；
2. 东和钢铁企业股份有限公司，台北 104；
3. 成功大学建筑研究所，台南 台湾 700)

摘　要：钢结构之高强度与高韧性颇适合于高层建筑之发展；其轻量化与施工快速亦造就许多大跨距之桥梁与重型机具设施，在近代工程已成为建设项目最主要之建材。由于钢结构之造价较混凝土结构略高，在金融风暴与财务吃紧之环境，其发展自然受到阻碍。本文谨就台湾钢结构产业之今昔汇整归纳，供相关单位规划之参考，并期望更美好之未来。

关键词：高强度；高韧性

THE STEEL STRUCTURE DEVELOPMENT AT PRESENT AND IN THE NEAR FUTURE OF TAIWAN

W. K. LIN[1], C. C. HO[2], C. S. CHEN[3]

(1. Industrial Analyst, Metal Industries Research & Development Center, Kaohsiung 800, Taiwan;
2. Special Assistant to Board Director, Ton Ho Steel Enterprise Co., Taipei 104, Taiwan;
3. Associate Professor Expert, Cheng-Kung University Architecture Dept., Kaohsiung 700, Taiwan)

Abstract: The high strength and toughness performance of structural steel has always been used for high-rise buildings, other advantages of rapid construction schedule and lighter weight for steels are also provided for many long span bridges and heavy construction equipments. Actually, steel materials have become one of the most important construction materials. On the contrary, due to higher cost to compare with concrete, the steel structure industry development also presents some extent of difficulties. This paper collects some statistics information of steel structure industry in Taiwan for the time being and expects to develop more construction jobs in the near future.

Keywords: high strength; toughness

第一作者：林伟凯（1971—），男，产业分析师，主要从事钢结构产业研究，E-mail: weikai@mail.mirdc.org.tw。
第二作者：何长庆（1956—），男，董事长特别助理，主要从事钢结构材料质量营业管理；E-mail: davidho@tung-hosteel.com。
通讯作者：陈纯森（1948—），男，副教授专家，主要从事钢结构技术推广与训练，E-mail: cscgrace@ms8.hinet.net。

1. 引言

所谓钢结构业,依经济部工业产品最新分类(2006年8月),归类于金属结构制造业下的金属结构,SIC code 为 25210,其中又细分为厂房钢结构、大楼钢结构、桥梁钢结构、铁塔钢结构、其他金属建筑结构及组件等五大类产品。因钢结构是应用型钢、钢管、钢板等钢材,经加工、焊接、组立及安装后建造成之工程结构,故属钢铁下游加工产业之一。

钢结构具有高强度、高韧性、耐震性佳、工业化程度高、环境污染少,及施工速度快等特性,百年来发展非常快速,尤其是从 20 世纪下半叶起,随着世界钢铁产量的大幅增加,钢结构更加扩展了应用领域。钢结构成为高层建筑构造的主流,已是不可避免的趋势。

2. 产品应用鱼骨图

钢结构应用范围非常广泛,举凡桥梁、大楼、中低层住宅、厂房、造船、焚化炉、吊车、集尘设备、仓库、停机棚、高架电塔、汽车车体、火车车厢、储藏架、排水设备等,都会使用到钢结构,详细之产品应用范围整理如图1。

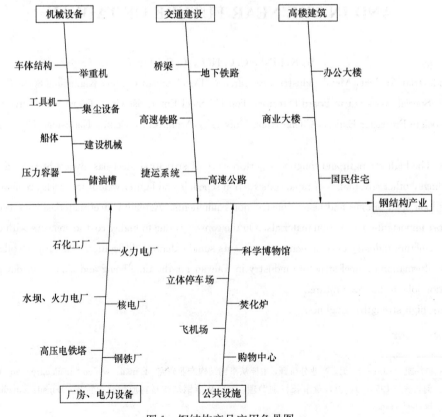

图 1 钢结构产品应用鱼骨图

3. 产业特质

在工业先进国家，钢结构产业均列为重工业的一环，其产业特质包括"劳力/资本密集产业"、"技术密集产业"、"内需市场为主"、"采订货生产，资金成本压力大"、"投入成本高，影响获利空间"、"上下游产业关联性高"等，说明如表1。

钢结构产业特质 表1

产业特质	说　　　　明
劳力/资本密集产业	钢结构体积庞大，需要大量土地、厂房、设备供制造及原料储存，且因生产流程繁杂，需较多人力
技术密集产业	钢结构生产流程主要为设计/制造/生产管理三大环节，每一生产流程都必须有专门的技术人才及专用设备，才能确保钢构质量并提升生产效率
内需市场为主	钢结构多为重厚长大型式，不适合长途运输，除了工厂制作之外，也需要现场进行实地安装，故外销比例不高，出口比例始终维持在5%以下
采订货生产，资金成本压力大	钢结构无法做一标准规格的整合，因此钢结构业者必须对各种不同的结构设计施工，无法生产固定产品。且钢结构多为大型建物，工程期间长，工程费用庞大，一般钢构厂必须事先投入相当多的工程资金
投入成本高，影响获利空间	钢结构产业固定成本高，且客户来源大部分是营造厂，此种发包制度使得钢构厂轮为丙方，因此在获利上自然受到严重压缩。 钢材成本高，比重约占总成本60%，钢材的买卖常需现金交易，因此利息成本高，影响获利，且钢材料价格不稳定，也会影响业者利润
上下游产业关联性高	钢结构上游产业主要为钢板及热轧H型钢制造商；钢板方面，中钢几乎皆可供应；热轧H型钢方面，东和钢铁、中龙已能满足国内需求。 在接单方面，受限于钢构业无法独立承包的缘故，均得透过营造厂商转包而来。所以，钢构业与其他产业关联性甚大

资料来源：金属中心产研组。

4. 钢结构产销分析

4.1 内部产销概况

目前参加钢铁公会钢结构小组的厂商有28家，年产能约120万吨；未参加钢铁公会之年产能合计约100万吨。因此台湾地区钢结构厂总产能约220万吨，但年需求量不足100万吨，呈现产能过剩现象。台湾钢结构产业的年产值占钢铁产业中的前五大，在金属工业中占有举足轻重的地位。然而，台湾虽有能力建造出全球最高的钢结构大楼，但因钢结构产业属于内需型的产业，大型工程除由大型厂商承揽外，转包或分包方式之加工情形亦十分普遍。在中小型与低技术性厂商进出抢标下，钢构厂普遍面临彼此削价竞争的局面，使得钢构业者之获利空间降低，加上2000年开始面临建筑业不景气，此状况持续了

2～3年，钢结构产业深受影响，使得2003年钢结构产值衰退成不到新台币300亿元，是自1991年来的最低点。

2004年起由于台湾地区多家TFT-LCD光电大厂纷纷投入扩厂计划，钢构市场的需求量明显激增，加上钢料价格大幅上涨，使得当年钢结构产值首次突破新台币500亿元，达到522亿元之历史新高，较2003年一举成长77％。

2005年钢结构产值回落到新台币408亿元，较2004年衰退22％，2006年起政府推动"加速推动都市更新"、"便捷生活设施"等公共建设计划，钢铁结构产值为新台币440亿元，较95年产值小幅成长8％。

此外，为了改善钢结构产业体质，提升竞争力，达到永续经营之目标，中国钢铁公司与财团法人金属工业研究发展中心在经济部技术处的支持下，共同进行钢结构产业调查，发现钢结构建筑比钢筋混凝土（RC）建筑建造成本高20％，此为钢结构未能全面普及化的最大障碍。因此，中钢公司于2007年5月邀集相关厂商及学研机构，共同筹组"钢结构高值化研发联盟"，期能提升台湾地区钢结构的使用比率与扩大市场需求。

4.1.1 供给面

2010年钢结构产值为新台币367.5亿元，较2009年衰退8.9％，2009年钢结构产值为新台币403.2亿元，约略与2005年产值相当，钢结构近十年产值变化如图2所示。

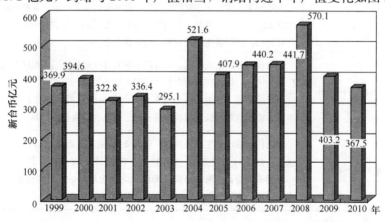

图2 钢结构产值变化（1999～2010年）

2010年钢结构产值367.5亿元，较2009小幅衰退35.7亿元，原因为营建业需求未大幅提升，钢价水平适中，2008年钢结构产值570.1亿元是有史以来最高，最主要原因是钢结构原物料钢材价格大幅飙涨。以中厚板为例，2007年每吨平均价格为23,583元，2008年飙涨至35,317元，足足涨了50％；再以H型钢为例，2007年每吨平均价格为22,350元，2008年为31,067元，也涨了近40％。如果从钢材生产量来看，2007年钢板生产量约139万吨，2008年约130万吨；H型钢2007年生产量约137万吨，2008年约118万吨，两者生产量2008年都是呈现衰退的现象。因此2008年钢结构产值虽然创下新高，较2007年多出128亿元，但在钢价大幅上涨及钢材减产状况下，实际上其产量应比2007年还少。

4.1.2 需求面

钢结构业之荣枯与下游房地产景气有着密切的关系，政府自1992年宣布全面实施容积率管制，造成建商大量抢建，1992年建造执照核准面积创下7,644万平方米的新高记录，而建物在1994年陆续申请使用执照，使核发使用执照面积也达到高峰5,816万平方米。之后建造执照与使用执照核准面积便开始逐年减少，加上2000年开始的建筑业不景气，使得2001年建造执照核准面积创下历史新低，仅2,170万平方米；紧接着2002年核发之建物使用执照为2,369万平方米，亦是最低记录。

2004年在电子厂增建效应下，建造执照核准面积回升至4,250万平方米，2005年仍维持在4,320万平方米，此后三年连续下滑，至2008年仅有2,616万平方米。至于使用执照核准面积在电子厂陆续完成建厂后，2006年及2007年都回升在3,600万平方米左右，2008年再下滑到3,275万平方米，在金融海啸的影响下，2009年再下滑到2,798万平方米，2010上升至3,005万平方米。参见图3。

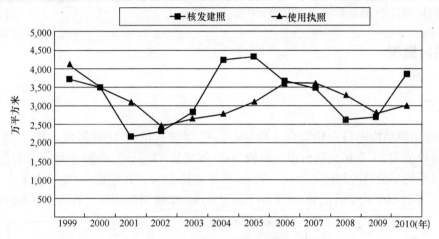

图3 核发建造执照及使用执照总面积变化（1999～2010年）
【资料来源：台湾/"内政部"营建署】

近10年钢结构市场供需分析汇整如表2。由数据可以看出，2008年钢结构在产值与需求都创下历史的新高，主要原因是钢结构原物料钢材价格大幅飙涨所造成，使得需求成长率高达26.6%。至于自给率近年皆在90%～95%左右，出口比例自2004年逐渐成长，2008年已逼近5%。种种数据显示钢结构业为高度内需型产业。

钢结构市场供需分析（1998～2010年） 表2

单位：亿元新台币

项目	产值	出口值	进口值	台湾总需求	需求成长率	出口比例	进口依存度	台湾自给率
年	A	B	C	D=A-B+C	E	F=B/A	G=C/D	1-G
1998	397.2	11.2	22.2	408.2	-1.50%	2.82%	5.44%	94.56%
1999	369.9	8.4	28.1	389.6	-4.56%	2.27%	7.21%	92.79%
2000	394.6	8.2	19.9	406.3	4.29%	2.08%	4.90%	95.10%
2001	322.8	10.4	24.6	337.0	-17.06%	3.22%	7.30%	92.70%
2002	336.4	10.3	24.3	350.4	3.98%	3.06%	6.93%	93.07%
2003	295.1	11.5	29.5	313.1	-10.64%	3.90%	9.42%	90.58%
2004	521.6	11.6	27.3	537.3	71.61%	2.22%	5.08%	94.92%

续表

项目 年	产值 A	出口值 B	进口值 C	台湾总需求 $D=A-B+C$	需求成长率 E	出口比例 $F=B/A$	进口依存度 $G=C/D$	台湾自给率 $1-G$
2005	407.9	11.7	22.8	419.0	-22.02%	2.87%	5.44%	94.56%
2006	440.2	13.3	32.4	459.3	9.62%	3.02%	7.05%	92.95%
2007	441.7	16.9	43.9	468.7	2.05%	3.83%	9.37%	90.63%
2008	570.1	28.2	51.5	593.4	26.61%	4.95%	8.68%	91.32%
2009	403.2	22.4	33.6	414.2	-30.18%	5.45%	7.97%	92.03%
2010	367.5	18.2	26.3	375.3	-9.42%	4.90%	6.93%	93.07%

资料来源：ITIS产销数据库、海关进出口统计月报。

5. 进出口概况

5.1 进口

近10年钢结构进口变化如图4所示，2010年进口量为6.85万吨，较2009年成长15%，进口值为41.2亿元新台币，也较2009年成长41.5%。2010年受国际钢价上扬影响，进口平均单价每吨超过6万元，较2009年成长24%，几乎比2009年多了2万元/公吨，2008年进口平均单价虽较2007年下滑9.8%，单价仍高达每吨近6.4万元新台币。

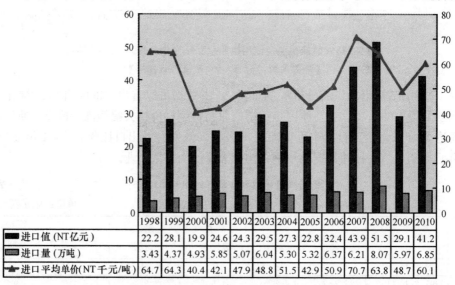

图4 台湾钢结构进口变化分析（1998～2010年）【资料来源：台经院】

在进口排名方面，中国内地、越南、日本名列前三，进口量占有率合计刚好九成，其中中国内地一枝独秀，占有率高达71%。中国内地的进口量虽大，但其平均单价最低，每公吨约3.9万元，远不及整体进口平均单价的6.4万元/吨。从平均单价的数据来看，自美国进口的钢结构产品平均单价最高，每吨平均单价高达209570元；日本次之，每吨为184261元，详见表3。

2010年钢结构前五大进口国家贸易表现　　　表3

单位：吨、百万新台币、元/t

排名	进口国家	进口量	占有率	进口值	平均单价
1	中国大陆	56,468	71.3%	2,237.3	39,615
2	越南	5,628	8.5%	254.8	45,273
3	日本	3,844	5.4%	708.3	184,261
4	韩国	2,671	3.9%	317.7	118,944
5	美国	1,024	1.7%	214.6	209,570

资料来源：台经院。

5.2 出口

2010年钢结构出口量较2009年成长24%，达2.51万吨，出口值也达到12.8亿元新台币，较2009年成长33%，2008年出口量值都是历年之最高。2008年出口平单价则与2007年相当，每吨出口平均单价为5.6万元，见图5。

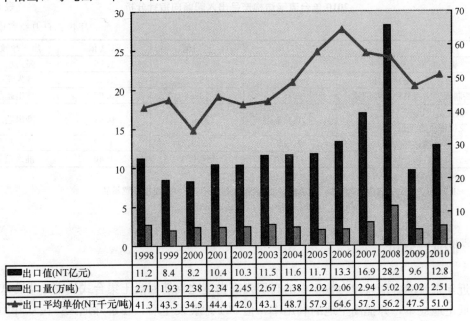

	1998	1999	2000	2001	2002	2003	2004	2005	2006	2007	2008	2009	2010
出口值（NT亿元）	11.2	8.4	8.2	10.4	10.3	11.5	11.6	11.7	13.3	16.9	28.2	9.6	12.8
出口量（万吨）	2.71	1.93	2.38	2.34	2.45	2.67	2.38	2.02	2.06	2.94	5.02	2.02	2.51
出口平均单价（NT千元/吨）	41.3	43.5	34.5	44.4	42.0	43.1	48.7	57.9	64.6	57.5	56.2	47.5	51.0

图5　钢结构出口变化分析（1998～2010年）【资料来源：台经院】

出口国排名方面，美国为台湾钢结构外销最主要的市场，2010年出口量占有率达29.5%，出口值达531.6百万新台币。新加坡则为第二大出口市场，出口量占有率为9.2%，出口值为179.8百万新台币。其中，值得注意的是新加坡出口量逐年升高，排名已跃居第二名，反观日本以往在2～3名徘徊，2010年落至第5名。若以出口平均单价来比较，2010年出口至日本的平均单价最高，每吨出口平均单价达59286元，参见表4。

2010 年钢结构前五大出口国家贸易表现 表 4

单位：吨、百万新台币、元/t

排　名	出口国家	出口量	占有率	出口值	平均单价
1	美国	12,986	29.5%	531.6	40,936
2	新加坡	4832	9.2%	179.8	37,210
3	阿拉伯	2980	6.7%	117.6	39,463
4	越南	2017	5.7%	81.5	40,406
5	日本	1906	5.4%	113.0	59,286

资料来源：台经院、本研究。

5.3 出入超指标

台湾钢结构产品 2010 年出口比例及进口依存度仅为 3.98% 及 5.85%，虽然贸易量并不大，但就外贸逆顺差的角度而言，钢结构产品仍以入超倾向为主，两大强入超产品分别为"其他钢铁结构体"与"钢铁制桥及桥体段"，RCA 值分别为 1.52 及 1.96。属于准出超品产品则有"钢铁制供鹰架、窗套、支柱或坑道支持用设备"与"钢铁制门、窗及其框架及门坎"，2010 年钢结构产品出入超倾向如表 5 所示。

2010 年台湾钢结构产品出入超倾向指针分析 表 5

单位：百万新台币

项　目	出口值	进口值	RCA 值	出入超倾向
钢铁制桥及桥体段	2.5	109.7	1.96	强入超品
钢铁制塔及格状桅杆	6.5	8.5	1.13	准入超
钢铁制门、窗及其框架及门坎	251.7	133.6	0.69	准出超品
钢铁制供鹰架、窗套、支柱或坑道支持用设备	144.8	143.7	0.99	准出超品
其他钢铁结构体	736.2	2330.2	1.52	强入超品
钢结构合计	1162.0	2729.4	1.40	准入超品

注：RCA 值 $=1-(E-I)/(E+I)$，E 为出口值，I 为进口值
　　$1.5 < RCA \leq 2$：表强入超；$1 < RCA \leq 1.5$：表准入超；$RCA = 1$：表水平贸易品
　　$0.5 \leq RCA < 1$：表准出超品；$0 \leq RCA < 0.5$：表强出超品

6. 产业前景分析

近年来台湾各地兴建为数不少的高层及高密度之钢筋混凝土集合式住宅，集合式住宅可降低土地成本，并提高土地的利用率，但 921 地震发生后，在市郊地区，民众开始青睐低层数的轻钢构建筑，在高地价之市区，则兴建钢结构高层建筑，主要是希望以钢材之韧性来承担作用力。另外，由于劳动人口减少、工资高涨、工地安全与环保的重视、建材轻量化的趋势、建材防火性的要求、工程质量的提升以及工期的缩短，传统厚重的建材（钢筋混凝土）与需要众多劳力的施工方式已逐渐式微，取而代之的是高质量、轻量化的营建材料，以及短工期与低劳动量的施工方法，因此造成钢构建筑逐渐受到重视。因此从以上因素得知，不论未来的建筑是往高楼或低层建筑方向发展，在钢构渐被采用的同时，国内营建业面对生存环境的改变与转型，已是不争之事实，也显示出营建业转型的必要性。

钢结构产业为一内需型产业，其产业前景深受营建业的荣枯与政府公共工程释出量的

影响，对于未来影响钢构业发展的正负面因素整理如表6。其中较近期的正面因素主要在于日本大地震，未来东北地区重建需求可观，加上大地震后，将更强化钢结构建筑的重要性与安全性，再者，政府推动扩大内需方案，包括桥梁、快速道路等工程，以提振经济。此外政府积极推动绿建筑钢构造的观念，将有利于钢结构市场之推广。再者，未来二氧化碳排放量的管制，水泥在制造生产与用于营建时所产生的环保问题，以及RC建筑物在拆除重建时再利用有其困难，采用钢结构似乎是建筑发展趋势。因此世界各国皆不断地发展以钢材为建筑物的基本原料，针对市场的需求之下，近20年来钢结构已渐渐地在世界各国成为重要建筑与桥梁所使用的主流建材。

负面因素为自2008年下半年起，金融风暴席卷全球，营建业亦遭受波及，连带影响钢结构市场之正常发展。除此之外，钢结构专业人才不足，证照、检验制度未落实，钢结构管理规范不完备及营造业体制不健全，钢结构成本仍偏高、绿建筑观念仍不普及等。

有鉴于欧美先进国家广泛采用钢结构为主要建筑材料，对于国家资源及人民安全多了份保障，台湾地区也亟须政府法令规范以及消费者和建商建立共识。在生命财产重于成本考虑的观念导正下，政府如能正视问题所在，修改营建法规相关法令，缩短钢结构与钢筋混凝土之成本价格差距，钢结构建筑的推动才得以具体落实。

未来钢结构市场发展正负面因素分析　　　　　表6

因素项目	正面影响	负面影响	说明分析
日本大地震，未来东北地区重建需求可观	✌		2011年3月11日13时46分，日本东北地区和关东地区受地震影响最大，而这两个地区是日本工业重地，未来东北地区重建需求可观
政府扩大公共工程建设	✌		2009年台湾推动扩大内需方案，全台有近60座桥梁将优先展开维修兴建，这与钢结构较为相关
环保问题与建筑发展趋势	✌		由于水泥的开采设限与设厂制造对环境的冲击，以及砂石的短缺，环保意识的不断提高的时代，使用具环保且可回收的钢材作为建筑材料，应是台湾未来发展的一个重要趋势
政府积极推动"绿建筑"观念	✌		钢结构具有维护生态环境之绿建筑指标，将有利于钢结构市场之推广
钢结构专业人才不足，证照、检验制度未落实		✌	1.学校教育师资不足，建筑、土木、营建科系学生对钢结构相关的素养普遍不足。2.设计、施工相关教材不足。3.施工专业技术人员（放样、吊装、焊接、检验）缺乏。4.专业技术人员能力之重要性未被肯定，造成薪资结构不合理
全球金融风暴		✌	自2008年下半年起，金融风暴席卷全球，营建业亦遭受波及，2009年市场景气依旧处于低迷状态，连带将影响我国钢结构市场之正常发展
部分钢构设施投资高，维护不易		✌	台湾地区地属潮湿多雨之海岛型气候，钢铁结构之防锈蚀乃极为重要，如何提高钢结构之耐久性为努力目标
营造业体制之加强		✌	营造业贵为钢构业之业主，体制有待加强，主管建筑机关专业能力及人力不足，无法有效监督
尚未建立完整的现场监督机制		✌	钢构组装，须现场人工焊接之部位甚多，而施工人员良莠不齐，容易产生人为的疏误及瑕疵，因此有必要建立完整的现场监督与质量检测机制，以确保钢构建筑的耐震安全

7. 结论

就技术观点言，钢结构确实具备许多优点。如何使钢结构能更进一步之发展，依赖产业界、学术界及政府机关之正视与共同努力。

参考文献

［1］ Ho Ming Kam，Steel construction building science and technology R & D and industrial development，Republic of China Steel Structure Association，the sixth General Assembly，April 27，2007
［2］ Republic of China Customs import and export statistics，Directorate General of Customs of the Ministry of Finance
［3］ Industrial Production Statistics Monthly，Ministry of Economic Affairs and Statistics Department

COLD-FORMED STEEL-AN ALTERNATIVE FOR BUILDING DESIGN AND CONSTRUCTION

Ir Prof Paul PANG[1], Ir Dr CHAN Wai Tai[2]

(1. Honorary Fellow, Hong Kong Institute of Steel Construction
2. Vice-President and Membership Officer, Hong Kong Institute of Steel Construction)

Abstract: Cold-formed steel is made from quality steel plate or coil that is formed into shape either through press-braking (for heavy gauge metal plate) or bend braking (for light gauge metal sheet) (see Figure 1), or more commonly, by roll-forming through a series of dies at ambient temperature (see Photo 7). No heat is required to form the section and thus the name "cold-formed steel" is acquired.

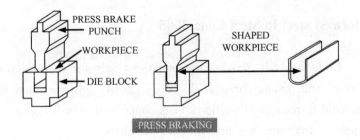

Fig 1　Press braking process
(Bend braking process is similar but its punching rate is faster) (Wikipedia)

　　The use of cold-formed steel as primary structural members in building design and construction has been prevailing in countries such as Australia, Mainland China, Japan, European countries and Northern American countries, etc.

　　In the Code of Practice for the Structural Use of Steel 2005 (i. e. Steel Code 2005), it gave recommendations only for the design of cold-formed sheet profiles (i. e. thin gauge corrugated open sections) with nominal thickness up to 4mm and other cold-formed steel sections (both open sections and closed hollow sections) with nominal thickness up to 8mm. Welding was not permitted and this hampers the wide application of cold-formed steel in building design and construction.

　　In the advent of the Code of Practice for the Structural Use of Steel 2011 (i. e. Steel Code 2011), design guidelines are substantially formulated for the use of cold-formed steel hollow sections with nominal thickness up to 22mm and cold-formed steel sheet pile sections with nominal thickness up to 16mm. Their requirements on tensile strength, notch toughness and ductility should comply with the essential requirements for the parent hot-rolled steel.

　　Another revamp in the Steel Code 2011 is the permission of welding within a length of 5 times the section thickness on either side of cold-formed zone. Occasionally, sections may be cold-formed with tight corners which violate the minimum radius-thickness ratio and no pre-normalizing treatment is done. In these circumstances, the Registered Structural Engineer should submit a valid Welding Procedure Specification for the approval of the Building Authority before the carrying out of welding works in cold-formed zones.

During cold-forming of steel sections, high residual stresses are created at the bent locations. This may render cracking/peeling at bent corners when the sections undergo hot-dip galvanizing bath. Special consideration should be made for this particular phenomenon at the time of order placement.

As the Steel Code 2011 promulgates the use of cold-formed steel, it offers the construction industry an alternative to choose. Now that special steel grades to BS 10149 Part 2 and Part 3 are stipulated in the Steel Code 2011, very tight corners of cold-formed sections with inside bend radius equal to the plate thickness are permitted. It is anticipated that peculiar section shapes generally not available from hot-rolled products can be tailor-made using the cold-forming technologies so as to produce an aesthetic outlook that tallies with innovative design conceived by the architects.

Keywords : Cold-formed steel, hollow section, sheet pile section, radius-thickness ratio, external corner profile, inside radius, Welding Procedure Specification, Welding Procedure Approval Record, hydrogen embitterment, circular hollow section (CHS), rectangular hollow section (RHS), square hollow section (SHS), tight corner, cracking.

Use of Cold-formed steel in Steel Code 2005

In the Steel Code 2005, it gave recommendations only for the design of cold-formed sheet profiles (i.e. thin gauge corrugated open sections) with nominal thickness up to 4mm and other cold-formed steel sections (i.e. open and closed hollow sections) with nominal thickness up to 8mm. Welding was not permitted.

Mechanical properties of cold-formed steel

Cold forming is a process whereby the main forming of metal section is done at ambient temperature. It changes the material properties of steel and impairs ductility as well as notch toughness but enhances strength. These changes may also limit the ability to weld in cold-formed zones. The extent to which the properties are changed depends upon the type of steel, the forming temperature and the degree of deformation. The basic requirements on strength, notch toughness and ductility shall comply with parent material. Figure 2 shows the stress-strain relationship of cold-formed steel.

Cold-formed thin gauge corrugated open sections

Cold-formed thin gauge corrugated sheet profiles are commonly used, such as Lysaght Bondek sections and Lysaght Spandek sections, etc.

For instance, Lysaght Bondek section is of re-entrant sheet profile which is well known for its excellent capacities for greater strength and less deflection and is usually used as permanent formwork for composite slab construction. It has re-entrant ribs of 54mm indented into the concrete slab at about 200mm centre to centre spacing. It is availa-

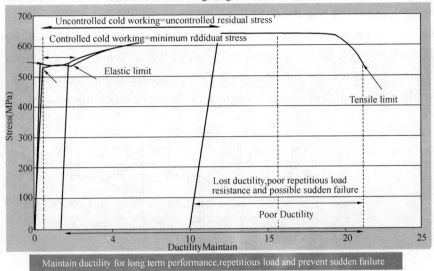

Figure 2 Stress-strain curve for cold-formed steel

ble in base metal thicknesses (BMT) of 0.6mm, 0.75mm and 0.9mm and 1.0mm respectively.

Lysaght Spandek section is of trapezoidal sheet profile which is ideal for use as roofing material and vertical building envelope cladding. The base metal thicknesses (BMT) are of 0.42mm and 0.48 mm respectively.

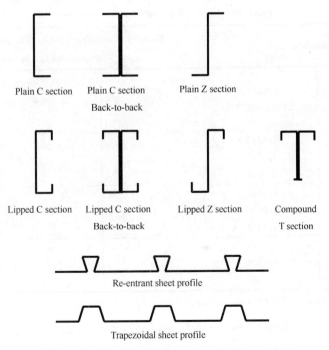

Figure 3 Typical cold-formed steel open sections

The base metal thickness (BMT) is the thickness of the bare metal inclusive of the galvanized coatings. For the usual Z275 zinc coated sheet metal, which means there are 275 grams of zinc metal coated over 1 m², the thickness of zinc coating on every square metre of the surface would be 0.0385mm (or $275/7.14/10^6 \times 10^3$ mm) say 0.04mm, given the density of zinc metal is 7.14 gram per cm³. Therefore, the design thickness of corrugated sheet profiles should deduct the zinc coating of 0.04mm from the base metal thickness (BMT).

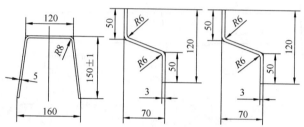

Figure 4 Examples of cold-formed hollow sections

Cold-formed open sections and closed hollow sections

The Steel Code 2005 stipulated the use of open sections and closed hollow sections with nominal thickness up to 8mm. The Code tallied with the thickness limits as stipulated in BS EN 10025-2, and Table 1 is extracted below for reference.

EN 10025-2 : 2004 (E)

Table 13—Cold roll forming of flat products

Designation		Minimum recommended inside bend radii[a] for nominal thicknesses (t) in mm		
According EN 10027-1 and CR 10260	According EN 10027-2	$t \leqslant 4$	$4 < t \leqslant 6$	$6 < t \leqslant 8$
S235JRC	1.0122			
S235J0C	1.0115	$1t$	$1t$	$1,5t$
S235J2C	1.0119			
S275JRC	1.0128			
S275J0C	1.0140	$1t$	$1t$	$1,5t$
S275J2C	1.0142			
S355J0C	1.0054			
S355J2C	1.0579	$1t$	$1,5t$	$1,5t$
S355K2C	1.0594			

a The values are applicable for bend angles⩽90°.

Table 1 Extract of Table 13 from BS EN 10025-2

For cold-formed hollow sections, they are manufactured by cold-forming the plate or coil into circular hollow section (CHS) (see Photo 1) or rectangular (RHS) / square (SHS) hollow section followed by longitudinal welding (see Photos 2 and 3). Occasionally, circular hollow section (CHS) is manufactured by cold-forming the coil into circular shape followed by spiral welding. RHS/SHS may alternatively be manufactured by cold-forming the as-built CHS into the required shape. All cold-forming works are carried out at ambient temperature.

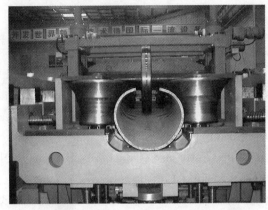

Photo 1 Cold-forming hollow section from steel plate prior to longitudinal welding

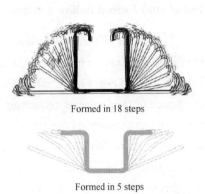

Figure 5 Roll-forming of sections

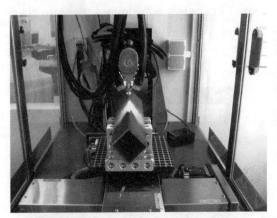

Photo 2 Longitudinal weld is made along an edge of SHS.

Photo 3 Longitudinal weld is made on a flat surface of SHS.

Welding in cold-formed steel

In general, welding in cold-formed steel was not permitted due to difficult control on welding procedure and quality as stipulated in the Steel Code 2005. It did not elaborate the degree of difficulties and the factors that caused such difficulties. Furthermore, it did not stipulate the extent of cold-formed zone in which welding was not permitted.

Use of Cold-Formed steel in Steel Code 2011

In the Steel Code 2011, the scope has been extended for the use of cold-formed hollow sections and cold-formed sheet pile sections. As a conservative design, there is no enhancement in the design strength of cold-formed sections.

Use of cold-formed hollow sections

The major difference between cold-formed hollow section and hot-finished hollow section is the corner profile and the corresponding residual stresses so induced at the cold-formed zones (See Figure 6). Cold-formed hollow sections are available worldwide and the nominal thickness as recommended in the Steel Code 2011 is set up to 22mm. Such sections are recommended for use as primary structural members in trusses and portal frames of modest span. When sufficient experience is gained in the use of thicker cold-formed hollow sections, the lifting of 22mm limit would be considered in the future.

From submission point of view, mill certificates of hollow sections compliant with BS EN 10219 or equivalent are considered acceptable to the Building Authority. The RSE should ensure that the external corner profile complies with the requirements in Table 2. He should also ensure that the inside bend radius complies with the requirements in Table 5 if there are welding works in the hollow sections. Otherwise, a Welding Procedure Specification (WPS) should be submitted to the Building Authority for prior approval. The WPS should include the Welding Procedure Approval Record (WPAR), which should have contained satisfactory material test results at cold-formed zones.

Photo 4 Stockpiles of SHS/RHS.

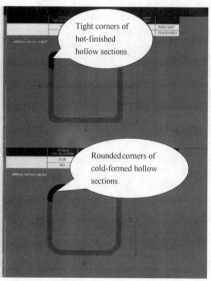

Figure 6 Difference in the corner profile between cold-formed hollow section and hot-finished hollow section.

Control of external corner profile of cold-formed hollow sections

In order to prevent corner bend cracking, control of dimensions of external corner profile should conform to the minimum requirements as given in Table 2. Most of the SHS/RHS available in the market generally comply with the requirements in Table 2.

Thickness t (mm)	External corner profile C_1, C_2 or R_{ext}
$t \leqslant 6$	$1.6t$ to $2.4t$
$6 < t \leqslant 10$	$2.0t$ to $3.0t$
$10 < t$	$2.4t$ to $3.6t$

N.B.: The sides need not be tangential to the corner arcs.

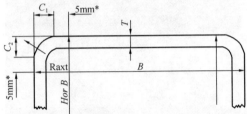

*This dimension is a mensuring when mensuring B or H and a minimum when mensuring T

Table 2 External corner profile of SHS/RHS (this is applicable to CHS)

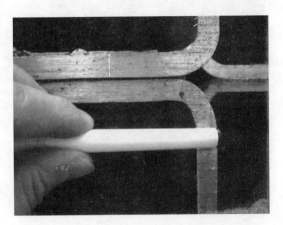

Photo 5 Square Hollow Section (SHS)

Use of Cold-Formed sheet pile sections

Cold-formed sheet pile sections conforming to BS 10249 or equivalent in various forms are available worldwide and the nominal thickness as recommended in the Steel Code 2011 is up to 16mm, which is generally the limit manufactured in the industry (See photos 6 to 10). Such sections are recommended for temporary use in steel sheet piling designed as vertical retaining elements in excavation and lateral support works. If the parent materials conforming to BS EN 10025-2 is used, the thickness can only be used up to 8mm with the minimum inside bend radius is limited to 1.5 times the thickness of the sheet pile sections.

From submission point of view, mill certificates of parent sheet pile material should be submitted to the Building Authority. The RSE should ensure that the inside bend radius complies with the requirements in Table 3 or 4 and even Table 5 if there are welding works in the sheet pile sections. Otherwise, a Welding Procedure Specification (WPS) should be submitted to the Building Authority for prior approval. The WPS should include the Welding Procedure Approval Record (WPAR), which should have contained satisfactory material test results at cold-formed zones.

Photo 6 Cold-formed sheet pile sections

Photo 7 Roll-forming using a series of dies at ambient temperature

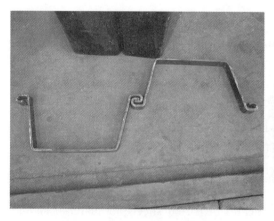

Photo 8 U-type cold-formed sheet pile sections

Photo 9 Z-type cold-formed sheet pile sections

As can be seen from photo 11, the inside bend radius is obviously less than 1.5 times the thickness of the sheet pile sections (about one time the thickness of the section). In the advent of special steel grades as stipulated in BS EN 10249, BS EN 10149-2, BS EN 10149-3 or equivalent standard, tighter bends are permitted.

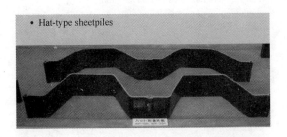

Photo 10 Hat-type sheet pile sections

Photo 11 End clipping, the inside bend radius of which is about one time the plate thickness

The material specifications of parent hot rolled strip or sheet used for cold-forming are:

i) S275JRC/S355J0C stipulated in BS EN 10025-2;

ii) S315MC/S355MC stipulated in BS EN 10149-2 (the steels are thermo-mechanically treated); and

iii) S260NC/S315NC/S355NC stipulated in BS EN 10149-3 (the steels are normalized) respectively.

The strip thickness ranges from 1.5mm to 16mm for sheet pile sections, which has specified minimum yield strength of $260N/mm^2$ up to and including $355N/mm^2$. The available steel grades of alloy quality steels are given in Tables 3 and 4.

Minimum inside radii for cold-formed sheet pile sections

When cold-formed sheet pile profile is manufactured using JC steel grade, the nominal thickness is limited to 8mm and the minimum inside radii should conform to Table 3 below.

Table 3 Minimum inside radii for JC steel grade to BS EN 10025-2

Grade designation	Minimum inside radii for nominal thickness (t) in mm		
	$t \leqslant 4$	$4 < t \leqslant 6$	$6 < t \leqslant 8$
S275JRC	$1.0t$	$1.0t$	$1.5t$
S355J0C	$1.0t$	$1.5t$	$1.5t$

The above minimum inside radii shall apply to JC steel grade to BS EN 10025-2 only as shown in Table 3. For tolerances on shape and dimensions, they are specified in BS EN 10249-2 or equivalent standard or equivalent standard. The inside radius to thickness ratio at bent corner of the interlocking crimped end should be limited to 1.5.

When cold-formed sheet pile profile is manufactured using MC or NC steel grade, the nominal thickness is limited to 16mm and the minimum internal radii should conform to

Table 4 below.

Table 4 Minimum inside radii for MC / NC steel grades to BS EN 10149

Grade designation	Minimum inside radii for nominal thickness (t) in mm		
	$t \leqslant 3$	$3 < t \leqslant 6$	$6 < t$
S315MC	$0.25t$	$0.5t$	$1.0t$
S355MC	$0.25t$	$0.5t$	$1.0t$
S260NC	$0.25t$	$0.5t$	$1.0t$
S315NC	$0.25t$	$0.5t$	$1.0t$
S355NC	$0.25t$	$0.5t$	$1.0t$

The above minimum inside radii shall apply to MC/NC steel grades only. For tolerances on shape and dimensions, they are specified in BS EN 10249-2 or equivalent standard. The inside radius to thickness ratio at bent corner of the interlocking crimped end is now reduced to 1.0.

Welding at cold-formed zones

Welding is sometimes inevitable at SHS and sheet piles. (see Figure 7 and Photo 12).

Figure 7 Welding carried out at SHS   Photo 12 Crimped/Welded sheet pile sections

Welding may be carried out within a length 5t either side of the cold-formed zone (see Table 5), provided that one of the following conditions is satisfied:

(i) cold formed areas are normalized after cold forming but before welding;

(ii) internal radius-to-thickness r/t ratio satisfies the relevant value given in Table 5; or

(iii) Registered Structural Engineer shall submit a Welding Procedure Specification (WPS) for the approval of the Building Authority prior to the commencement and carrying out of welding works in cold-formed hollow sections.

Welding Procedure Specification (WPS)

A preliminary Welding Procedure Specification (pWPS) is drafted by a qualified welding personnel of the contractor. The pWPS should be justified with the Welding Procedure Approval Test and witnessed by a qualified welding inspector of an accredited inspection body. The test results, if found satisfactory, should be endorsed as Welding Procedure Approval Record (WPAR) by the qualified welding inspector of the accredited inspection body. Upon endorsement of the WPAR, the pWPS is then qualified as a valid Welding Procedure Specification (WPS) and can be used by the contractor.

Table 5 Conditions for welding cold-formed areas and adjacent materials

Minimum internal radius/ thickness (r/t) ratio	Strain due to cold forming (%)	Maximum thickness (mm)		Fully killed Aluminium-killed steel (AL≥0.02%)
		Generally		
		Predominantly static loading	Where fatigue predominates	
≥3.0	≤14	22	12	22
≥2.0	≤20	12	10	12
≥1.5	≤25	8	8	10
≥1.0	≤33	4	4	6

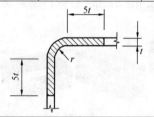

Structural design of cold-formed sections

In general, cold-formed hollow sections may be manufactured by forming the metal at ambient temperature followed by longitudinal weld or spiral weld. The design of bending moment and shear for cold-formed hollow section of this thickness range subjected to various modes of loading may follow the general design provisions stipulated in the Steel Code 2011. For cold-formed sections subjected to compression, the buckling curve 'c' with initial imperfection of 1/300 is used in second order analysis. Deflections should be calculated using elastic analysis. Due allowance shall be made for the effects of non-uniform loading. Design recommendations on connections and fastenings with the bolt and screw should refer to the Steel Code 2011.

Cracking at corners of hot-dip galvanized cold-formed RHS/SHS

Cracking at corners of hot-dip galvanized cold-formed RHS/SHS was reported (see

Photo 13 Corner cracking in cold-formed SHS

Photo 13). The exact causes are yet to be investigated. It is postulated that the cracking may be caused by hydrogen embitterment at the corner of RHS/SHS. The phenomenon is probably a great release of energy due to the formation of hydrogen molecules from free hydrogen atoms, which migrate into the highly strained metal lattice of the RHS/SHS during acid pickling just before hot-dip galvanizing bath.

Should hot-dip galvanizing be required for cold-formed RHS/SHS, a test sample should be arranged by the manufacturer for hot-dip galvanizing to demonstrate its galvanized performance for a given bath.

Current application in the construction industry

The use of cold-formed sections is becoming popular as the Steel Code 2011 provides an alternative for the RSE and the RGBC to choose in lieu of hot-rolled steel sections. A variety of cold-formed sections are now available in the market (see Photos 14、15 and Figure 8).

Photo 14 A variety of cold-formed sections available in the market

Applications of cold-formed steel can be found in many projects in the mainland (see Photos 16 to 21) and in other countries (see Photos 22 to 26).

Trends in the use of cold-formed steel

In the advent of the codification of structural cold-formed sections in the Steel Code 2011, it starts to show promise in the local construction industry that cold-formed steel products are becoming popular. They are comparatively cost-effective as no heat treatment is required. Photos 27 to 33 show the possible application of cold-formed steel in construction industry.

There are occasions whereby certain sizes of hot-rolled sections may not always be available in the manufacturers' stock, and extra time may be required for the mill to hot-roll the sections before they can be delivered to the site. The extra time is beyond expectation and would adversely affect not only the construction programme and also bring about unfavourable financial implications.

Cold-formed steel is considered more efficient as it utilizes the exact plate thickness to form the required section instead of ordering oversized hot-rolled sections. By this, there is economical savings in steel materials. Common structural cold-formed sections may comprise unequal angles, U channels and Z-shaped sections.

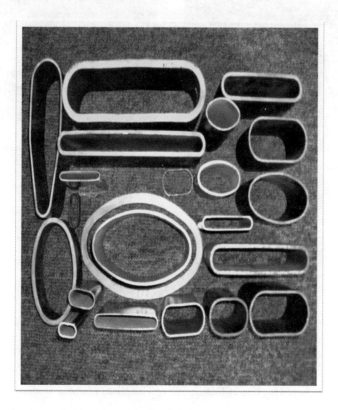

Photo 15　A variety of cold-formed shapes available in the market

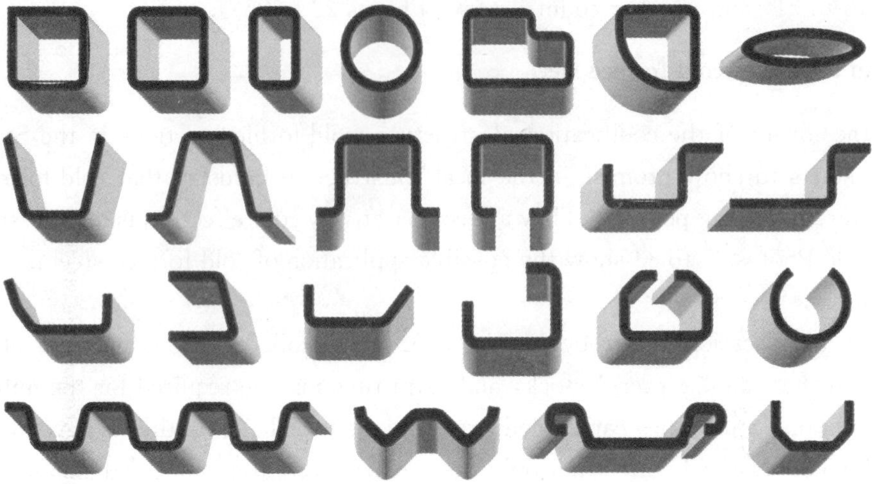

Figure 8 Common cold-formed steel sections available in the market

Photo 16 Use in Shanghai Southern Railway Station

Photo 17 Use in Guangzhou Exhibition Centre

Photo 18 Use in balustrabes of Hangzhou Harbour Bridge

Photo 19　Use of cold-formed steel in residential buildings in seismic areas, Sichuan

Photo 20　Use of cold-formed steel in residential house in Sichuan

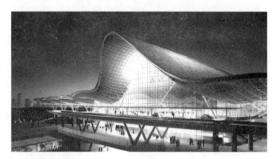

Photo 21　Wuhan Railway Station

Photo 22　Cold-formed bridge in New South Wales, Australia

Photo 23　Cold-formed hoarding used in Australia

Photo 24　Cold-formed car park in Japan

Photo 25　Cold-formed steel office building

Photo 26 Cold-formed warehouse in USA

Photo 27 Use as permanent retaining structure

Photo 28 Use as permanent retaining wall for river training

Photo 29 Use as permanent retaining structure for port works

Photo 30 Use in pylon steel structures

Photo 31 Cold-formed balustrades

Photo 32　Cold-formed balustrades used in Beijing Railway System

Photo 33　Cold-formed oval-shaped section

　　Meanwhile, sections may be cold-formed with thicknesses as large as 12mm, while some factories in Hong Kong can even cold-form the section with thicknesses up to 16mm. It is anticipated that the introduction of cold-formed structural steel would bring about flexibility to both the RSE and the contractor in the steel construction industry.

References

[1]　Code of Practice for the Structural Use of Steel 2011, published by Buildings Department of HKSAR Government.

[2]　BS EN 10025: 2004 Hot-rolled products of non-alloy structural steels. Technical delivery conditions.

[3]　BS 5950-7: 1992 Structural use of steelwork in building. Specification for materials and workmanship: cold formed sections

[4]　BS EN 10149-1: 1996 Specification for hot-rolled flat products made of high yield strength steels for cold forming. Part 1: General delivery conditions

[5]　BS EN 10149-2: 1996 Specification for hot-rolled flat products made of high yield strength steels for cold forming. Part 2: Delivery conditions for thermomechanically rolled steels

[6]　BS EN 10149-3: 1996 Specification for hot-rolled flat products made of high yield strength steels for cold forming. Part 3: Delivery conditions for normalized or normalized rolled steels

[7]　BS EN 10219-1: 2006 Cold formed welded structural hollow sections of non-alloy and fine grain steels. Part 1: Technical delivery requirements

[8]　BS EN 10249-1: 1996 Cold formed sheet piling of non alloy steels. Part 1: Technical delivery conditions

Acknowledgment

　　Ir Prof Paul PANG and Ir Dr W T CHAN hereby acknowledge the financial support by the Research Grant Council of HKSARG on the projects "Collapse Analysis of Steel Tower Cranes and Tower Structures (PolyU 5119/10E)" and "Second-order and Advanced Analysis and Design of Steel Towers made of Members with Angle cross-section (PolyU 5115/08E)".

Biographies of Ir Prof Paul PANG and Ir Dr W T CHAN

Ir Professor Paul Pang is an Honorary Fellow of the Hong Kong Institute of Steel Construction, the Chairman of the Fire Division; a Council Member; a Founding Member of the Fire Discipline Advisory Panel; the Immediate Past Chairman of the Structural Discipline Advisory Panel; and a Past Chairman of the Joint Structural Division of the HKIE. Ir Professor Pang serves the HKSAR Government as an Assistant Director of the Buildings Department, responsible for the control and enforcement policies of existing private buildings in Hong Kong. He is also the Chairman of the Technical Committees on the Codes of Practice for the Structural Use of Steel and Concrete under the Buildings Department. He acquires vast experience including planning, design, construction, forensic and control of building and civil engineering works through his professional career. Currently, he is an adjunct professor at the Hong Kong Polytechnic University and the Hong Kong City University.

Ir Dr W T Chan is the Vice President and the Membership Officer of the Hong Kong Institute of Steel Construction (HKISC). Dr CHAN was the past Chairman of the Jointing, Welding and Cold-Formed Steel Group. He is a Committee Member of the Fire Division of the Hong Kong Institution of Engineers. Dr Chan works in the HKSAR Government as a Senior Structural Engineer in the Buildings Department, and serves as the Secretary to the Technical Committee on the Code of Practice for the Structural Use of Steel under the Buildings Department. His interests cover welding of structural steel, fire engineering, façade engineering, the use of high strength steel and timber construction.

国产钢结构钢材质量现状及质量控制措施

侯兆新,何文汇

(国家钢结构工程技术研究中心;中冶建筑研究总院有限公司,北京 100088)

摘 要:本文通过对近几年国产钢结构钢材性能及质量专项调研情况的介绍,得出了国产钢结构钢材质量的现状及存在的问题,对目前钢结构用钢材标准的相互协调、钢板厚度偏差对结构安全影响,以及钢材质量现场检验等关键共性问题进行了分析研究,提出了有建设性的改进意见和应对措施。可供从事钢结构工程技术人员和相关管理部门参考。

关键词:国产钢结构钢材　钢材标准　钢材质量复验　钢板厚度允许偏差　建筑钢结构

中图分类号:TU391

QUALITY CURRENT CONDITION AND QUALITY CONTROL MEASURES OF DOMESTIC STEEL FOR STEEL STRUCTURE

Z. X. Hou, W. H. He

(National Engineering Research Center for Steel Construction, Beijing 100088, China.)

Abstract: This article is based on the special research work of the recent years' regarding domestic steel structure rolled steel, and summarizes the quality current conditions and outstanding common problems of domestic steel structure rolled steel. It analyses the current existing key common problems such as standard unification of structural steel, the influence of the steel plate thickness deviation to the structural safety and the on-site inspection of steel products quality. It also proposes constructive improvements and solutions of these problems. This can be used as reference for engineers and technicians and other related managerial staff of steel structure.

Keywords: domestic steel structure rolled steel; standard of steel products; quality inspection of steel products; permissible deviation of steel plate; steel structural building

1. 国产钢结构钢材性能及质量的调研情况

1.1 专项调研活动历程

在钢结构工程技术标准特别是国家标准《钢结构设计规范》的编制历程中,针对国产建筑钢结构钢材先后进行过五次规模较大的专项调研活动:

(1)第一次较为全面的统计在 20 世纪 80 年代,用于编制《钢结构设计规范》GBJ

17—88，其主要调研对象为当时的 A3 钢、A3F 钢和 16Mn 钢。

(2) 第二次调研是针对国产建筑结构钢厚板（A3F 钢和 16Mn 钢）进行的，成果用于编制《高层民用建筑钢结构技术规程》JGJ 99—98。

(3) 第三次在修订《钢结构设计规范》GB 50017—2003 前，主要是对第一次调研数据的再分析及补充调研。

(4) 第四次是在 2007 年，对舞阳钢厂按《建筑结构用钢板》GB/T 19879—2005 生产的 Q345GJ 钢材数据进行统计分析。

(5) 第五次是自 2008 年以来针对正在修编国家标准《钢结构设计规范》GB 50017 而开展的调研工作。

1.2 钢材调研工作的主要内容

为配合国家标准《钢结构设计规范》修编工作，修正钢材抗力分项系数和强度设计值，确定新的钢材品种的设计参数等，自 2008 年以来，中冶建筑研究总院（国家钢结构工程技术研究中心）组织开展了较系统的调研工作，主要内容包括以下五个方面：

(1) 收集整理大型工程如中央电视台新址工程、国贸三期、国家游泳馆、深圳证券大楼、石家庄开元环球中心、锦州会展中心、新加坡圣淘沙名胜世界等所用钢材的质检报告和复检报告，其中包括 Q235 钢、Q345 钢、Q390 钢、Q420 钢、Q460 钢和 Q345GJ 钢。钢材生产年限从 2004 年到 2009 年，厚度范围 5~100mm（少量为 100~135mm），数据既包括力学性能，还包括化学元素含量等。总计为 14608 组。

(2) 从钢材生产厂舞钢、湘钢、首钢、武钢、太钢、鞍钢、安阳、新余、济钢、宝钢征集指定钢材牌号、规定钢板厚度的拉伸试件，板厚范围为 16~100mm，牌号为 Q345、Q390、Q420、Q460 和 Q345GJ 钢。集中后统一由独立的第三方进行试验，在人员、设备和方法一致的条件下，获得实验数据。总计为 557 组。

(3) 对影响材性不定性的试验因素（如加载速度和试验机柔度）进行系统的测试分析，以 3 种牌号钢材，3 种板厚，3 种加载速度，2 种刚度的试验机为试验参数，共进行 245 件试验。

(4) 通过十家钢结构制造厂：安徽鸿路、安徽富煌、江苏沪宁、上海宝冶、浙江恒达、东南网架、杭萧钢构、二十二冶、鞍钢建设、中建阳光，测定钢厂生产的钢板、型钢和钢结构厂制作构件的厚度和几何尺寸偏差，共计 25578 组，进行统计分析。

(5) 其他试验及统计分析，如延伸率、屈强比、裂纹敏感性指数和碳当量，硫含量及厚度方向断面收缩率等。

独立的第三方试验数据和工程调研数据相互印证，反映我国钢材生产质量的真实水平，在各钢材牌号，厚度组别一致时，二者的屈服强度平均值、标准差、统计标准值接近，可以以工程调研和独立试验的组合数据作为《钢结构设计规范》确定抗力分项系数和强度设计指标的基础。本次取得数据的对象涵盖广泛，钢材规格品种增加，从 Q235 到 Q460 钢及 Q345GJ 钢。

2. 国产钢结构钢材质量的现状及问题

从 2008 年开始的第五次全国建筑钢结构钢材性能及质量调研结果上看，统计数据分析说明中国国产结构钢材的各项力学性能指标（屈服强度、抗拉强度、延伸率、冷弯、冲击韧性等）基本可达到现行中国钢材标准，也符合中国《钢结构设计规范》对钢材的性能要求，具体情况归纳概述如下：

2.1 Q235 钢材屈服强度值变异较大，整体质量有所下降

Q235 钢材的统计屈服强度平均值较 20 世纪 80 年代统计有明显增加，但其标准差成倍增加，屈服强度波动严重，而计算标准值变化不大，整体质量较以前稍有下降，这主要是由 Q235 的生产企业良莠不齐造成的。

部分 Q235 钢材是由于设备条件差、管理不规范的小型钢厂生产，造成 Q235 钢整体质量不稳定，淘汰部分落后产能势在必行。

2.2 Q345GJ 钢材比 Q345 钢材性能更稳定

在《低合金高强度结构钢》GB/T 1591—2008Q345 钢材屈服强度分布中，屈服强度平均值虽略有增加，但标准差有所增大，波动区间较 20 世纪 80 年代统计也有所增加，屈服强度的计算标准值略有下降，但 Q345 钢材整体质量变化不大，还是能达标的。

《建筑结构用钢板》GB/T 19879—2005 中，Q345GJ 钢的屈服强度分布区间较几年前统计有所增大，标准值略有提高，和 Q345 钢的实际统计表现相比，性能更稳定。

目前有些钢厂误认为 Q345GJ 钢就是高一个质量等级的 Q345 钢（如 Q345C 就相当于 Q345GJB），在提交的材质单中存在屈服强度区间超限、屈强比和碳当量超标等现象。

2.3 Q390 钢材质量总体达标

Q390 钢屈服强度平均值普遍较高，强度波动较小，各项指标均满足《低合金高强度结构钢》GB/T 1591—2008 的要求，将来可作为钢结构钢材的主要品种之一。

2.4 Q420、Q460 钢材质量有待稳定

Q420 和 Q460 高强钢种已能生产，现有数据表明，多数屈服强度值偏向下限值一边，平均值较小。厚度为 $35<t\leqslant50$ 和 $50<t\leqslant100$ 的 Q420 钢和 Q460 钢中少量屈服强度实验值还不能完全达到《低合金高强度结构钢》GB/T 1591—2008 规定的屈服强度标准值。可以说，Q420 和 Q460 高强度合金钢材已可基本满足国内重大工程关键部位的需要，但钢材总体质量水平还有待提高，在工程应用中需要加强现场复验力度。

2.5 钢材截面尺寸以负偏差为主

（1）公称厚度小于 50mm 的钢板，基本以负偏差轧制，负偏差板材数量占到总钢板数量的 90% 以上，且随钢板的厚度变薄，负偏差所占比重增加，负偏差程度也增大。

（2）在钢板厚度大于 50mm 时，情况好转，尺寸质量有保证，钢板整体合格率在

99.5%以上,大型钢厂的厚板生产水平和质量较好。

(3) 热轧H型钢翼缘厚度偏差情况和板材厚度类似,大多为负偏差;截面宽度和高度数据正负偏差大致对称,表现良好。

(4) 角钢的角肢厚度偏差情况同样与板材厚度类似,且角钢角肢高度和厚度不合格的现象比较多,这是由于角钢大多由管理不规范的小型钢厂生产造成的。

2.6 同时并存两个钢材标准问题

《建筑结构用钢板》GB/T 19879—2005 标准在一定程度上能和《低合金高强度结构钢》(GB/T 1591—2008) 互补,但在发布后对钢材力学性能的规定有很大的重叠;目前两个标准对可焊性指标(碳当量CE、裂纹敏感性指数 P_{cm})的规定尚有一些差别,但实际两种钢材表现出的可焊性相差却不大。

《建筑结构用钢板》GB/T 19879—2005 是为了改善《低合金高强度结构钢》(GB/T 1591—1994) 而制定的,有其历史背景,对提高建筑钢材性能和质量起到积极作用。但《建筑结构用钢板》GB/T 19879—2005 采用上屈服强度作为屈服强度进行取值,及在钢板厚度分组等方面,存在着改进的地方,目前技术人员特别是设计人员对使用《低合金高强度结构钢》GB/T 1591—2008 和《建筑结构用钢板》GB/T 19879—2005 标准方面存在诸多模糊和不便。

3. 热轧钢板厚度允许偏差对结构安全影响

3.1 钢板厚度允许偏差类别

现行国家标准《热轧钢板和钢带的尺寸、外形、重量及允许偏差》GB/T 709—2006 中对钢板厚度允许偏差规定五个类别,具体规定见表1。

热轧钢板厚度允许偏差类别　　　　　　　　　　表1

厚度偏差类别	允许偏差规定	备注
N类偏差	正偏差和负偏差相等	不特别注明的情况下,按N类供货
A类偏差	在N类公差值不变的情况下,加大正偏差,减少负偏差绝对值	需要合同注明
B类偏差	在N类公差值不变的情况下,固定负偏差值—0.3mm	需要合同注明
C类偏差	在N类公差值不变的情况下,固定负偏差值为零	需要合同注明
特殊类别	正负偏差值供需双方协商规定	需要合同注明

从表1可知,钢板采购时,应该提出板厚允许偏差类别要求,如果合同中没有注明偏差类别,钢厂原则上按照N类偏差供货。

3.2 钢板厚度允许偏差值

现行国家标准《热轧钢板和钢带的尺寸、外形、重量及允许偏差》GB/T 709—2006 中分别给出N、A、B、C类钢板厚度允许偏差值,仅以常用的钢板公称宽度2500~4000mm,公称厚度在4~100mm情况为例,将允许偏差汇总于图1中。

从图1可知,钢板厚度偏差值是按照在一定公称厚度范围内,固定公差值的方法来确

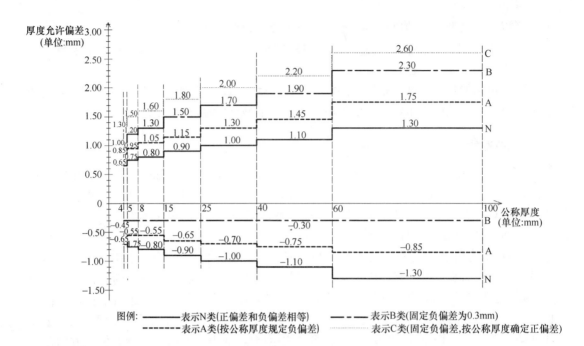

图例： —— 表示N类(正偏差和负偏差相等)   ——— 表示B类(固定负偏差为0.3mm)
       ---- 表示A类(按公称厚度规定负偏差) ······ 表示C类(固定负偏差,按公称厚度确定正偏差)

图1 钢板厚度允许偏差值（板宽2500~4000mm）

定的，通过限制负偏差值，分别得到相应的N、A、B、C类钢板厚度允许偏差值。

本次调研中对钢板、型钢和钢结构厂制作构件的厚度和几何尺寸偏差，共计25578组，进行统计分析，统计结果见图2。

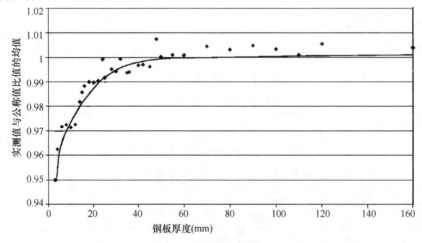

图2 钢板厚度实测统计结果

从图2可以看出，厚度在40mm以上的厚钢板，厚度偏差较小，且正偏差为主；厚度在40mm以下的中厚板及薄板，随板厚变薄，实测值与公称厚度比值（偏差率）的均值呈急剧增大之势，且厚度偏差几乎都是负偏差，似有人为控制之嫌。

3.3 钢板厚度偏差率

钢板厚度允许偏差值与其公称厚度的比值称为偏差率，图3绘出钢板公称宽度2500~

4000mm，公称厚度在100mm情况下的偏差率。

从图3可以明显看出，钢板厚度偏差率随着板厚由厚变薄，呈喇叭状增大，当板厚为5mm时，正偏差最大超过+30%，负偏差低于-15%。当正偏差偏大时（+10%以上），结构自重相应加大较多，增加结构负荷；当负偏差绝对值偏大时（-5%以下），构件截面相应地减小，对结构安全有直接的影响。

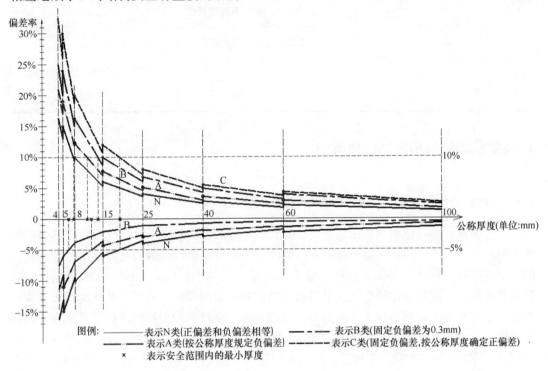

图3 钢板厚度允许偏差率（板宽2500～4000mm）

3.4 钢材厚度偏差类别选用规则

现行国家标准《热轧钢板和钢带的尺寸、外形、重量及允许偏差》GB/T 709—2006是针对热轧钢板的通用标准，适用于各种用途，在有些情况下，特别是板厚较薄时，虽然板厚偏差值满足其产品标准的要求，属于合格产品，但偏差率远远超过建筑结构安全所能允许的范围，因此，结构用钢板特别是薄钢板，仅仅满足其产品标准是不够的，还应对尺寸偏差特别是板厚度偏差进行限制，依据图3及现行国家标准《建筑结构可靠度设计统一标准》GB 50068，具体建议见表2。

热轧钢板厚度偏差选用规则　　　　　　　　　　表2

建筑钢结构及构件类型	钢板厚度偏差类别选用	备 注
安全等级一级主要受力构件（梁、柱、支撑、桁架弦杆等）	厚度≥60mm：N类偏差 >40～60mm：A类偏差 >15～40mm：B类偏差 ≤15mm：C类偏差	负偏差率控制在-2%以内

35

续表

建筑钢结构及构件类型	钢板厚度偏差类别选用	备 注
安全等级二级主要受力构件（梁、柱、支撑、桁架弦杆等）及安全等级一级的次要构件	厚度≥40mm：N类偏差 ＞25～40mm：A类偏差 ＞10～25mm：B类偏差 ≤10mm：C类偏差	负偏差率控制在－4％以内
安全二级的次要构件及安全等级三级的构件	厚度≥15mm：N类偏差 ＞10～15mm：A类偏差 ＞5～10mm：B类偏差 ≤5mm：C类偏差	负偏差率控制在－6％以内

4. 建筑结构用钢板产品标准协调

4.1 《建筑结构用钢板》GB/T 19879—2005 特点

现行国家标准《建筑结构用钢板》GB/T 19879—2005 是在原冶金行业标准《高层建筑用钢》YB 4104—2000 基础上修订并升级为国家标准，可以理解为建筑钢结构量身定做的一个专用钢材标准，与国家标准《低合金高强度结构钢》GB/T 1591—1994 相比，有其显著的特点，其很多性能更能满足建筑钢结构对钢材的要求，特别是高强度厚板确实解决了前些年我国大型钢结构工程所急需，效果显著。修订后的现行国家标准《低合金高强度结构钢》GB/T 1591—2008 其性能有了较大提高和完善，很多性能指标与《建筑结构用钢板》GB/T 19879—2005 相近，在结构用钢板方面形成了两个标准共存的局面。表3 列举了两个钢材标准之间存在的主要不同点。

建筑结构用钢板标准与低合金高强度结构钢标准的主要差异点　　表3

项 目	《低合金高强度结构钢》GB/T 1591—2008	《建筑结构用钢板》GB/T 19879—2005
适用范围	低合金高强度结构钢含钢板、钢带、型钢、钢棒	碳素结构钢和低合金高强度结构钢仅含钢板（6～100mm）
钢材牌号及质量等级	牌号：Q345、Q390、Q420、Q460、Q500、Q550、Q620、Q690 质量等级：A、B、C、D、E	牌号：Q235GJ、Q345GJ、Q390GJ、Q420GJ、Q460GJ 质量等级：B、C、D、E
厚度偏差类别	N、A、B、C 或协商	B（负偏差限定为－0.3mm）或协商
屈服点取值	下屈服点	上屈服点（比下屈服点高3％～5％）
厚度分组	≤16mm ＞16～40mm ＞40～63mm ＞63～80mm ＞80～100mm	6～16mm ＞16～35mm ＞35～50mm ＞50～100mm
屈强比	没有具体列出	Q235GJ：≤0.80 Q345GJ：≤0.83 Q390GJ 及以上：≤0.85
化学成分	硫、磷含量相对较高碳当量相对较高	硫、磷含量相对较低碳当量相对较低

4.2 对《建筑结构用钢板》GB/T 19879—200 改进建议

为了与现行国家标准《低合金高强度结构钢》GB/T 1591—2008 协调，同时考虑与国际标准的接轨，突出建筑结构用高性能钢板的特点，对今后国家标准《建筑结构用钢板》GB/T19879—2005 修订提出以下建议：

（1）屈服强度取值由上屈服点改为下屈服点。
（2）钢板厚度分组与现行国家标准《低合金高强度结构钢》GB/T 1591—2008 保持一致，即：

$$\leqslant 16mm$$
$$>16\sim 40mm$$
$$>40\sim 63mm$$
$$>63\sim 80mm$$
$$>80\sim 100mm$$

（3）增加 Q275GJ 和 Q355GJ 两个牌号，其性能与欧洲相应钢材相匹配，以利于国产结构钢材进入国际市场以及等效代换。
（4）增加低屈服强度高伸长率钢材牌号，例如 Q160GJ 等。
（5）增加高强钢材 Q550GJ 和 Q690GJ 牌号。

5. 国产钢结构钢材质量现场复验规定

5.1 钢材质量现场复验的依据

钢材作为生产环节中各批间质量特征的统计参数差异较大的材料，根据《建筑结构可靠度设计统一标准》GB 50068—2001，结构钢材的质量控制流程如图 4 所示：

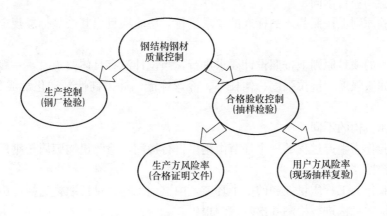

图 4 钢材质量控制流程示意图

从图 4 可以看出，钢材质量除了钢厂控制外，需要增加合格验收控制的环节，这个环节是通过现行国家标准《钢结构工程施工质量验收规范》GB 50205 来实施的。

5.2 建筑钢结构钢材合格验收的规定

(1) 全数检查钢材的质量合格证明文件、中文标志及检验报告等,检查钢材的品种、规格、性能等符合现行国家产品标准和设计要求。

(2) 对属于下列情况之一的钢材,应进行抽样复检,其复验结果应符合现行国家产品标准和设计要求。

1) 结构安全等级一级重要建筑的主体钢结构用钢材;

2) 结构安全等级二级一般建筑,当其结构跨度或高度超过30m时的主体钢结构用钢材;

3) 板厚等于或大于40mm,且设计有Z向性能要求的厚板;

4) 进口钢材及强度等级超过390MPa高强度钢材;

5) 钢材混批,或质量证明文件不齐全的钢材;

6) 设计文件或合同文件要求复验的钢材。

5.3 钢材出厂检验与工程现场复验的差别

钢材现场复验与钢厂出厂检验有所不同,主要体现以下两个方面:

(1) 检验批量不同

出厂检验按照生产组批进行,每批应由同一牌号、同一质量等级、同一炉罐号、同一规格、同一交货条件的钢材组成,每批重量不大于60t。

现场复验属于抽样检验,根据建筑结构及构件的重要性、钢材的品种及数量等因素,确定最小抽样样本数量。实际操作中可以把钢材出厂检验批的次数作为样本容量,例如,某一钢材品种用量600t,相当于出厂检验批为10次(600/60=10),按照样本容量为10,确定最小抽样样本数量为3,意味着600t钢材抽3次,通俗地讲就是每200t抽一次样本,检验批量不大于300t。

(2) 检验的项目不同

出厂检验原则上按照其产品标准的要求进行,检验项目由其产品标准规定,一般是逐项检验。

现场复验的项目原则上按照设计要求进行,当设计没有具体要求时,一般情况只进行力学性能(屈服强度、抗拉强度、伸长率、冷弯性能、冲击韧性)和化学成分(硫、磷、碳)检验。

(3) 检测机构的不同

出厂检验作为生产过程的一个环节由钢厂自行检测,检测机构归属于钢厂,但其资质和认证要符合国家的规定。

现场复验作为工程质量控制的一个环节,由监理工程师见证取样送样,由独立法人资格和资质的检测机构进行检测并出具检测报告。

5.4 钢材复验检验批量的确定

(1) 检验批量标准值

根据现行国家标准《建筑结构检测技术标准》GB 50344—2004对建筑结构抽样检测

的最小样本容量的要求，制定钢材复验检验批量标准值见表4。

钢材复验检验批量标准值确定 表4

检验批容量 （60t/个）	相应钢材用量 （t）	最少抽样样本数	每个样本所代表的 最大钢材量 （t）	检验批量标准值 （t）
2～8	≤500	3	160	180
9～15	501～900	4	225	220
16～25	901～1500	6	250	240
26～50	1501～3000	10	300	300
51～90	3001～5400	15	360	340
91～150	5401～9000	24	375	360
大于150	大于9000	1个/400t	400	400

注：① 同批钢材由同一牌号、同一质量等级、同一规格、同一交货条件的钢材组成；
② 对于钢板厚度，同一规格可参照厚度分组（≤16mm；>16～40mm；>40～63mm；>63～80mm；>80～100mm）执行。

（2）检验批量

根据建筑结构的重要性及钢材品种不同，对检验批量标准值进行修正，即可得到检验批量值，修正系数见表5。经过修正的检验批量值，按照4舍5入原则取整。

钢材复验检验批量修正系数 表5

项　　目	修正系数	备　　注
建筑结构安全等级一级，且设计使用年限100年重要建筑用钢材	0.50	
Q420、Q460及以上高强度钢材	0.75	
板厚40mm以上Z15、Z25、Z35钢材	0.50	不含超声波探伤项目
同批钢材用量超过3000t时，同一厂家首次进场前600t钢材	0.50	
其他情况	1.00	

注：当同时出现2种或2种以上情况时，修正系数取较低者。

5.5 钢材复验项目的确定

钢材复验项目原则上由结构设计工程师通过设计文件确定，当设计文件没有注明时，一般情况可按照表6执行。

每个检验批复验项目及取样数量 表6

序号	复验项目	取样数量	适用标准
1	屈服强度、抗拉强度、伸长率	1	GB/T 2975、GB/T 228
2	冷弯性能（设计要求时）	1	GB/T 2975、GB/T 232
3	冲击韧性（设计要求时）	3	GB/T 2975、GB/T 229
4	Z向钢厚度方向断面收缩率	3	GB/T 5313
5	化学成分 （碳当量CE、裂纹敏感性指数P_{cm}）	1	GB/T 20065、GB/T 223 GB/T 4336、GB/T 20125

6. 主要结论和建议

（1）国产钢结构钢材的各项力学性能指标（屈服强度、抗拉强度、延伸率、冷弯、冲击韧性等）基本可满足现行中国钢材标准的要求，整体上也符合现行国家标准《钢结构设计规范》GB 50017 对钢材的性能要求。

（2）Q345、Q390、Q345GJ 钢材质量已趋于稳定，其质量和性能已经接近并赶超世界先进水平，其性能技术指标能够满足大型钢结构工程的需要，在今后一段时间内将成为钢结构钢材的主打钢种。

（3）现行国家标准《热轧钢板和钢带的尺寸、外形、重量及允许偏差》GB/T 709—2006 规定的钢板厚度偏差特别是薄钢板厚度偏差率（C 类除外）超出建筑结构所允许范围，且市场上供应的中厚钢板和薄钢板基本上是负偏差，给结构安全带来隐患，建议根据不同建筑结构安全等级及不同板厚给出相应的偏差类别，参见表 2。

（4）建议对国家标准《建筑结构用钢板》GB/T 19879—2005 进行修编和完善，原则上与现行国家标准《低合金高强度结构钢》GB/T 1591—2008 协调一致。

（5）钢材现场复验是国产钢材质量控制的重要环节，根据建筑结构的重要性及钢材品种不同，分别确定钢材现场复验检验批量是必要的，根据被复验钢材用量的多少，采用抽样检验理论来确定检验批量标准值是可行的，检验批量标准值和修正系数分别见表 4 和表 5。

参考文献

[1] 国家标准《低合金高强度结构钢》GB/T 1591—2008
[2] 国家标准《建筑结构用钢板》GB/T 19879—2005
[3] 国家标准《热轧钢板和钢带的尺寸、外形、重量及允许偏差》GB/T 709—2006
[4] 国家标准《建筑结构可靠度设计统一标准》GB 50068—2001
[5] 国家标准《钢结构设计规范》GB 50017—2003
[6] 国家标准《钢结构工程施工质量验收规范》GB 50205—2001
[7] 国家标准《建筑结构检测技术标准》GB 50344—2004

冷弯成型钢管压弯构件抗震性能研究 *

沈祖炎[1,2]，温东辉[1]，李元齐[1,2]

（1. 同济大学 土木工程学院，上海 200092；

2. 同济大学 土木工程防灾国家重点实验室，上海 200092）

摘 要：本文对冷弯成型的方、矩形截面钢管压弯构件进行了常轴力、循环水平荷载作用下的拟静力试验，对比了试件的破坏形态、得到了试件的滞回曲线、骨架曲线、延性系数及耗能能力等抗震性能及指标，分析了宽厚比、轴压比和长细比对试件抗震性能的影响，并根据相关规范对试件的延性类别进行了初步判定，为冷弯成型钢管在抗震结构中的应用提供了初步依据。

关键词：冷弯厚壁型钢；压弯构件；抗震性能；滞回曲线；延性系数

中图分类号：TU392.3

STUDY ON COLD-FORMED THICK-WALLED STEEL MEMBERS UNDER COMPRESSION AND CYCLIC BENDING

Z. Y. Shen[1,2], D. H. Wen[1], Y. Q. Li[1,2]

(1. College of Civil Engineering, Tongji University, Shanghai 200092, China;

2. State Key Laboratory for Disaster Reduction in Civil Engineering,

Tongji University, Shanghai 200092, China.)

Abstract: In this paper, quasi-static tests on cold-formed thick-walled steel box members under constant axial force and cyclic bending were reported. The failure modes were discussed and compared, and the hysteretic loops, skeleton curves, ductility and energy-dissipation capacity were investigated. The effect of width-thickness ratio, axial compression ratio and slenderness ratio on seismic performance of specimens was also analyzed, by which the ductility category of specimens was simply considered in accordance with the relevant standard for the reference of further using cold-formed thick-walled steel box members in seismic structures.

Keywords: cold-formed thick-walled steel; members under compression and cyclic bending; seismic performance; hysteresis curve; ductility index

---

* 基金项目：国家自然科学基金资助项目（51178330）。

第一作者：沈祖炎（1935—），男，院士，教授，主要从事钢结构方面的研究，E-mail：zyshen@tongji.edu.cn.

通讯作者：李元齐（1971—），男，博士，教授，博导，主要从事大跨度空间钢结构非线性分析和抗风理论、冷弯型钢结构设计理论、钢结构检测与鉴定技术等方面的研究，E-mail：liyq@tongji.edu.cn.

1. 引言

冷弯型钢作为建筑钢结构主要用材之一,具有截面高效、经济性好和制作工业化程度高、绿色环保等特点,在发达国家应用广泛,适用厚度已达 25.4mm。但我国现行国家标准《冷弯薄壁型钢结构技术规范》[1]的适用厚度不超过 6mm,导致其应用范围相对非常有限。近年来我国冷弯型钢行业发展迅猛,已能生产加工壁厚 20mm 左右的冷弯型钢,并在低多层和小高层住宅等建筑领域呈现广阔应用前景。目前,国内外学者对冷弯型钢构件的抗震性能研究主要集中在冷弯薄壁构件($t \leqslant 6mm$)[2~8]抗震性能的研究,对 6mm 以上冷弯厚壁构件的滞回性能研究几乎是空白。而我国建筑物普遍需要考虑抗震设防要求。因此,冷弯型钢作为结构构件,其抗震性能及相应设计方法的研究至关重要。本文基于滞回性能试验,对 6mm 以上冷弯成型的方、矩形截面钢管压弯构件的抗震性能进行研究,为冷弯成型钢管在抗震结构中的应用提供初步依据和建议。

2. 试验研究

2.1 材性试验

为研究冷加工对钢材性能的影响,在冷弯成型的方、矩形钢管的平板、焊缝和角部分别取样,根据《金属材料室温拉伸试验方法》GB/T 228—2002)[9]进行材性试件设计,材性试验研究结果表明:(1)位于钢管截面焊缝两邻边的屈服强度和极限强度差异都很小,可用各个截面两邻边的平均值作为平板部位的代表值;(2)焊缝两邻边强屈比平均值对 Q235 和 Q345 钢分别是 1.383 和 1.389,达不到规范[1]给定的 1.58 和 1.48 值;(3)焊缝对面平板屈服强度有明显提高,且提高系数随宽厚比的增大而减小;(4)角部的屈服强度和极限抗拉强度比焊缝邻边平板处均有较大提高,平均提高率分别为 42% 和 15%;而伸长率明显降低,平均降低率为 38%。另外,由于角部屈服强度提高,导致角部强屈比平均减小为 1.12。

2.2 压弯构件反复试验

2.2.1 试件设计

考虑宽厚比 b/t、轴压比 n 和长细比 l 三个因素选择钢管截面,各个试件的设计参数见表1。

方、矩形截面压弯试件设计试验参数 表1

序号	试件编号	截面规格(mm)	试件设计参数			来源
			b/t	n	l	
1	Q2-S-250-8-MC-1-A	250×250×8	30	0.2	70	宝钢
2	Q2-S-350-12-MC-2	350×350×12	30	0.4	50	宝钢
3	Q2-S-350-12-MC-3	350×350×12	30	0.6	30	宝钢
4	Q2-S-220-10-MC-1	220×220×10	20	0.2	50	宝钢
5	Q2-S-350-16-MC-2	350×350×16	20	0.4	30	宝钢
6	Q2-S-220-10-MC-3	220×220×10	20	0.6	70	宝钢

续表

序号	试件编号	截面规格（mm）	b/t	n	l	来源
7	Q2-S-135-10-MC-1	135×135×10	10	0.2	30	宝钢
8	Q2-S-108-10-MC-2	108×108×10	10	0.4	70	宝钢
9	Q2-S-120-10-MC-3	120×120×10	10	0.6	50	宝钢
10	Q2-S-250-8-MC-1-B	250×250×8	30	0.2	70	武钢
11	Q1-S-220-10-MC-1-A	220×220×10	20	0.2	50	宝钢
12	Q1-S-220-10-MC-1-B	220×220×10	20	0.2	50	武钢
13	Q1-S-108-10-MC-2	108×108×10	10	0.4	70	宝钢
14	Q1-S-350-14-MC-2	350×350×14	25	0.4	35	宝钢
15	Q1-S-250-16-MC-3	250×250×16	15	0.6	60	宝钢
16	Q2-R-350-12-MC-1	350×250×12	30	0.2	75	宝钢
17	Q2-R-350-12-MC-2	350×250×12	30	0.4	50	宝钢
18	Q2-R-350-12-MC-3	350×250×12	30	0.6	25	宝钢
19	Q2-R-300-8-MC-1	300×200×8	45	0.2	50	宝钢
20	Q2-R-300-8-MC-2	300×200×8	45	0.4	25	宝钢
21	Q2-R-300-8-MC-3	300×200×8	45	0.6	75	宝钢
22	Q2-R-400-10-MC-1	400×200×10	40	0.2	25	宝钢
23	Q2-R-400-10-MC-2	400×200×10	40	0.4	75	宝钢
24	Q2-R-400-10-MC-3	400×200×10	40	0.6	50	宝钢

表1中，试件编号为：钢材等级-截面类型-长边尺寸-壁厚-受力类型-轴压比-重复试件编号。其中钢材等级Q1，Q2分别表示Q235，Q345；截面类型S、R分别表示方管、矩形管；受力类型MC表示往复水平加载；1，2，3分别表示设计轴压比为0.2，0.4和0.6；字母A和B表示重复试件，但钢管来源不同。

2.2.2 试验装置

本试验采用同济大学建工系试验室1000t大型多功能结构试验机系统。加载方式采用悬臂柱加载方法。整个加载装置包括竖向加载系统和水平加载系统。竖向加载系统可施加的最大压力为1000t；两个不同高度处的水平150t和300t的电液伺服作动器可用来施加水平往复荷载。压弯反复试件的试验装置和试件加载全景见图1。

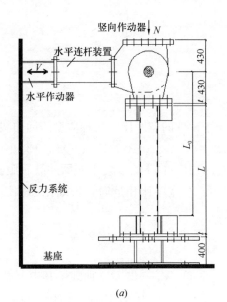

(a) (b)

图1 压弯反复试件加载装置

(a) 试验装置示意图；(b) 试验加载全景

2.2.3 加载制度

本试验正式加载前先进行竖向荷载和水平荷载预加载，目的一是确保试验机及采集系统工作正常，二是进行物理对中。正式加载先以荷载控制在柱顶部施加轴力到预定值，后保持不变；以位移控制施加反复水平荷载。所有构件均采用先推后拉循环加载，加载制度按照规范 ATC-24[10] 中选用，见图 2。此处 D_y 是试件受力最大纤维出现屈服时的柱顶侧向位移，采用名义屈服强度计算得到。具体如下：1） $0.25D_y$、$0.5D_y$、$0.75D_y$、D_y、$2D_y$ 和 $3D_y$ 时，每级循环 3 圈；2） $4D_y$、$5D_y$、$6D_y$……时，每级循环 2 圈。

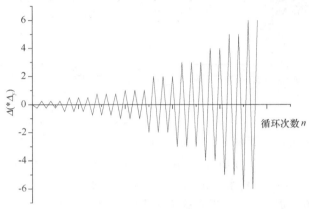

图 2 压弯反复试件水平位移加载制度

2.2.4 试验测量内容

试验测试数据有：柱顶加载点处的水平荷载和位移；靠近柱底截面塑性铰区和上端弹性段截面的应变；靠近柱底截面塑性铰区的转角。为考察底座是否刚性，另布置位移计监测支座的位移。

3. 试验结果与分析

3.1 试验现象

方、矩形钢管截面试件的破坏模式主要取决于试件的宽厚比。四种典型滞回曲线对应的试件破坏模式见图 3。其中，宽厚比较小的试件 Q2-S-220-10-MC-1、Q2-S-135-10-MC-1、Q2-S-108-10-MC-2、Q2-S-120-10-MC-3 和 Q1-S-108-10-MC-2 发生四周外鼓屈服破坏，其他试件发生两对边内凹或外鼓的局部屈曲破坏。在宽厚比相近时，长细比和轴压比越小，局部屈曲出现越晚，且构件承载力降低越慢。

3.2 滞回曲线

图 4 给出典型的无量纲化的水平荷载-侧移关系（$V/V_y - D/D_y$）曲线，其中 V 为水平荷载；V_y 为最大受力纤维处屈服时对应的水平承载力；D 为实测销轴处侧向位移（向西为正，向东为负）；D_y 为各试件的屈服位移。水平屈服承载力和屈服位移均按照实测截面尺寸和材性结果计算得到。

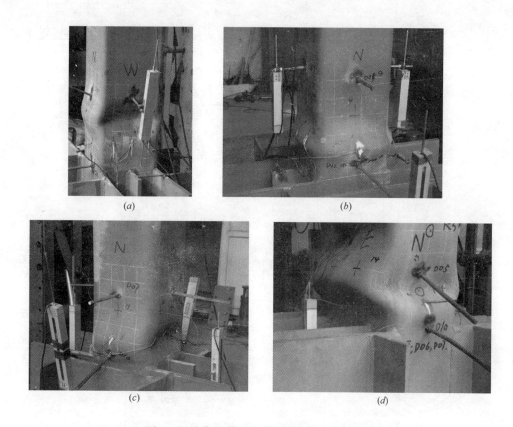

图 3 四种典型滞回曲线对应的试件破坏模式
(a) 典型形状 1 (Q2-S-250-8-MC-1-A); (b) 典型形状 2 (Q2-S-220-10-MC-1);
(c) 典型形状 3 (Q1-S-250-16-MC-3); (d) 典型形状 4 (Q2-R-400-10-MC-1)

滞回曲线的形状和饱满程度与试件的宽厚比 b/t、长细比 l 和轴压比 n 等参数密切相关。根据各参数范围，可分为四种典型情况：

(1) 当参数 b/t、l 和 n 取值适中，b/t 和 l 较大、n 较小，或 b/t 和 n 较大、l 较小时，滞回曲线形状如图 4（a）所示。此类试件局部屈曲和整体失稳相互耦合影响，试件的承载力达到峰值荷载以后，试件强度和刚度明显下降。特别是当参数 n、l 和 b/t 均较大时，此类试件达到峰值荷载后随即破坏，试件的延性和耗能较差。

(2) 当参数 b/t 和 n 较小，同时 l 取值适中时，滞回曲线形状如图 4（b）所示。试件的承载力达到峰值荷载以后，强度和刚度几乎不退化，经过若干循环，构件根部接近全截面屈服以后，承载力开始缓慢降低。

(3) 当参数 n 和 l 较大、b/t 较小时，滞回曲线形状如图 4（c）所示。此类试件整体失稳起控制作用，由于轴压比较大，试件受到侧向荷载后，承载力迅速下降，同时由于试件的宽厚比较小，试件不发生局部屈曲，破坏时柱脚根部出现四面外鼓现象。

(4) 当参数 b/t 较大，n 和 l 较小时，滞回曲线形状如图 4（d）所示。此类试件局部屈曲起控制作用，达到峰值荷载后，试件的强度和刚度随着局部屈曲的逐步发展缓慢下降。

试验得到的四种典型滞回曲线的各参数范围见表 2。各试件的滞回曲线参数见表 3。

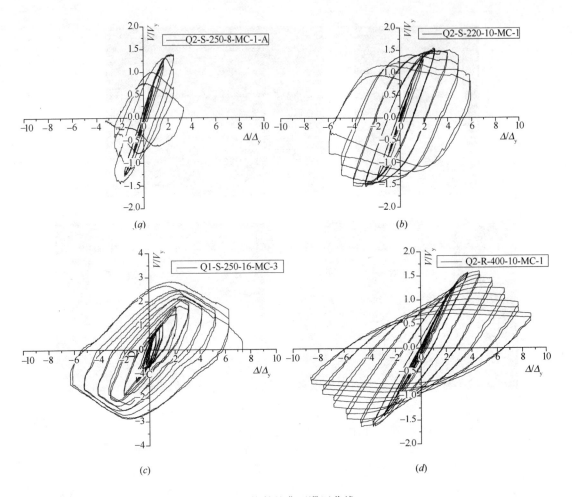

图 4 构件的典型滞回曲线
(a) 典型形状1（Q2-S-250-8-MC-1-A）；(b) 典型形状2（Q2-S-220-10-MC-1）；
(c) 典型形状3（Q1-S-250-16-MC-3）；(d) 典型形状4（Q2-R-400-10-MC-1）

压弯试件四种典型滞回曲线的参数范围　　　　　表 2

参数	典型形状1				典型形状2	典型形状3	典型形状4
	情况1	情况2	情况3	情况4			
轴压比 n_e	0.33—0.47	0.18—0.24	0.40—0.71	0.41—0.67	0.19—0.23	0.43—0.71	0.20
长细比 l	37.36—61.28	58.99—90.13	29.64—42.98	59.83—89.89	42.51—61.35	63.49—95.26	29.48
宽厚比 b/t	28.74—37.10	34.54—46.76	34.54—46.76	46.76—49.35	18.43—28.75	14.64—28.75	49.35
试验现象	局部失稳与整体失稳耦合				强度破坏为主	整体失稳为主	局部失稳为主

注：轴压比 n_e 按有效截面计算得到。

3.3 骨架曲线

图5给出了四种典型滞回曲线的水平荷载-侧移骨架曲线。由图可知，所有骨架曲线正、反向基本对称，走势相似，从弹性变形到屈服点，达最大荷载后，开始下降直至塑性

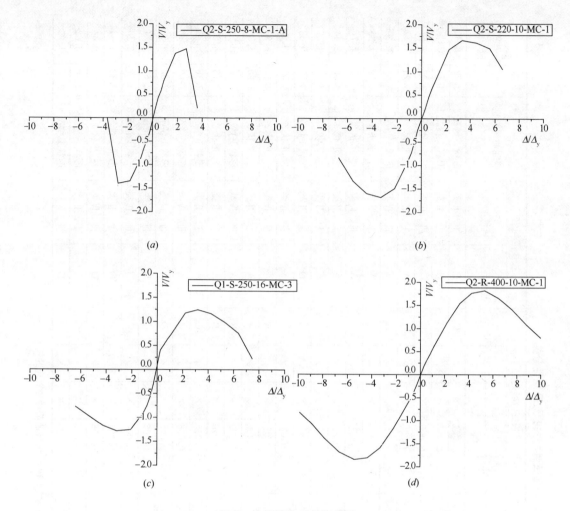

图 5 构件的典型骨架曲线
（a）典型形状 1（Q2-S-250-8-MC-1-A）；（b）典型形状 2（Q2-S-220-10-MC-1）；
（c）典型形状 3（Q1-S-250-16-MC-3）；（d）典型形状 4（Q2-R-400-10-MC-1）

破坏。长细比、宽厚比和轴压比越大，试件的峰值荷载、延性越小，加载后期试件的承载力及刚度退化越严重。

4. 延性性能

本文采用位移延性系数 m 表达构件的延性，计算公式为：

$$m = D_u/D_y \tag{1}$$

式中，D_u 为构件的极限位移，取骨架曲线峰值荷载下降 15% 时对应的位移；D_y 为构件的屈服位移，采用能量法[11]由骨架曲线计算得到。

图 6（a）-（c）从正交试验结果的角度分别给出各参数对试件延性的影响。由图可见，位移延性系数随长细比、宽厚比和轴压比的增大而呈现降低趋势；三个参数相互耦合，共同影响压弯试件的延性性能。各试件的延性系数及延性分类见表 3。

表 3　压弯试件相关参数及延性分类

试件编号	试件实际参数				根据试验判定				按规范判定		滞回曲线类型	m	h_e [3]
	n	n_e	$b/t\sqrt{f_y/235}$	$l/\sqrt{f_y/235}$	$\geq M_p$	q/rad 值	≥ 0.02	$\geq M_y$	结论	截面类别[1]	延性类别[2]			
Q2-S-250-8-MC-1-A	0.20	0.22	41.25	87.05	×	0.004	×	√	C	D	Ⅲ、Ⅳ类	形状1-情况2	2.62	8.89
Q2-S-350-12-MC-2	0.36	0.38	37.10	61.28	×	0.007	×	√	C	C	Ⅲ、Ⅳ类	形状1-情况1	3.77	16.40
Q2-S-350-12-MC-3	0.54	0.58	37.10	35.60	√	0.006	×	×	D	C	Ⅲ、Ⅳ类	形状1-情况3	3.38	3.67
Q2-S-220-10-MC-1	0.22		28.75	61.35	√	0.02	√	√	A	A	Ⅰ、Ⅱ类	形状2	3.73	7.42
Q2-S-350-16-MC-2	0.33		30.70	37.36	√	0.02	√	√	A	A	Ⅲ、Ⅳ类	形状1-情况1	3.67	32.46
Q2-S-220-10-MC-3	0.65		28.75	90.05	×	0.005	×	√	C	A	Ⅰ、Ⅱ类	形状3	2.79	1.25
Q2-S-135-10-MC-1	0.23		18.43	42.51	√	0.03	√	√	A	A	Ⅰ、Ⅱ类	形状2	6.75	7.76
Q2-S-108-10-MC-2	0.46		14.94	95.26	√	0.02	√	√	A	A	Ⅲ、Ⅳ类	形状3	3.36	2.44
Q2-S-120-10-MC-3	0.71		16.27	63.49	√	0.02	√	√	A	A	Ⅲ、Ⅳ类	形状3	4.11	1.27
Q2-S-250-8-MC-1-B	0.18	0.20	40.43	90.13	×	0.005	×	√	C	D	Ⅲ、Ⅳ类	形状1-情况2	2.89	9.68
Q1-S-220-10-MC-1-A	0.23		24.70	50.27	√	0.02	√	√	A	A	Ⅲ、Ⅳ类	形状2	4.51	11.24
Q1-S-220-10-MC-1-B	0.19		25.93	57.15	√	0.02	√	√	A	A	Ⅲ、Ⅳ类	形状2	4.30	8.93
Q1-S-108-10-MC-2	0.43		14.64	91.97	√	0.02	√	√	A	A	Ⅲ、Ⅳ类	形状3	3.20	1.86
Q2-S-350-14-MC-2	0.44		28.74	37.57	√	0.009	×	√	B	A	Ⅲ、Ⅳ类	形状1-情况1	3.39	10.49
Q2-S-250-16-MC-3	0.65		20.15	71.68	√	0.010	×	√	B	A	Ⅲ、Ⅳ类	形状3	3.88	6.27
Q2-R-350-12-MC-1	0.24		34.54	76.91	×	0.005	×	√	C	A	Ⅲ、Ⅳ类	形状1-情况2	3.38	15.18
Q2-R-350-12-MC-2	0.47		34.54	57.62	×	0.006	×	√	C	A	Ⅲ、Ⅳ类	形状1-情况1	3.00	5.76
Q2-R-350-12-MC-3	0.71		34.54	42.98	×	0.005	×	×	D	A	Ⅲ、Ⅳ类	形状1-情况3	2.82	3.50
Q2-R-300-8-MC-1	0.20	0.22	46.76	58.99	×	0.005	×	√	C	D	Ⅲ、Ⅳ类	形状1-情况2	2.63	1.24
Q2-R-300-8-MC-2	0.39	0.45	46.76	29.64	×	0.004	×	√	C	D	Ⅲ、Ⅳ类	形状1-情况3	2.87	2.15
Q2-R-300-8-MC-3	0.59	0.67	46.76	88.34	×	0.001	×	×	D	D	Ⅲ、Ⅳ类	形状4	2.18	1.44
Q2-R-400-10-MC-1	0.20	0.23	49.35	29.48	×	0.008	×	√	C	D	Ⅰ、Ⅱ类	形状1-情况4	3.43	9.04
Q2-R-400-10-MC-2	0.40	0.45	49.35	89.89	×	0.006	×	√	C	D	Ⅲ、Ⅳ类	形状1-情况4	2.36	6.32
Q2-R-400-10-MC-3	0.60	0.68	49.35	59.83	×	0.005	×	×	D	D	Ⅲ、Ⅳ类	形状1-情况4	2.58	1.95

注：①依规范[12] 8.2.4 条判定；②依规范[12] 8.2.3 条判定；③耗能系数 h_e 为峰值荷载处对应的耗能。

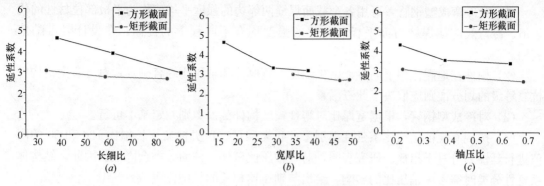

图 6 各参数对延性系数的影响
(*a*) 长细比；(*b*) 宽厚比；(*c*) 轴压比

5. 耗能能力

采用等效黏滞阻尼 h_e（滞回环面积）来描述试件在屈服阶段、峰值荷载处及极限阶段的耗能能力。各试件的等效黏滞阻尼 h_e 见表 3。根据文中定义的四种典型滞回曲线形状，图 7 给出各种类型的曲线在各阶段的耗能系数。由图可知：

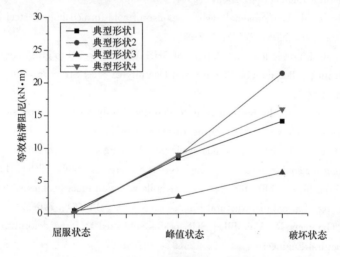

图 7 构件的典型滞回曲线的耗能

（1）各试件在屈服状态的耗能较小，且比较接近，此时试件基本上处于弹性范围。

（2）构件达峰值荷载时，除破坏形式以整体失稳为主的构件耗能较小外，其他三类破坏形式的耗能能力差异不大。

（3）在破坏状态，各试件的耗能能力均明显提高，且以强度破坏为主的试件耗能最好，局部失稳为主和局部失稳与整体失稳耦合破坏的试件耗能次之，整体失稳为主的试件耗能最差。

6. 结论

通过材性和压弯构件抗震试验研究，可得到以下结论：

（1）对冷弯成型钢管，可用各个截面焊缝两邻边的强度平均值作为平板部位材性的代表值。冷弯效应使焊缝对面平板和角部的强度均有不同程度的提高，在设计时可酌情利用。

（2）构件在地震作用下的性能由长细比、宽厚比和轴压比综合确定，对冷弯型钢构件抗震等级的划分准则应取决于此三因素。

（3）对冷成型钢管，根据宽厚比对塑性铰区构件截面类别判定基本可行。

（4）按现有规范结构体系延性类别Ⅰ、Ⅱ类和Ⅲ、Ⅳ类框架柱长细比120和60的限值进行延性划分过于粗糙。研究表明，对大轴压比试件，截面分类会降低。因此，结构体系延性分类时需考虑轴压比的影响，给出与轴压比相关的长细比限值。

（5）冷成型压弯构件破坏前可经历充分的塑性变形，其延性性能和耗能能力等抗震性能指标较好，在满足相关参数限值条件下，能够在强震区的多、高层结构中使用。具体的参数限值需结合考虑冷弯效应和残余应力影响的参数分析结果确定。

参考文献

[1] 冷弯薄壁型钢结构技术规范. GB 50018—2002. 北京：中国计划出版社，2002
[2] Mohamed Elchalakani, et al. Tests of cold-formed circular tubular braces under cyclic axial loading. Journal of Structural Engineering, 2003, 129(4): 507-514
[3] J. M. Goggins. et al. Experimental cyclic response of cold-formed hollow steel bracing members. Engineering Structures, 2005, 27(7): 977-989
[4] Grzebieta R, et al. Multiple low cycle fatigue of SHS tubes subjected to gross pure bending deformation. Proc. 5th Int. Colloquium on Stability and Ductility of Steel Structure, University of Nayoya, Nayoya, Japan, 847-854
[5] Elchalakani, M. et al. Cyclic bending tests to determine fully ductile section slenderness limits for cold-formed circular hollow sections. Journal of Structural Engineering, 2004, 130(7): 1001-1010
[6] Elchalakani, M. et al. Variable amplitude cyclic pure bending tests to determine fully ductile section slenderness limits for cold-formed CHS. Engineering Structures, 2006, 28(9): 1223-1235
[7] Zhao X. L., Void-filled cold-formed rectangular hollow section braces subjected to large deformation cyclic axial loading. Journal of Structural Engineering, 2002, 128(6): 746-753
[8] Zhao X. L., Grzebieta R. Void-filled SHS beams subjected to large deformation cyclic bending. Journal of Structural Engineering, 1999, 125(9): 1020-1027
[9] 金属材料室温拉伸试验方法. GB/T 228—2002. 北京：中国标准出版社，2002
[10] Guidelines for cyclic seismic testing of components of steel structures. ATC-24. Applied Technology Council, Redwood City, Calif, 1992. 8-13
[11] 罗金辉. L形钢管混凝土柱-H型钢梁框架节点抗震性能研究[博士学位论文]. 同济大学，2011. 60-61
[12] 高层建筑钢结构设计规程. DG/T J08—32—2008, J11195—2008. 上海市工程建设规范，2008. 17-19, 58-59

高温下冷弯不锈钢螺栓连接结构的试验研究*

蔡炎城，杨立伟

（香港大学 土木工程系，中国香港）

摘 要：目前冷弯不锈钢螺栓连接结构的规范条文只适用于螺栓连接结构在室温下的设计，高温条件下单剪螺栓连接结构的研究并不常见。本研究通过稳态测试方法（恒温加载）对三种不同等级的不锈钢螺栓连接结构进行试验研究，试验构件包括25个材料力学性能试验以及24个单剪螺栓连接结构试验，螺栓连接结构由两个螺栓顺着加载方向排列，试验温度范围从室温22℃到950℃高温。三种不同等级的不锈钢包括奥氏体不锈钢EN1.4301(AISI 304)和EN1.4571(AISI 316Ti)以及节约型双相不锈钢EN1.4162 (AISI S32101)，其中EN1.4571(AISI 316Ti)含有少量钛元素。单剪螺栓连接结构的试验结果主要以承压破坏模式为主。试验的极限承载力值与美国、澳大利亚/新西兰以及欧洲的冷弯不锈钢结构设计规范中的名义强度值进行了比较。在计算螺栓连接结构的名义强度值时，应用了相应的冷弯不锈钢在高温下的材料力学性能。比较结果显示，运用现行规范得到单剪螺栓连接结构高温下的名义强度值通常低估了试验结果的极限承载力值。高温条件下本研究中的奥氏体不锈钢EN1.4571(AISI 316Ti)通常比其他两种等级不锈钢抗火性能更好。

关键词：承压破坏；螺栓连接；试验研究；材料性能；不锈钢；极限强度

EXPERIMENTAL INVESTIGATION OFCOLD-FORMED STAINLESS STEEL BOLTED CONNECTIONS AT ELEVATED TEMPERATURES*

Yancheng CAI, Ben YOUNG

(Department of Civil Engineering, The University of Hong Kong,
Pokfulam Road, Hong Kong, China)

Abstract: The current design rules on bolted connections of cold-formed stainless steel structures are allows for room (ambient) temperature condition only. Research on structural behavior of single shear bolted connections at elevated temperatures is limited. In this study, 25 coupon specimens and 24 single shear two-parallel bolted connection specimens involving three different grades of stainless steel were conducted by using steady state test method in the temperature ranged from 22to 950. The three different grades of stainless steel are austenitic stainless steel EN1.4301 (AISI 304) and EN1.4571 (AISI 316Ti having small

---

* 基金项目：中国香港特别行政区研究基金(HKU718612E)。
第一作者：蔡炎城，男，博士生，主要从事高温下不锈钢螺栓连接结构的性能研究。E-mail：caiyc@hku.hk。
通讯作者：杨立伟，男，教授，主要从事冷弯薄壁钢结构、不锈钢结构以及铝合金结构的性能研究。E-mail：young@hku.hk。

amount of titanium) as well as lean duplex stainless steel EN1. 4162 (AISI S32101). The bearing failure mode was mainly observed in the stainless steel single shear two-parallel bolted connection tests. The test results were compared with the nominal strengths calculated from the American Specification, Australian/New Zealand Standard and European Codes for cold-formed stainless steel structures. In calculating the nominal strengths of the connections, the material properties of stainless steel obtained at elevated temperatures were used. It is shown that the strengths of the single shear two-parallel bolted connections predicted by the specifications are generally conservative at elevated temperatures. The austenitic stainless steel type EN 1. 4571 (AISI 316Ti) generally performed better than the other two stainless steel types at elevated temperatures.

Keywords: bearing failure; bolted connection; experimental investigation; material properties; stainless steel; ultimate strength

1. INTRODUCTION

In recent years, significant progress has been made in developing design rules for stainless steel structures at room temperature, but the performance of fire resistance has received less attention[1]. Bolted connections are one of the common connection types in cold-formed steel structures construction. The design rules of cold-formed stainless steel bolted connections are available in current specifications, i. e. the American Society of Civil Engineers Specification (ASCE)[2], Australian/New Zealand Standard (AS/NZS)[3] and European Code 3 Part 1. 4 (EC3-1. 4)[4]. Tests of cold-formed steel bolted connections have been conducted at ambient temperature by several researchers, such as Zadanfarrokh[5], Rogers and Hancock[6~8], and Bouchaïr et al. [9]. However, investigation of stainless steel bolted connections at elevated temperatures is limited. Yan and Young[10~11] recently studied the structural behavior of single shear bolted connections of thin sheet carbon steels at elevated temperatures by steady state and transient state test methods. Previous research has shown that the strength and stiffness retention of austenitic stainless steel at elevated temperatures is superior to those of carbon steel[12]. Itshould be noted that the bolted connection design rules in the current specifications[2~4] are applicable at ambient temperature condition only.

In this study, the material properties of three different types of stainless steels, namely the austenitic stainless steel EN1. 4301 (AISI 304) and EN1. 4571 (AISI 316Ti having small amount of titanium) as well as lean duplex stainless steel EN1. 4162 (AISI S32101)were firstly determined by tensile coupon tests using the steady state test method for the temperature ranged from 22 to 950℃. Based on the coupon test results at elevated temperatures obtained from this study, seven critical high temperature levels were selected for the single shear bolted connection tests. The bearing failure mode was mainly observed from the tests at elevated temperatures. The ultimate strength of single shear two-parallel bolted connections of different stainless steel types at elevated temperatures were compared, and it was shown that the type EN 1. 4571 (AISI 316Ti) stainless steel generally performed better than the other two types of stainless steels at elevated temperatures.

2. COUPON TESTS

Steady state test method was used for both the coupon and bolted connection tests in this study. The tensile coupon tests at elevated temperatures were conducted using an MTS810 Universal testing machine. The MTS model 653.04 high temperature furnace was used to heat up the specimen to the specified temperature. An external thermal couple was used to measure the actual temperature of the coupon specimen. The external thermal couple was inserted inside the furnace and contacted on the surface of the coupon specimen at mid-length. The temperature obtained from the external thermal couple was recorded as the specimen temperature in this study. The MTS model 632.54F-11 high temperature axial extensometer was used to measure the strain of the middle section of the coupon specimen.

The coupon test specimens were designed according to the Australian Standard AS-2291 [13]. A total of 25 specimens were conducted to obtain the material properties of the stainless steels at elevated temperatures. The coupon specimens involved three different grades of stainless steel, namely the austenitic stainless steel EN1.4301 (AISI 304) and EN1.4571 (AISI 316Ti having small amount of titanium) as well as the lean duplex stainless steel EN1.4162 (AISI S32101). The lean duplex stainless steel EN1.4162 (AISI S32101) is a high strength material and it is a relatively new kind of material in civil engineering construction, thus it is not covered in any current design specifications; while the austenitic stainless steels EN1.4301 (AISI 304) and EN1.4571 (AISI 316Ti) have a lower strength than lean duplex material. The type EN1.4571 (AISI 316Ti) contains titanium (element Ti) and has good resistance at high temperature. For simplicity, the three types of stainless steels, EN1.4301 (AISI 304), EN1.4571 (AISI 316Ti) and EN1.4162 (AISI S32101) are labeled as types A, T and L, respectively, in the context of this paper.

The mechanical properties of the three types of cold-formed stainless steels obtained at room (ambient) temperature are presented in Table 1. The reduction factors $f_{0.2,T}/f_{0.2,N}$ and $f_{u,T}/f_{u,N}$ of the three types of stainless steels versus the specimen temperatures are plotted in Fig. 1a and 1b, respectively. The vertical axis of the graphs is the normalized reduction factors $f_{0.2,T}/f_{0.2,N}$ and $f_{u,T}/f_{u,N}$, while the horizontal axis plotted against the actual specimen temperatures. The symbols employed in Tables 1 and Fig. 1 are defined as follows: E_N is elastic modulus at room temperature; $f_{0.2,N}$ and $f_{0.2,T}$ is longitudinal 0.2% tensile proof stress at room temperature and elevated temperatures, respectively; $f_{u,N}$ and $f_{u,T}$ is longitudinal tensile strength at room temperature and elevated temperatures, respectively; $\varepsilon_{u,N}$ is ultimate strain at room temperature; $\varepsilon_{f,N}$ is elongation (longitudinal tensile strain) at fracture at room temperature and n is exponent in the Ramberg-Osgood expression. It is shown that the reduction factor $f_{u,T}/f_{u,N}$ dropped rapidly in the temperature ranged from 500 to 950℃. Furthermore, the reduction factors $f_{0.2,T}/f_{0.2,N}$ and $f_{u,T}/f_{u,N}$ were compared with those calculated using the equations proposed by Chen and Young [14] and the factors obtained from EC3-1.2 [15]. It is shown that similar trend of deterioration at elevated temperatures were obtained. The deterioration of different types of stainless steels at elevated temperatures was also compared. It is shown that the austenitic stainless steel type T having a small amount of titanium has better performance in 0.2% proof stress than the lean duplex stainless steel type L and austenitic stainless steel type A in the temperature ranged from 200 - 950℃. The stainless steel type T also has a better performance in ultimate strength than the stainless steel types A and L when the temperature exceeded 500℃.

Table 1 Coupon test results at room temperature

Series	Type	E_N GPa	$f_{0.2,N}$ MPa	$f_{u,N}$ MPa	$\varepsilon_{u,N}$ %	$\varepsilon_{f,N}$ %	n
A	EN1.4301 (AISI 304)	199	474	759	45.2	52.6	5
T	EN1.4571 (AISI 316Ti)	199	463	677	38.8	46.8	7
L	EN1.4162 (AISI S32101)	200	724	862	19.7	36.8	7

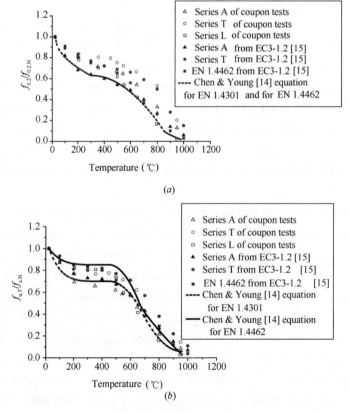

Fig. 1 Comparison of test results with results predicted by Chen and Young [14] and EC-3 1.2 [15]
(a) 0.2% proof stress; (b) Tensile strength

3. BOLTED CONNECTION TESTS

3.1 Bolted connection specimen design

The single shear two-parallel bolted connection specimens were designed by varying the type of stainless steel, namely, type A, T and L, respectively. The detailed dimensions of the specimens are illustrated in Fig. 2. The specimens were cut from stainless steel rectangular hollow sections with a specified length. The stainless steel tubes were supplied from STALA Tube Finland in uncut lengths of 3000 mm and nominal section size $20 \times 50 \times 1.5$ mm (with $\times$ depth $\times$ thickness). The A4 stainless steel bolts M8 [16] with grade 8.8 were used in this study. The corresponding size of stainless steel washers and nuts were used. The stainless steel washers were assembled in both sides of the bolt.

Lips were designed in each bolted connection specimens as illustrated in Fig. 2. Previous researchers found that standard flat specimens curled out of plane affecting the mode of failure[8] and may not accurately represent the true behavior of profiled structural members, for example channel sections. Therefore, lips of 10mm height were used to prevent the out-of-plane curling at the overlapped connection part. All bolts were hand-tightened to a torque of approximately 10Nm, which allowed for slip of the connection after applied a small loading. The specimens are separated into three series according to the stainless steel types, namely A, T and

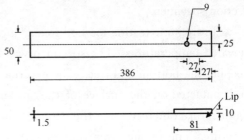

Fig. 2 Nominal dimension of two-parallel bolted connection specimen

L, respectively. Each specimen was labeled by four segments, for example "A-S-2Pa-8". The first letter "A" indicates the type of stainless steel of which the bolted connection specimen is assembled. The second letter "S" shorts for the single shear bolted connection. The third segment of the label "2Pa" means two bolts are used in the connection specimen with the arrangement parallel to the loading direction. The fourth part of the label "8" means the nominal diameter (8mm) of the bolt used in the connection.

3.2 Test set-up and procedure

The test set-up of stainless steel single shear bolted connections at elevated temperatures is shown in Fig. 3. The bolted connection tests were conducted by the same MTS Universal testing machine and furnace as the coupon tests. A total of 24 single shear two-parallel bolted connection specimens including the repeated test specimens were tested in this study under seven different elevated temperature levels. In general, it was found that the reduction factors of 0.2% proof stress dropped regularly at elevated temperatures,

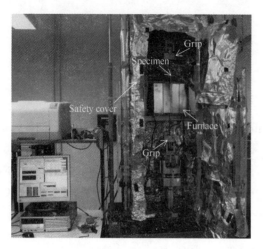

Fig. 3 Test set-up of single shear bolted connection at elevated temperatures

while the ultimate strength of the [...] types of stainless steels reduced rapidly when temperature goes beyond 500 1 as shown in Fig. 1. Hence, the nominal temperatures for the connection tests were chosen as 22 (room temperature), 200, 350, 500, 650, 800 and 9501. The test specimen was assembled on a pair of gripping apparatus, which was specially fabricated in order to provide the pin end boundary condition of the test. Two special gaskets were inserted in both grips such that the shear surface of the single shear bolted connection specimen was in-line to the loading direction. The details of the gripping apparatus are shown in Yan and Young [10]. Similar to the coupon tests, an external thermal couple was used to measure the actual temperature of the connection specimen.

Steady state testing method was adopted for the single shear bolted connection tests at elevated temperatures. The specimen was firstly set-up with clamping the top end, while keeping the bottom end free. The external thermal couple was inserted inside the furnace and contacted on the surface of the specimen in the middle of the overlapped part. The temperature obtained from the external thermal couple was recorded as the specimen temperature. The furnace was then closed and the temperature was raised to a pre-selected level. The thermal expansion of the specimen was allowed by the free bottom end of the specimen during the heating process. Once the pre-selected temperature was reached, the temperature was hold for a period of 8 to 15 minutes, such that allows the temperature to stabilize and the heat to transform uniformly in the specimen, and then the bottom end of the specimen was gripped. The bolted connection tests were conducted by displacement control with the loading rate of 1.5mm/min. A data acquisition system was used to record the furnace air temperature, the specimen temperature and the applied load at regular intervals during the test.

3.3 Test results

The test strengths ($P_{u,N}$ and $P_{u,T}$) of the single shear two-parallel bolted connection specimens at room and elevated temperatures are given in Table 2. The deterioration of the connection strengths at elevated temperatures was plotted in Fig. 4. The vertical axis of the graphs show the test strengths normalized with the test strength at room temperature ($P_{u,T}/P_{u,N}$) for each test series, while the horizontal axis plotted against the actual specimen temperatures. The $P_{u,N}$ is ultimate load of bolted connection test at room temperature, while the $P_{u,T}$ is ultimate load of bolted connection test at elevated temperatures. Repeated test was conducted in each series of connection specimens. It was found that the ultimate strengths of the connections $P_{u,T}$ dropped rapidly when the temperature exceeded 500℃. It was also found that the stainless steel type T (EN1.4571 or AISI 316Ti) generally performed better at elevated temperatures compared with the other two types A (EN1.4301 or AISI 304) and L (EN1.4162 or AISI S32101). Furthermore, the stainless steel type L has a better performance than type A for the temperature ranged from 22 to 500℃, but type A has a slightly better performance than type L for the temperature ranged from 650 to 950℃ in this study.

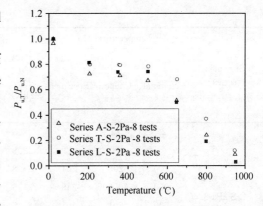

Fig. 4 Comparison of single shear bolted connection test results by steady state method

Table 2 Comparison of test results with nominal strengths for single shear bolted connections

| Specimen series | Temperature(℃) | | | $P_{u,N}$ or $P_{u,T}$ (kN) | $P_{u,T}/P_{u,N}$ | $P_{u,N}/P_{ASCE}$ or $P_{u,T}/P_{ASCE}$ | $P_{u,N}/P_{EC}$ or $P_{u,T}/P_{EC}$ | Failure mode | | |
	Nominal	Coupon	Specimen					ASCE and AS/NZS	EC	Test
A-S-2Pa-8	22	22	22	38.9	1.00	1.39	1.14	NS	NS	B
			22	37.5	0.96	1.33	1.09	NS	NS	B
	200	205	206	28.1	0.72	1.27	1.09	—	—	B
	350	351	359	27.6	0.71	1.36	1.15	—	—	B
	500	496	502	26.1	0.67	1.38	1.16	—	—	B
	650	648	648	20.0	0.51	1.36	1.16	—	—	B
	800	800	800	9.4	0.24	1.38	1.18	—	—	B
	950	950	948	3.5	0.09	1.29	1.10	—	—	B
					Mean	1.35	1.13			
					COV	0.033	0.031			
T-S-2Pa-8	22	22	22	36.7	1.00	1.33	1.13	NS	B	B
	200	206	208	29.3	0.80	1.34	1.15	—	—	B
	350	356	354	29.2	0.80	1.34	1.15	—	—	B
			359	29.1	0.79	1.34	1.14	—	—	B
	500	498	504	28.7	0.78	1.38	1.18	—	—	B
	650	645	651	25.0	0.68	1.42	1.21	—	—	B
	800	800	798	13.6	0.37	1.26	1.07	—	—	B+BS
	950	950	946	4.4	0.12	1.08	0.92	—	—	B+BS
					Mean	1.31	1.12			
					COV	0.079	0.080			

续表

Specimen series	Temperature(℃)			$P_{u,N}$ or $P_{u,T}$ (kN)	$P_{u,T}/P_{u,N}$	$P_{u,N}/P_{ASCE}$ or $P_{u,T}/P_{ASCE}$	$P_{u,N}/P_{EC}$ or $P_{u,T}/P_{EC}$	Failure mode		
	Nominal	Coupon	Specimen					ASCE and AS/NZS	EC	Test
L-S-2Pa-8			22	42.8	1.00	1.18	1.00	NS	NS	B+BS
			22	42.8	1.00	1.20	1.03	NS	NS	B+BS
	200	206	204	34.8	0.81	1.17	1.00	—	—	B+BS
	350	356	348	31.6	0.74	1.08	0.93	—	—	B+BS
	500	501	502	31.7	0.74	1.21	1.03	—	—	B+BS
	650	652	647	21.4	0.50	1.43	1.22	—	—	B
	800	795	798	8.2	0.19	1.42	1.21	—	—	B
	950	948	951	1.3	0.03	1.24	1.06	—	—	B
					Mean	1.24	1.06			
					COV	0.099	0.097			

Note: B=Bearing failure, NS=Net section tension failure, BS=Bolt shear failure

The observed failure modes of each two-parallel bolted connection specimen are listed in Table 2. The characteristics of different failure modes of steel bolted connections were detailed in Yan and Young [10]. The specimens were mainly failed by bearing failure based on the experimental observation. The tear out failure (end pull out failure) mode was not observed in any tested specimens. The failure mode of specimens in Series A-S-2Pa-8 is bearing failure. The bolt shear failure was deliberately avoided in the design of the specimens in Series T-S-2Pa-8 at room temperature, but this failure mode was observed in the specimens for the temperature ranged from 800 to 950℃. It should be noted that bolt shear failure was found in the specimensL-S-2Pa-8 for temperature ranged from 22 to 500℃, but only bearing failure occurred at elevated temperatures from 650 to 950℃.

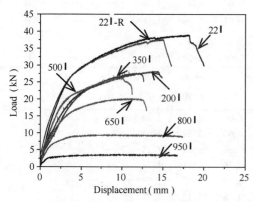

Fig. 5 Test curves of single shear bolted connections Series A-S-2Pa-8 at elevated temperatures

Fig. 5 exemplifies the test curves of two-parallel bolted connection specimens of Series A-S-2Pa-8 at different nominal temperatures, in which "R" represents the repeated test curve. The displacement of bolt slip during the initial loading stage was shifted in all the curves. The bearing failure mode of different specimens at different temperatures is shown in Fig. 6.

The nominal strengths (P_{ASCE} and P_{EC}) of the stainless steel single shear two-parallel bolted connections were calculated using the design equations in the current specifica-

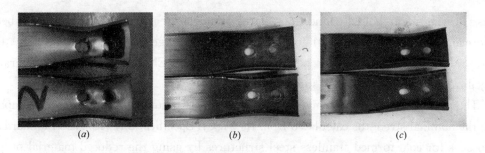

Fig. 6 Bearing failure mode of two-parallel bolted
connections at different temperatures
(a) A-S-2Pa-8 at room temperature; (b) T-S-2Pa-8 at 500°C; (c) L-S-2Pa-8 at 950°C

tions[2,4,17] with consideration of the deterioration of the material properties at elevated temperatures. In the design calculation, the reduced 0.2% proof stress and ultimate strength obtained from the coupon tests at elevated temperatures were used. The measured specimen dimensions were used to calculate the nominal strengths. The design rules for single shear bolted connections in the ASCE Specification[2] are identical to those in the AS/NZS Standard[3]. Therefore, the predicted values obtained from the two specifications are identical. The comparison of the test results with the predicted values calculated using the ASCE Specification[2], AS/NZS Standard[3] and Eurocodes[4,17] are shown in Table 2. It was found that the nominal strengths (P_{ASCE} and P_{EC}) of the two-parallel bolted connections calculated using the ASCE Specification[2] and Eurocodes[4,17] are generally conservative at elevated temperatures. The predictions P_{ASCE} were more conservative than the predictions P_{EC} as found in Table 2. The current design formulas in these three standards by substituting the reduced material properties at elevated temperatures generally underestimate the ultimate strength of the stainless steel single shear two-parallel bolted connections.

4. CONCLUSIONS

An experimental investigation of the stainless steel single shear two-parallel bolted connections at elevated temperatures has been presented. Three types of cold-formed stainless steels with nominal thickness 1.50mm were investigated. The three types of stainless steel are austenitic stainless steel EN1.4301 (AISI 304) and EN1.4571 (AISI 316Ti having small amount of titanium) as well as lean duplex stainless steel EN1.4162 (AISI S32101). A total of 25 coupon tests and 24 two-parallel bolted connection tests were conducted by steady state test method.

The reduction factor of different types of stainless steel single shear two-parallel bolted connections at elevated temperatures was compared. It is shown that the austenitic stainless steel type EN 1.4571 generally has better performance than the lean duplex stainless steel type EN1.4162 and austenitic stainless steel type EN1.4301 at elevated temperatures. Furthermore, bolted connections of the stainless steel type EN1.4162 has a better performance than type EN1.4301 for the temperature ranged from 22 to 500℃, but type EN1.4301 has a slightly better performance than type EN1.4162 for the temperature ranged from 650 to 950℃ in this study.

The test strengths of the single shear two-parallel bolted connections were compared with the nominal strengths calculated from the ASCE Specification, AS/NZS Standard and Eurocodes for cold-formed stainless steel structures by using the reduced material properties due to high temperatures. It was found that the nominal strengths predicted by the ASCE Specification and Eurocodes for the two-parallel bolted connections are generally conservative at elevated temperatures. The ASCE predictions are more conservative than the Eurocode predictions for the stainless steel single shear two-parallel bolted connections at elevated temperatures.

Acknowledgements

The authors are grateful to STALA Tube Finland for supplying the test specimens. The authors are also thankful to Mr. Ho-hin CHAU for his assistance in the experimental program as part of his final year undergraduate research project at The University of Hong Kong. The research work described in this paper was supported by a grant from the Research Grants Council of the Hong Kong Special Administrative Region, China (Project No. HKU718612E).

References

[1] Gardner, L. and Baddoo, N. R. Fire testing and design of stainless steel structures. Journal of Constructional Steel Research; 2006, 62(6): 532-543

[2] American Society of Civil Engineers (ASCE). Specification for the design of cold-formed stainless steel structural members. ASCE Standard, 2002, SEI/ASCE-8-02, Reston, Virginia

[3] Australian/New Zealand Standard (AS/NZS). Cold-formed stainless steel structures. AS/NZS 4673: 2001, Standards Australia, 2001, Sydney, Australia

[4] EC3-1.4. Eurocode 3. Design of steel structures-Part 1.4: General rules-Supplementary rules for stainless steels, European Committee for Standardization, 2006, BS EN 1993-1-4: 2006, CEN, Brussels

[5] Zadanfarrokh F. Analysis and design of bolted connections in cold formed steel members. PhD Thesis, 1991, The University of Salford

[6] Rogers, C. A. and Hancock, G. J. Bolted connection tests of thin G550 and G300 sheet steels. Journal of Structural Engineering ASCE, 1998, 124 (7): 798-808

[7] Rogers, C. A. and Hancock, G. J. Bolted connection design for sheet steels less than 1.0mm thick. Journal of Constructional Steel Research, 1999, 51(2): 123-146

[8] Rogers, C. A. and Hancock, G. J. Failure modes of bolted-sheet-steel connections loaded in shear. Journal of Structural Engineering ASCE, 2000, 126(3): 288-296

[9] Bouchair, A., Averseng, J., and Abidelah, A. Analysis of the behaviour of stainless steel bolted

connections. Journal of Constructional Steel Research, 2008, 64(11): 1264-1274

[10] Yan, S. and Young, B. Tests of single shear bolted connections of thin sheet steels at elevated temperatures - Part I: Steady state tests. Thin-Walled Structures 2011, 49(10): 1320-1333

[11] Yan, S. and Young, B. Tests of single shear bolted connections of thin sheet steels at elevated temperatures - Part II: Transient state tests. Thin-Walled Structures, 2011, 49(10): 1334-1340

[12] Gardner, L., and Ng, K. T. Temperature development in structural stainless steel sections exposed to fire. Fire Safety Journal, 2006, Vol. 41(3), 185-203

[13] AS-2291: 1979. Methods for the tensile testing of metals at elevated temperatures. Sydney, Australia, 1979, Standards Australia

[14] Chen, J., and Young, B. Stress-strain curves for stainless steel at elevated temperatures. Engineering Structures, 2006, Vol. 28(2), 229-239

[15] EC3-1. 2. Eurocode 3: Design of steel structures-Part 1. 2: General rules-Structural fire design, European Committee for Standardization, 2005, BS EN 1993-1-2: 2005, CEN, Brussels

[16] BS EN ISO 3506-1. Mechanical properties of corrosion-resistant stainless steel fasteners-Part 1: Bolts, screws and studs; BS EN ISO 3506-1: 1998

[17] EC3-1. 8. Eurocode 3: Design of steel structures-Part 1. 8: Design of joints. European Committee for Standardization, 2005, BS EN 1993-1-8: 2005, CEN, Brussels

应用反应谱分析法之阻尼器最佳化配置

黄婉婷，吕良正

(台湾大学土木工程学系，台北 10617)

摘　要：结构在进行阻尼器最佳化配置时常产生非古典阻尼之动力系统，一般都以直接积分法来求得此种结构系统的动力反应，进而进行最佳化，由于直接积分法费时，因此影响阻尼器最佳化之计算效率。为了改善此一问题，本文以两种实数域解耦法将非古典阻尼结构动力系统解耦，求得其模态方程式及各模态对应的模态频率及模态阻尼比，再透过非古典阻尼反应谱法，求得结构系统的最大反应值。最后将此法应用于阻尼器最佳化配置中，虽然会影响一些分析精度，但需要的分析时间可大为减少。

关键词：非古典阻尼系统；解耦；反应谱法；阻尼器最佳化配置

OPTIMAL PLACEMENT OF DAMPERS IN BUILDING STRUCTURES USING RESPONSE SPECTRUM ANALYSIS

W. T. Huang, L. J. Leu

(Department of Civil Engineering, National Taiwan University, Taipei 10617, Taiwan)

Abstract: when carrying out optimal allocation of supplemental dampers, the system will become non-classical damped, which is usually analyzed by a direct integration method. As the direct integration method is time consuming, such an optimal design process becomes computationally inefficient. In order to overcome the problem, two methods are used in this paper to decouple the non-classically damped system in real field. Modal equations, modal damping ratios and modal frequencies can be obtained in such a decoupling process. Response spectrum analysis is then performed to obtain the maximum response of the system. Finally application of the above procedure to optimal placement of dampers is presented. As expected, the proposed strategy reduces the computational time significantly while the accuracy is not much affected.

Keywords: non-classically damped system; decouple; response spectrum method; optimal placement of dampers

1. 引言

近二十年来随着结构分析软件的快速发展与进步，结构控制（structural control）理论也随之兴起，研究者利用各类主、被动式消能装置（active or passive energy dissipation device）装设于结构上，用以减轻结构物受外力时的反应。目前结构设计采用外加阻尼器的方式来消能在多数地震频繁之国家已经相当普遍如美国、日本及我国。在各类学术期刊中，各种不同形式阻尼器的设计理论与应用蓬勃发展。虽然如此，在各相关规范中，仅仅纳入关于阻尼器之设计与分析，但是有关于阻尼器的配置位置与方式的准则，目前仍未有规范提出一既定之准则。因此，有关于阻尼器配置的最佳化方法也屡屡见于学术期刊中，但是大多数文章在求解阻尼器最佳配置时，经常所对应的结构动力系统为非古典阻尼结构动力系统，此时在求取结构最大反应值时仅能采用直接积分法，当分析的次数增多所需的分析时间也会增多。为了改善此一问题，本研究应用最近一些学者所提之方法将非古典结构动力系统解耦，并进而使用反应谱分析法来求得结构最大反应，如此便可以节省大量的分析时间。

2. 复数解耦法

2.1 二次特征值问题

在此将依含线性黏性阻尼器之多自由度系统的运动方程式表示如下：

$$[M]\{\ddot{u}(t)\} + [C]\{\dot{u}(t)\} + [K]\{u(t)\} = \{f(t)\} \tag{1}$$

其中，$[M]$、$[C]$ 及 $[K]$ 分别表示 $n \times n$ 质量矩阵、含线性黏性阻尼的阻尼矩阵及结构劲度矩阵，$\{f(t)\}$ 表示 n 维随时间变化的外力向量，$\{u(t)\}$ 为 n 维相对于地表的位移向量；令 (1) 式中 $\{f(t)\}$ 为零，可得到系统之自由振动（free vibration）的运动方程式：

$$[M]\{\ddot{u}(t)\} + [C]\{\dot{u}(t)\} + [K]\{u(t)\} = \{0\} \tag{2}$$

令 (2) 式的解为 $\{u(t)\} = \{v\}e^{\lambda t}$ 则

$$(\lambda^2[M] + \lambda[C] + [K])\{v\} = \{0\} \tag{3}$$

为包含线性黏性阻尼器结构动力系统之二次特征问题（Quadratic eigenvalue problem, QEP）。由 (3) 式解得之特征值，在模态阻尼为次阻尼（under-damped）情况下，会有两两成对之特性，因此可以得到 n 对共轭复数特征值及其对应的共轭特征向量：

$$\begin{cases} \lambda_j = \alpha_j + i\omega_{dj} = -\xi_j\omega_{nj} + i\omega_{nj}\sqrt{1-\xi_j^2} \\ \bar{\lambda}_j = \alpha_j - i\omega_{dj} = -\xi_j\omega_{nj} - i\omega_{nj}\sqrt{1-\xi_j^2} \end{cases} \tag{4}$$

$$\{v\}_j, \{\bar{v}\}_j = \{r_{j1}e^{\mp i\varphi_{j1}} \quad r_{j2}e^{\mp i\varphi_{j2}} \quad \cdots \quad r_{jn}e^{\mp i\varphi_{jn}}\}^T \tag{5}$$

其中 $j = 1, 2, \cdots, n$，而各特征值及特征向量组成特征值矩阵与特征向量矩阵，分列如下：

$$\{\Lambda\} = diag(\lambda_1, \lambda_2, \cdots, \lambda_n), \{\bar{\Lambda}\} = diag(\bar{\lambda}_1, \bar{\lambda}_2, \cdots, \bar{\lambda}_n) \tag{6}$$

$$[V] = [\{v\}_1, \{v\}_2, \cdots, \{v\}_n], [\bar{V}] = [\{\bar{v}\}_1, \{\bar{v}\}_2, \cdots, \{\bar{v}\}_n] \tag{7}$$

由 (4) 式成对之特征值，可以求得模态频率（modal frequency）及模态阻尼（modal damping ratio），分别定义为

$$\omega_{nj} = \sqrt{\lambda_j \bar{\lambda}_j} \tag{8}$$

$$\xi_j = -\frac{(\lambda_i + \bar{\lambda}_j)}{2\omega_{nj}} = -\frac{(\lambda_i + \bar{\lambda}_j)}{2\sqrt{\lambda_j \bar{\lambda}_j}} \tag{9}$$

但当模态阻尼渐渐增大，直至其值大于 1（$\xi > 1$）时，是为过阻尼（over-damped）的情况，特征值会由复数形态转变为实数形态，此时 (4) 式要改写为

$$\lambda_j = -\xi_j\omega_{nj} + \omega_{nj}\sqrt{\xi_j^2 - 1}, \bar{\lambda}_j = -\xi_j\omega_{nj} - \omega_{nj}\sqrt{\xi_j^2 - 1} \tag{10}$$

而对应之特征向量也是实数的形态；而特征值矩阵及特征向量矩阵同样也可表达成如 (6)

及（7）式之形式，此时系统之模态阻尼及模态频率依然可由（8）及（9）式求得。

2.2 状态空间特征值问题

若考虑结构为非古典阻尼结构动力系统的情况，则无法用传统的模态分析法将系统的阻尼矩阵对角化，而不能将（1）式的耦合解除，因此将系统转换到状态空间中，在此进行解耦的动作。式在状态空间中可写成

$$[A]\{\dot{y}(t)\} + [B]\{y(t)\} = \{f_s(t)\} \tag{11}$$

其中，$[A] = \begin{bmatrix} [C] & [M] \\ [M] & [0] \end{bmatrix}$、$[B] = \begin{bmatrix} [K] & [0] \\ \{0\} & -[M] \end{bmatrix}$ 及 $\{f_s(t)\} = \begin{Bmatrix} \{f(t)\} \\ \{0\} \end{Bmatrix}$ 为状态向量，一样可令（11）式中的外力为零，可求得对应的状态空间特征值问题如下

$$(\lambda[A] + [B])\{\psi\} = \{0\} \tag{12}$$

由（12）式可求得特征值及特征向量为

$$\lambda_j = -\xi_j \omega_{nj} + i\omega_{nj}\sqrt{1-\xi_j^2}, \bar{\lambda}_j = -\xi_j \omega_{nj} - i\omega_{nj}\sqrt{1-\xi_j^2} \tag{13}$$

$$\{\psi\}_j = \begin{Bmatrix} \{v\}_j \\ \lambda_j \{v\}_j \end{Bmatrix}, \{\bar{\psi}\}_j = \begin{Bmatrix} \{\bar{v}\}_j \\ \bar{\lambda}_j \{\bar{v}\}_j \end{Bmatrix} \tag{14}$$

将其特征向量以矩阵之型式表达成一复数特征向量矩阵

$$[\Psi] = [\{\psi\}_1, \cdots, \{\psi\}_n, \{\bar{\psi}\}_1, \cdots, \{\bar{\psi}\}_n] = \begin{bmatrix} [V] & [\bar{V}] \\ [V][\Lambda] & [\bar{V}][\bar{\Lambda}] \end{bmatrix} \tag{15}$$

由状态空间解得之复数模态特征向量分别对 $[A]$ 及 $[B]$ 矩阵有正交性存在；复数特征向量分别可以对 $[A]$ 及 $[B]$ 矩阵展开如下

$$\{\psi\}_j^T [A] \{\psi\}_j = \{v\}_j^T (2\lambda_j [M] + [C]) \{v\}_j = a_j \tag{16}$$

$$\{\bar{\psi}\}_j^T [A] \{\bar{\psi}\}_j = \{\bar{v}\}_j^T (2\bar{\lambda}_j [M] + [C]) \{\bar{v}\}_j = \bar{a}_j$$

$$\{\psi\}_j^T [B] \{\psi\}_j = \{v\}_j^T (-\lambda_j^2 [M] + [K]) \{v\}_j = b_j = -\lambda_j a_j \tag{17}$$

$$\{\bar{\psi}\}_j^T [B] \{\bar{\psi}\}_j = \{\bar{v}\}_j^T (-\bar{\lambda}_j^2 [M] + [K]) \{\bar{v}\}_j = \bar{b}_j = -\bar{\lambda}_j \bar{a}_j$$

因此广义的正交性质可表示如下，并且定义状态空间中对角化的复数模态矩阵 $[a]$ 及 $[b]$ 矩阵

$$[a] = [\Psi]^T [A] [\Psi] = \mathrm{diag}[a_1, a_2, \cdots, a_n, \bar{a}_1, \bar{a}_2, \cdots, \bar{a}_n] \tag{18}$$

$$[b] = [\Psi]^T [B] [\Psi] = \mathrm{diag}[b_1, b_2, \cdots, b_n, \bar{b}_1, \bar{b}_2, \cdots, \bar{b}_n]$$

而特征向量依照下式来进行正规化，

$$\{v\}_j^T (2\lambda_j [M] + [C]) \{v\}_j = \lambda_j - \lambda_{n+j} \tag{19}$$

$$\{\bar{v}\}_j^T (2\bar{\lambda}_j [M] + [C]) \{\bar{v}\}_j = \lambda_{n+j} - \lambda_j$$

2.3 复数解耦法

首先我们定义复数模态的转换关系

$$\{y(t)\} = \begin{Bmatrix} \{u(t)\} \\ \{\dot{u}(t)\} \end{Bmatrix} = [\Psi]\{x(t)\} = \begin{bmatrix} [V] & [\bar{V}] \\ [V][\Lambda] & [\bar{V}][\bar{\Lambda}] \end{bmatrix} \{x(t)\} \tag{20}$$

上式中的 $[V]$ 及 $[\Lambda]$ 矩阵皆可由（6）式及（7）式得到，而 $\{x(t)\}$ 为 $2n \times 1$ 之复数函数向量

$$\{x(t)\} = \{x_1(t), x_2(t), \cdots, x_n(t), \bar{x}_1(t), \bar{x}_2(t), \cdots, \bar{x}_n(t)\}^T \quad (21)$$

将（20）式代入（11）式中，并且左右前乘 $[\Psi]^T$ 得

$$[\Psi]^T [A] [\Psi] \{\dot{x}(t)\} + [\Psi]^T [B] [\Psi] \{x(t)\} = [\Psi]^T \{f_s(t)\} \Rightarrow$$

$$\begin{bmatrix} [\Lambda] - [\bar{\Lambda}] & [0] \\ [0] & [\bar{\Lambda}] - [\Lambda] \end{bmatrix} \{\dot{x}(t)\} + \begin{bmatrix} ([\bar{\Lambda}] - [\Lambda])[\Lambda] & [0] \\ [0] & ([\Lambda] - [\bar{\Lambda}])[\bar{\Lambda}] \end{bmatrix} \{x(t)\}$$

$$= \begin{Bmatrix} [V]^T \{f(t)\} \\ [\bar{V}]^T \{f(t)\} \end{Bmatrix}$$

$$(22)$$

（22）式为状态空间解耦后之复数模态方程式。虽然在状态空间中可以将含线性黏性阻尼器的结构系统解耦，但是此过程为复数运算，在执行过程中比起实数运算需要消耗较多的分析时间，因此在实数域中的解耦法对于后续在阻尼器最佳化配置中是非常必要的。

3. 实数解耦法

在本研究中采用两种实数解耦法，第一种为由 Ma et al.（2010）所提出利用相位同步法的原理来解耦非古典阻尼结构动力系统。第二种为由 Song et al.（2008）提出运用拉氏转换法的运算，此法先将结构系统在频率域中解耦，并透过反拉氏转换得到非古典阻尼结构动力系统在时间域解耦的摩态方程式，在此法中也推导出非古典阻尼反应谱分析法。此两种方法之运算皆建立在复数模态分析之上，尔后再透过不同的方法将非古典阻尼结构动力系统解耦，因为皆为实数解耦法，之后在本文称由 Ma et al.（2010）提出的实数解耦法为相位同步法，而由 Song et al.（2008）提出应用拉氏转换的实数解耦法为拉氏转换法，以下分别简述此两种实数解耦方法。

3.1 相位同步法

在此先简述相位同步法的原理，假设由二次特征方程式解得 $2n$ 个相异的复数特征值及其对应的特征向量，如（4）式及（5）式。特征解 $\{v\}_j e^{\lambda_j t}$ 由特征值及特征向量组合而成，与互为共轭的特征解 $\{\bar{v}\}_j e^{\bar{\lambda}_j t}$ 依线性组合产生一振动模态（damped mode）

$$\{s(t)\}_j = d_j \{v\}_j e^{\lambda_j t} + \bar{d}_j \{\bar{v}\}_j e^{\bar{\lambda}_j t} \quad (23)$$

上式的 d_j 为任意常数，以极坐标表示为 $2d_j = h_j e^{-i\theta_j}$，$h_j$ 及 θ_j 可由初始条件求得，代入（23）式中

$$\{s(t)\}_j = 2\text{Re}\left[d_j \{v\}_j e^{(\alpha_j + i\omega_{dj})t}\right] = h_j e^{\alpha_j t} \begin{Bmatrix} r_{j1} \cos(\omega_{dj} t - \theta_j - \varphi_{j1}) \\ r_{j2} \cos(\omega_{dj} t - \theta_j - \varphi_{j2}) \\ \vdots \\ r_{jn} \cos(\omega_{dj} t - \theta_j - \varphi_{jn}) \end{Bmatrix} \quad (24)$$

系统之反应可表为

$$\{u(t)\} = \sum_{j=1}^n (d_j \{v\}_j e^{\lambda_j t} + \bar{d}_j \{\bar{v}\}_j e^{\bar{\lambda}_j t}) = \sum_{j=1}^n \{s_j(t)\}$$

$$= \sum_{j=1}^{n} h_j e^{\alpha_j t} \begin{Bmatrix} r_{j1}\cos(\omega_{dj}t - \theta_j - \varphi_{j1}) \\ r_{j2}\cos(\omega_{dj}t - \theta_j - \varphi_{j2}) \\ \vdots \\ r_{jn}\cos(\omega_{dj}t - \theta_j - \varphi_{jn}) \end{Bmatrix} \quad (25)$$

(25) 式中的 φ_{jk} 即表示在第 j 个振动模态第 k 个分量所造成之相位差（phase shift），在非古典阻尼结构动力系统时其值不为零，此时每个振态的分量通过平衡位置的时间点会不一样，而形成相位不同步的情况，导致系统耦合在一起无法解耦，因此在每个振态中导入一适当的相位差，使得振态 $\{s_j(t)\}$ 中 $\varphi_{j1} = \varphi_{j2} = \cdots = \varphi_{jn} = 0$，此时振态中的每个分量通过平衡位置的时间点就会一样，达到相位同步的状态。这时系统便被转换到具有古典阻尼性质的结构动力系统之下，即可进行模态分析来解耦结构动力系统。

在应用此概念前，先讨论当由（3）式解出来之特征值为混合特征值的情况，假设有 $2c$ 个复数特征值及 $2r$ 个实数特征值，分别对应的为次阻尼模态及过阻尼模态。依照（26）式对特征值进行排列并且配对

$$\{2c\text{complex}\lambda\} = \{\lambda_1, \cdots, \lambda_c, \lambda_{n+1} = \bar{\lambda}_1, \cdots, \lambda_{n+c} = \bar{\lambda}_c\}$$
$$\{2r\text{real}\lambda\} = \{\lambda_{c+1} < \cdots < \lambda_n < \lambda_{n+c+1} = \tilde{\lambda}_{c+1} < \cdots < \lambda_{2n} = \tilde{\lambda}_n\} \quad (26)$$

此配对及排序方式，组合成的特征值矩阵及特征向量矩阵可表示为

$$[\Lambda] = \text{diag}[\lambda_1, \lambda_2, \cdots, \lambda_n], [\Lambda^*] = \text{diag}[\lambda_{n+1}, \lambda_{n+2}, \cdots, \lambda_{2n}]$$
$$[V] = [\{v\}_1 | \{v\}_2 | \cdots | \{v\}_n], [V^*] = [\{v\}_{n+1} | \{v\}_{n+2} | \cdots | \{v\}_{2n}] \quad (27)$$

以（27）为基本模态参数，可以应用前述相位同步法的概念，将非古典阻尼结构动力系统解耦，求得模态方程式为

$$\{\ddot{q}(t)\} + [D_z]\{\dot{q}(t)\} + [\Omega_z]\{q(t)\} = \{g(t)\} = [\alpha]^T\{f(t)\} + [\beta]^T\{\dot{f}(t)\} \quad (28)$$

其中，

$$[\alpha] = ([V][\Lambda^*] - [V^*][\Lambda])([\Lambda^*] - [\Lambda])^{-1}, \quad (29)$$
$$[\beta] = ([V^*] - [V])([\Lambda^*] - [\Lambda])^{-1}$$
$$[D_z] = -\text{diag}[\lambda_j + \lambda_{n+j}], [\Omega_z] = \text{diag}[\lambda_j \lambda_{n+j}] \quad (30)$$

其初始条件为

$$\begin{Bmatrix} \{q(0)\} \\ \{\dot{q}(0)\} \end{Bmatrix} = \begin{bmatrix} [I] & [I] \\ [\Lambda] & [\Lambda^*] \end{bmatrix} \begin{bmatrix} [V] & [V^*] \\ [V][\Lambda] & [V^*][\Lambda^*] \end{bmatrix}^{-1} \begin{Bmatrix} \{u(0)\} \\ \{\dot{u}(0)\} \end{Bmatrix} + \begin{Bmatrix} [0] \\ [\beta]^T\{f(0)\} \end{Bmatrix} \quad (31)$$

当从解耦系统（28）求得模态位移 $\{q(t)\}$ 时，代入下式可求得系统位移 $\{u(t)\}$

$$\{u(t)\} = [\alpha]\{q(t)\} + [\beta]\{\dot{q}(t)\} - [\beta][\beta]^T\{f(t)\} \quad (32)$$

当结构系统可透过传统模态分析法来解耦，则此即为古典阻尼结构动力系统，为了维持相位同步法之广义性，假设所有特征向量皆依照（19）式来做正规化，并假设所有由二次特征值问题解得之特征值皆为相异复数根；则 $[V] = [V] = [\Phi]$，$[\Phi]$ 为由无阻尼或是古典阻尼矩阵所对应之特征值问题（$[K] - \omega^2[M]$）$[\Phi] = [0]$ 求得之模态阻尼矩阵。（29）式中 $[\alpha] = [\Phi]$ 及 $[\beta] = [0]$，模态方程式则为

$$\{\ddot{q}(t)\}+[D_z]\{\dot{q}(t)\}+[\Omega_z]\{q(t)\}=\{g(t)\}=[\alpha]^T\{f(t)\} \quad (33)$$

模态叠加公式则会简化成

$$\{u(t)\}=[\Phi]\{q(t)\} \quad (34)$$

3.2 拉氏转换法

此法为以复数模态分析法为基础来解耦非古典阻尼结构动力系统。如同相位同步法，在拉氏转换法中同样考虑由二次特征值解得之所有特征值为混合特征值的情况，但是不同于3.1节所采用之配对方法，在此假设有 $2n_c$ 个复数特征值及 $n_p\,[=2(n-n_c)]$ 个实数特征值（过阻尼），在此仅对复数特征值依照其自然成对的特性来进行配对，而不将实数特征值进行配对的动作，因此每一个实数特征值皆对应到一个各自独立的过阻尼模态，令实数特征值为下列之形式

$$\lambda_{pi}=-\omega_{pi}\quad(i=1,2,\cdots,n_p) \quad (35)$$

定义（35）式中 ω_{pi} 为第 i 个过阻尼模态频率（ith over-damped modal circular natural frequency），其中 $\omega_{pi}>0$，单位为 rad/sec。而每个过阻尼模态皆会对应到其实数特征向量 $\{\psi\}_{pi}$

$$\{\psi\}_{pi}=\begin{Bmatrix}\{v\}_{pi}\\ \lambda_{pi}\{v\}_{pi}\end{Bmatrix}\quad(i=1,2,\cdots,n_p) \quad (36)$$

所有特征值及特征向量依照下式组成特征值及特征向量矩阵

$$\begin{aligned}[\Lambda_g]&=\mathrm{diag}(\lambda_1,\lambda_2,\cdots,\lambda_{n_c},\bar{\lambda}_1,\bar{\lambda}_2,\cdots,\bar{\lambda}_{n_c},\lambda_{p1},\lambda_{p2},\cdots,\lambda_{pn_p})\\ [V_g]&=[\{v\}_1,\{v\}_2,\cdots,\{v\}_{n_c},\{\bar{v}\}_1,\{\bar{v}\}_2,\cdots,\{\bar{v}\}_{n_c},\{v\}_{p1},\{v\}_{p2},\cdots,\{v\}_{pn_p}]\end{aligned} \quad (37)$$

由（37）式组成模态矩阵 $[\Psi]$

$$[\Psi]=\begin{bmatrix}[V_g]\\ [V_g][\Lambda_g]\end{bmatrix} \quad (38)$$

以上述之模态参数为基础，首先在初始条件为零的情况下，对（20）式及（22）式做拉氏转换，将其表示为

$$\{Y(s)\}=\begin{Bmatrix}\{U(s)\}\\ \{\dot{U}(s)\}\end{Bmatrix}=[\Psi]\{X(s)\} \quad (39)$$

及

$$\begin{cases}(a_js+b_j)X_j(s)=\{v\}_j^T\{F(s)\}(j=1,2,\cdots,n_c)\\ (\bar{a}_js+\bar{b}_j)\bar{X}_j(s)=\{\bar{v}\}_j^T\{F(s)\}(j=1,2,\cdots,n_c)\\ (a_{pi}s+b_{pi})X_{pi}(s)=\{v\}_{pi}^T\{F(s)\}(i=1,2,\cdots,n_p)\end{cases} \quad (40)$$

其中 s 为拉式参数（Laplace parameter）以及

$$\{X(s)\}=\{X_1(s),X_2(s),\cdots,X_{n_c}(s),\bar{X}_1(s),\bar{X}_2(s),\cdots,\bar{X}_{n_c}(s),X_{p1}(s),X_{p2}(s),\cdots,X_{pn_p}(s)\}^T \quad (41)$$

$\{X(s)\}$ 为在频率（s-domain）之模态向量。令外力 $\{F(s)\}=-[M]\{l\}\ddot{U}_g(s)$，$\{l\}$ 为影响向量（influence vector），而 $U\dot{U}_g(s)$ 为地表加速度 $\ddot{u}_g(t)$ 之拉氏转换，将外力向量代入（40）式中，再将（40）式代入（39）式，可得到在频率域之模态位移向量

$$\{U(s)\} = \sum_{j=1}^{n_c} \{U(s)\}_j + \sum_{i=1}^{n_p} \{U(s)\}_{pi} \qquad (42)$$

在（42）式中位移向量 $\{U(s)\}_j$ 及 $\{U(s)\}_{pi}$ 分别可表示成

$$\{U(s)\}_j = -\left(\frac{\{A_D\}_j s}{s^2+2\xi_j\omega_{nj}s+\omega_{nj}^2} + \frac{\{B_D\}_j}{s^2+2\xi_j\omega_{nj}s+\omega_{nj}^2}\right)\dot{U}_g(s), \{U(s)\}_{pi} = -\frac{\{A_{pD}\}_i}{s+\omega_{pi}}\dot{U}_g(s) \qquad (43)$$

而

$$\{A_D\}_j = 2\mathrm{Re}([R]_j)[M]\{l\}, \{A_{pD}\}_i = \mathrm{Re}([R]_{pi})[M]\{l\}$$
$$\{B_D\}_j = 2\omega_{nj}(\xi_j\mathrm{Re}([R]_j) - \sqrt{1-\xi_j^2}\mathrm{Im}([R]_j))[M]\{l\} \qquad (44)$$
$$[R]_j = \mathrm{Re}([R]_j) + i\mathrm{Im}([R]_j) = \frac{\{v\}_j\{v\}_j^T}{a_j}, [R]_{pi} = \frac{\{v\}_{pi}\{v\}_{pi}^T}{a_{pi}}$$

再对（43）式做反拉氏转换可得模态方程式

$$\ddot{q}_j(t) + 2\xi_j\omega_{nj}\dot{q}_j(t) + \omega_{nj}^2 q_j(t) = -\ddot{u}_g(t) \qquad (45)$$
$$\dot{q}_{pi}(t) + \omega_{pi}q_{pi}(t) = -\ddot{u}_g(t) \qquad (46)$$

因为在拉氏转换法中，并不将实数特征值配对，因而每一个实数特征值所对应的模态方程式为一阶常微分方程式，如（46）式所示。模态方程式之初始条件可由（47）式求得

$$\begin{Bmatrix}\{q(0)\}\\\{q(0)\}\\\{q(0)\}_p\end{Bmatrix} = \begin{bmatrix}[A_D] & [B_D] & [A_{pD}]\\[A_V] & [B_V] & [A_{pV}]\end{bmatrix}^{-1}\begin{Bmatrix}\{u(0)\}\\\{\dot{u}(0)\}\end{Bmatrix} \qquad (47)$$

最后，将求得之模态位移代入下列模态叠加式中

$$\{u(t)\} = \sum_{j=1}^{n_c}[\{A_D\}_j\dot{q}_j(t) + \{B_D\}_j q_j(t)] + \sum_{i=1}^{n_p}\{A_{pD}\}_i q_{pi}(t)(t) \qquad (48)$$

即可求得系统之位移。当结构系统为古典结构动力系统，则残差矩阵 $\mathrm{Re}([R]_j)=0$ 以及 $\mathrm{Im}([R]_j) = -\frac{\{v\}_j\{v\}_j^T}{2m_j\omega_{dj}}$，$m_j=\{v\}_j^T[M]\{v\}_j$ 为系统之模态质量。（44）式中 $\{A_D\}_j = \{0\}$ 和 $\{B_D\}_j = \frac{\{v\}_j\{v\}_j^T[M]\{l\}}{m_j} = \Gamma_j\{v\}_j$，在古典阻尼结构动力系统当中，依然会有过阻尼模态的情况出现，此时关于过阻尼模态的参数形式依然相同。因而位移向量可表示成

$$\{u(t)\} = \sum_{j=1}^{n_c}\Gamma_j\{v\}_j q_j(t) + \sum_{i=1}^{n_p}\Gamma_{pi}\{v\}_{pi}q_{pi}(t) \qquad (49)$$

4. 非古典阻尼反应谱法

在此将简单描述由前述两种非古典阻尼结构动力系统之实数解耦法应用于反应谱分析法中之公式及反应谱之绘制方法。

4.1 相位同步法之非古典阻尼反应谱法

在黄婉婷（2012）当中，探讨相位同步法式中 $[\beta]$ 矩阵对于系统反应的影响，并且从

后续的解析证明可得 $[\beta][\beta]^{\mathrm{T}}$ 该项为零，故（32）式在消去 $[\beta][\beta]^{\mathrm{T}}\{f(t)\}$ 的项次后为

$$\{u(t)\} = [\alpha]\{q(t)\} + [\beta]\{\dot{q}(t)\} \tag{50}$$

利用由 Zhou et al.（2004）提出之复数模态非古典阻尼结构动力系统之反应谱法，可推导出对应于（50）式之 CCQC（Complex Complete Quadratic Combination）及 CSRSS（Complex Square Root of the Sum of Squares）此两种模态组合法则。首先假设结构系统为 n 个自由度，则结构最大反应值 $|\{u(t)\}|_{\max}$ 可表示成

$$|\{u(t)\}|_{\max} = \Big[\sum_{i=1}^{n}\sum_{j=1}^{n}\rho_{ij}^{\mathrm{DD}}(\{\alpha\}_i\{\alpha\}_j + \mu_{ij}\{\beta\}_i\{\beta\}_j\omega_{\mathrm{n}i}\omega_{\mathrm{n}j} \\ + 2v_{ij}\{\beta\}_i\{\alpha\}_j\omega_i)|q_i(t)|_{\max}|q_j(t)|_{\max}\Big]^{1/2} \tag{51}$$

其中

$$\rho_{ij}^{\mathrm{DD}} = \frac{8\sqrt{\xi_i\xi_j}(\beta_{ij}\xi_i + \xi_j)\beta_{ij}^{3/2}}{(1-\beta_{ij}^2)^2 + 4\xi_i\xi_j\beta_{ij}(1+\beta_{ij}^2) + 4(\xi_i^2+\xi_j^2)\beta_{ij}^2} \quad (\beta_{ij} = \omega_{\mathrm{n}i}/\omega_{\mathrm{n}j},\, i,j=1,2,\cdots,n) \tag{52}$$

$$\rho_{ij}^{\mathrm{VV}} = \frac{8\sqrt{\xi_i\xi_j}(\xi_i + \xi_j\beta_{ij})\beta_{ij}^{3/2}}{(1-\beta_{ij}^2)^2 + 4\xi_i\xi_j\beta_{ij}(1+\beta_{ij}^2) + 4(\xi_i^2+\xi_j^2)\beta_{ij}^2} \quad (\beta_{ij} = \omega_{\mathrm{n}i}/\omega_{\mathrm{n}j},\, i,j=1,2,\cdots,n) \tag{53}$$

$$\rho_{ij}^{\mathrm{VD}} = \frac{4\sqrt{\xi_i\xi_j}(1-\beta_{ij})\beta_{ij}^{1/2}}{(1-\beta_{ij}^2)^2 + 4\xi_i\xi_j\beta_{ij}(1+\beta_{ij}^2) + 4(\xi_i^2+\xi_j^2)\beta_{ij}^2} \quad (\beta_{ij} = \omega_{\mathrm{n}i}/\omega_{\mathrm{n}j},\, i,j=1,2,\cdots,n) \tag{54}$$

$$\Rightarrow \mu_{ij} = \frac{\rho_{ij}^{\mathrm{VV}}}{\rho_{ij}^{\mathrm{DD}}},\, v_{ij} = \frac{\rho_{ij}^{\mathrm{VD}}}{\rho_{ij}^{\mathrm{DD}}} \quad (i,j=1,2,\cdots,n) \tag{55}$$

（51）式即为 CCQC 法则，此法则为假设地震外力为白噪音（white noise）形态下所推导出来的方法。$|q_i(t)|_{\max}$、$|q_j(t)|_{\max}$ 分别为第 i 个及第 j 个振态之最大反应值。ρ_{ij}^{DD} 为第 i 个及第 j 个振态间之位移相关系数；ρ_{ij}^{VV} 为第 i 个及第 j 个振态间之速度相关系数；ρ_{ij}^{VD} 为第 i 个及第 j 个振态间之位移与速度相关系数。ξ_i 与 ξ_j 为第 i 个及第 j 个振态之模态阻尼比。$\omega_{\mathrm{n}i}$ 与 $\omega_{\mathrm{n}j}$ 为第 i 个及第 j 个振态之模态频率。

若忽略每个模态间之相关性；则当 $i=j$ 时，对所有 i 及 j，$\rho_{ij}^{\mathrm{DD}}=0$ 以及 $v_{ij}=0$，则（51）式会简化为

$$|\{u(t)\}|_{\max} = \Big[\sum_{i=1}^{n}(\{\alpha\}_i^2 + \omega_i^2\{\beta\}_i^2)|q_i(t)|_{\max}|\dot{q}_i(t)|_{\max}\Big]^{1/2} \tag{56}$$

此即为 CSRSS 法则。

4.2 拉氏转换法之非古典阻尼反应谱法

在 Song et al.（2008）文献中，除了提出非古典阻尼结构动力系统在实数域之解耦法外，一并推导出对应（48）式之 GSRSS（General Complete Quadratic Combination）及 GSRSS（General Square Root of Sum Square）模态组合法则。同样假设结构系统有 n 个自由度，并且地震外力为白噪音的状态下，结构最大反应值 $|\{u(t)\}|_{\max}$ 可表示成

$$|\{u(t)\}|_{\max} = \left\{ \begin{array}{l} \sum_{i=1}^{n_c}\sum_{j=1}^{n_c}\rho_{ij}^{DD}[\mu_{ij}\omega_{ni}\omega_{nj}\{A_D\}_i\{A_D\}_j + \{B_D\}_i\{B_D\}_j \\ + 2v_{ij}\omega_{ni}\{A_D\}_i\{B_D\}_j]|q_i(t)|_{\max}|q_j(t)|_{\max} \\ + 2\sum_{i=1}^{n_c}\sum_{j=1}^{n_p}\rho_{ij}^{DP}[\omega_{pj}\{A_D\}_i\{A_{pD}\}_j + \{B_D\}_i\{A_{pD}\}_j]|q_i(t)|_{\max}|q_{pj}(t)|_{\max} \\ + \sum_{i=1}^{n_p}\sum_{j=1}^{n_p}[\rho_{ij}^{PP}\{A_{pD}\}_i\{A_{pD}\}_j]|q_{pi}(t)|_{\max}|q_{pj}(t)|_{\max} \end{array} \right\}^{1/2}$$

(57)

其中

$$\rho_{ij}^{DD} = \frac{8\sqrt{\xi_i\xi_j}(\beta_{ij}\xi_i + \xi_j)\beta_{ij}^{3/2}}{(1-\beta_{ij}^2)^2 + 4\xi_i\xi_j\beta_{ij}(1+\beta_{ij}^2) + 4(\xi_i^2+\xi_j^2)\beta_{ij}^2} \quad (\beta_{ij}=\omega_{ni}/\omega_{nj}, i,j=1,2,\cdots,n_c)$$

(58)

$$\rho_{ij}^{VV} = \frac{8\sqrt{\xi_i\xi_j}(\xi_i + \xi_j\beta_{ij})\beta_{ij}^{3/2}}{(1-\beta_{ij}^2)^2 + 4\xi_i\xi_j\beta_{ij}(1+\beta_{ij}^2) + 4(\xi_i^2+\xi_j^2)\beta_{ij}^2} \quad (\beta_{ij}=\omega_{ni}/\omega_{nj}, i,j=1,2,\cdots,n_c)$$

(59)

$$\rho_{ij}^{VD} = \frac{4\sqrt{\xi_i\xi_j}(1-\beta_{ij}^2)\beta_{ij}^{1/2}}{(1-\beta_{ij}^2)^2 + 4\xi_i\xi_j\beta_{ij}(1+\beta_{ij}^2) + 4(\xi_i^2+\xi_j^2)\beta_{ij}^2} \quad (\beta_{ij}=\omega_{ni}/\omega_{nj}, i,j=1,2,\cdots,n_c)$$

(60)

$$\Rightarrow \mu_{ij} = \frac{\rho_{ij}^{VV}}{\rho_{ij}^{DD}}, v_{ij} = \frac{\rho_{ij}^{VD}}{\rho_{ij}^{DD}} (i,j=1,2,\cdots,n_c)$$

(61)

$$\rho_{ij}^{PP} = \frac{2\sqrt{\omega_{pi}\omega_{pj}}}{\omega_{pi}+\omega_{pj}} (i,j=1,2,\cdots,n_p)$$

(62)

$$\rho_{ij}^{DP} = \frac{2\omega_{ni}\sqrt{2\xi_i\omega_{ni}\omega_{pj}}}{\omega_{ni}^2 + 2\xi_i\omega_{ni}\omega_{pj} + \omega_{pj}^2} \quad (i=1,2,\cdots,n_c, j=1,2,\cdots,n_p)$$

(63)

(57) 式即为 GCQC 法则。$|q_i(t)|_{\max}$、$|q_j(t)|_{\max}$ 分别为第 i 个及第 j 个次阻尼振态之最大反应值。$|q_{pi}(t)|_{\max}$、$|q_{pj}(t)|_{\max}$ 分别为第 i 个及第 j 个过阻尼振态之最大反应值。ρ_{ij}^{DD} 为第 i 个及第 j 个次阻尼振态间之位移相关系数；ρ_{ij}^{VV} 为第 i 个及第 j 个次阻尼振态间之速度相关系数；ρ_{ij}^{VD} 为第 i 个及第 j 个次阻尼振态间之位移与速度相关系数；ρ_{ij}^{VD} 为第 i 个及第 j 个振态间之位移与速度相关系数；ρ_{ij}^{PP} 为第 i 个及第 j 个过阻尼振态之相关系数；ρ_{ij}^{DP} 为第 i 个次阻尼振态及第 j 个过阻尼振态间的相关系数。ξ_i 与 ξ_j 为第 i 个及第 j 个振态之模态阻尼比。ω_{ni} 与 ω_{nj} 为第 i 个及第 j 个振态之模态频率。

若忽略每个模态间之相关性；则当 $i=j$ 时, 对所有 i 及 j, $\rho_{ij}^{DD}=0$、$\rho_{ij}^{PP}=0$ 以及 $v_{ij}=0$、$\rho_{ij}^{DP}=0$, 则 (57) 式会简化为

$$|\{u(t)\}|_{\max} = \left\{\sum_{i=1}^{n_c}(\omega_{ni}^2\{A_D\}_i^2 + \{B_D\}_i^2)|q_i(t)|_{\max}^2 + \sum_{i=1}^{n_p}\{A_{pD}\}_i^2|q_{pi}(t)|_{\max}^2\right\}^{1/2}$$

(64)

此即为 GSRSS 法则。

4.3 位移反应谱之建立

在此介绍传统反应谱之绘制方法，此位移反应谱可应用到 4.1 节中求取结构最大位

移。考虑一受地表运动 $\ddot{u}_g(t)$ 的单自由度系统其运动方程式可写成

$$\ddot{q}(t)+2\xi\omega_n\dot{q}(t)+\omega_n^2 q(t)=-\ddot{u}_g(t) \tag{65}$$

其中，$q(t)$、$\dot{q}(t)$ 及 $\ddot{q}(t)$ 为相对位移、速度及加速度反应；ξ 和 ω_n 为单自由度系统的阻尼比及自然振动频率；首先给定地表加速度历时 $\ddot{u}_g(t)$，选择单自由度系统之自然振动频率 ω_n 或是其对应之自然振动周期 T_n 以及一模态阻尼比，再计算单自由度系统受地表运动 $\ddot{u}_g(t)$ 时之位移反应，求其最大值 $|q(t)|_{max}$；再对各自然振动周期重复上述步骤。以 El Centro 为例，图 1 为其加速度历时，应用上述方法建立模态阻尼比为 5% 之位移反应谱见图 2。

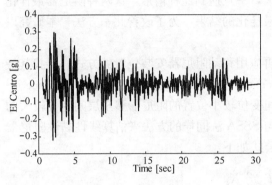

图 1　El Centro 地表加速度历时

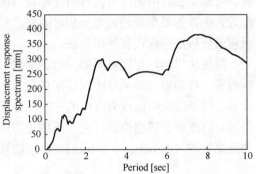

图 2　El Centro 地震位移反应谱

4.4　过阻尼模态反应谱之建立

过阻尼模态反应谱（Over-damped mode response spectrum）之定义与绘制方法与位移反应谱类似，但是因为在拉氏转换法中，不将实数特征值配对因此其对应之模态方程式为一阶常微分方程式，在此及针对此方程式简述其反应谱之绘制方法（Song et al.，2008）。考虑一单自由度过阻尼模态系统受地表振动 $\ddot{u}_g(t)$ 之运动方程式为

$$\dot{q}_{pj}(t)+\omega_{pj}q_{pj}(t)=-\ddot{u}_g(t) \tag{66}$$

其中，$q_p(t)$ 为过阻尼模态反应；$\dot{q}_p(t)$ 为对过阻尼模态反应 $q_p(t)$ 之微分，以及过阻尼模态频率 ω_p（over-damped modal natural frequency），单位为 rad/sec。与传统位移反应谱的绘制方法类似，首先给定地表加速度历时 $\ddot{u}_g(t)$，选择单自由度系统之过阻尼模态自然振动频率 ω_p 或是其对应之过阻尼模态自然振动周期 T_p，计算系统受地表运动 $\ddot{u}_g(t)$ 时之位移反应，求其最大值 $|q_p(t)|_{max}$；再对所有可能的过阻尼模态自然振动周期重复上述步骤，将其对应的最大值绘制于图中。以 El Centro 为例，图 1 为其加速度历时，应用上述方法建立过阻尼模态反应谱如图 3。图 3 可应用于（57）式及（64）式中求取结构系统过阻尼模的最大反应值。

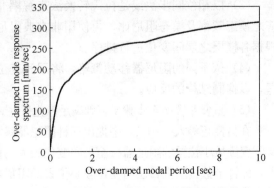

图 3　El Centro 地震过阻尼模态反应谱

5. 阻尼器最佳化配置方法

在此使用由吕良正等人（2011）发展出的阻尼器最佳化配置之简易法；此法之概念为将 Garcia and Soong（2002）的 SSSA 法改良而来，简易法较 SSSA 法更为省时且直观。但不管是简易法或是 SSSA 法，在每次改变阻尼器配置后皆须进行结构动力分析，然而，因为每一次的阻尼器配置都会让系统成为非古典阻尼结构动力系统，此系统在过去皆采用直接积分法来进行结构动力分析求得结构反应值；在自由度较少时，所花费的时间也较短，可是工程界中之结构物自由度都较为庞大，一旦遇到此种情形，这两种阻尼器最佳化配置方法皆需要花费较长之分析时间来求得最佳化的结果，为了改善此一方法，本研究以反应谱分析法来取代直接积分法。

简易法以均匀配置阻尼器为初始配置，再应用移动阻尼器安装位置的方式，寻找其最佳配置。移动阻尼器的原则以层间变位（inter-story drift）为性能指针（performance index），利用压抑结构较大反应的方式，将层间变位较小位置的阻尼器移至层间变位较大的位置，以改善受震结构的行为。简易法应用与 SSSA 法同样的方法来估算线性黏性阻尼器的阻尼常数（damping constant）c，其估算公式如下

$$c = \frac{\xi_{dj} T \sum_{i=1}^{i=n} K_i}{\pi n_d \cos^2\theta} \tag{67}$$

（67）式假设使用相同的阻尼器，且阻尼器摆放角度 θ 一样。式中结构的基本振动周期 T，K_i 为第 i 层的侧向劲度，n_d 为阻尼器总数，n 为楼层数，ξ_{dj} 为第 j 个模态的等阻尼比（equivalent damping ratio）；若固有阻尼比（inherent damping ratio）为 ξ_0，则结构的等效阻尼比（effective damping ratio）$\xi_T = \xi_0 + \xi_d$。假设结构系统的等效阻尼比 ξ_T（目标值）及固有阻尼比 ξ_0 已知，则可得阻尼器所要提供的等阻尼比 ξ_d，再将上述参数代入（67）式中，即可计算黏性阻尼器的阻尼常数。而阻尼器最佳化配置之简易法演算流程如下：

（1）选择加装阻尼器数 n_d 为楼层数的两倍（$n_d = 2n$），用（67）式来计算线性黏性阻尼器之阻尼常数 c。将 n_d 个阻尼器均匀配置（每层楼两个）于结构上，当作阻尼器最佳化的初始配置。

（2）调整地震力至初始配置时最大层间变位为 1.5cm。

（3）以相位同步法或是拉氏转换法来解耦非古典阻尼结构动力系统，并求得各模态对应之模态频率及模态阻尼比，再使用非古典阻尼反应谱分析法求得结构系统的最大反应。根据各楼层之层间变位进行排序。

（4）依不同的阻尼器移动策略，将层间变位较小位置的阻尼器移至层间变位较大的位置，以抑制结构的反应。

（5）重复步骤2至步骤4直到满足停止条件（stop criterion），即得到阻尼器最佳化配置。

在吕良正等人（2011）中提出三种阻尼器移动策略，（1）每一次移动一个阻尼器，即将层间变位最小位置的阻尼器移至层间变位最大的位置；（2）每一次移动半数楼层数目之阻尼器，即将半数楼层数目层间变位较小位置的阻尼器移至同个数层间变位较大的位置；（3）第一次将半数楼层层间变位较小位置的阻尼器移至另一半楼层层间变位较大的位置，然后依次

递减移动一半个数阻尼器。此三种移动策略，每一楼层每次只允许移动一个阻尼器。

6. 案例探讨

在此以十二层楼平面剪力屋架为例来探讨应用反应谱分析法之阻尼器最佳化配置的效益。十二层楼平面剪力屋架之楼层劲度如表1所示，每层楼的质量为52860kg，对应之结构第一周期为1.2sec。采用 El Centro 为输入外力。将每层楼均匀配置两根阻尼器为初始配置，并假设固有阻尼比为每个模态2％，阻尼器设计以第一模态之等阻尼比18％为标的，利用（67）式计算得每根阻尼器之阻尼常数为2950.73kN·sec/m。调整地震力至初始配置时最大层间变位为1.5cm。

楼 层 劲 度　　　　　　　　　　　　　　　　　　　　　表1

楼　层	楼层劲度（kN/cm）
1～4	1000
5～8	850
9～12	725

首先，讨论第一种移动策略搭配三种分析方法（直接积分法（Newmark's method）、CCQC及GCQC）。第一种移动策略为每一次移动一个阻尼器，即将层间变位最小位置的阻尼器移至层间变位最大的位置。配置过程详如表2、表3及表4所示。由实际分析发现最后阻尼器配置的形态会重复出现，因此可将配置重复出现作为本简易法的停止条件。于表中三种分析方法之次数依序为11次、17次及14次；整个最佳化过程所需分析时间依序为1617s、833秒及308s，虽然CCQC及GCQC较直接积分法的分析次数多，但是整体所花费的时间依分析方法的不同可省二分之一至五分之四的时间，尤其以GCQC法最为有效益。三种分析方法所得最佳配置下之最大层间变位分别为1.28cm、1.34cm、1.34cm，折减14.67％、10.67％及10.67％；其变化过程如图4所示，虽然CCQC及GCQC法对整体结构之最大层间变位的折减效果没有比直接积分法的效果好，但是亦可以有接近11％的抑制结构反应之效能。而三种方法之阻尼器最佳化配置如图5，从图中可看出阻尼器几乎都集中于低楼层，三种分析方法之最佳化配置差异性并不大。因此整体而言，以GCQC法之效益性最好。

以直接积分法配合第一种移动策略（一次移动一个）之阻尼器配置过程　　　表2

重排次数		0	1	2	3	4	5	6	7	8	9	10	11	PR(10)
最大层间变位(cm)		1.50	1.45	1.43	1.35	1.34	1.29	1.26	1.26	1.26	1.26	1.28	1.27	1.28
各楼层对应之阻尼器个数	12	2(−1)	1(−1)	0	0	0	0	0	0	0	0	0	0	0
	11	2	2	2(−1)	1(−1)	0	0	0	0	0	0	0	0	0
	10	2	2	2	2	2(−1)	1(−1)	0	0	0	0	0	0	0
	9	2	2	2	2	2	2	2(−1)	1	1(−1)	0	0	0	0
	8	2	2	2	2	2	2	2	2(−1)	1	1(−1)	0	0	0
	7	2	2	2	2	2	2	2	2	2	2	2	2	2
	6	2	2	2	2	2	2	2[+1]	3	3	3	3	3	3
	5	2	2[+1]	3	3	3	3	3	3	3	3[+1]	4(−1)	3[+1]	4
	4	2	2	2	2	2	2	2	2[+1]	3	3	3	3	3
	3	2	2	2	2	2[+1]	3	3	3	3	3	3[+1]	4(−1)	3
	2	2	2	2[+1]	3	3	3[+1]	4	4	4	4	4	4	4
	1	2	2[+1]	3	3	3[+1]	4	4	4	4[+1]	5	5	5	5

表3 以CCQC法配合第一种移动策略(一次移动一个)之阻尼器配置过程

重排次数		0	1	2	3	4	5	6	7	8	9	10	11	12	13	14	15	16	17	PR(16)
最大层间变位(cm)		1.76	1.67	1.61	1.58	1.62	1.55	1.40	2.32	1.37	1.87	1.38	1.47	1.68	1.43	1.87	1.64	1.62	1.45	1.62 (1.34)
各楼层对应之阻尼器个数	12	2(−1)	1(−1)	0	0	0	0	0	0	0	0	0	0	0	0	0	0	0	0	0
	11	2	2	2(−1)	1(−1)	0	0	0	0	0	0	0	0	0	0	0	0	0	0	0
	10	2	2	2	2	2(−1)	2	2(−1)	2(−1)	2(−1)	1[+1]	2(−1)	1	1	0	0	0	0	0	0
	9	2	2	2	2	2	2	2	2(−1)	1[+1]	1(−1)	0	0	0	0	0	0	0	0	0
	8	2	2	2	2	2	2	2	1	1	1(−1)	0	1	1	1	0	0	0	0	0
	7	2	2	2	2	2	2	2	2(−1)	1[+1]	1(−1)	0	1(−1)	1	1(−1)	0	0	0	0	0
	6	2	2	2	2	2	2	2	2	2	2	2	2(−1)	1	1	0	0	0	0	2
	5	2	2	2	2	2	2	2	2	2	2	2	2	2[+1]	2	2	2	2	2	3
	4	2	2	2	2	2	2[+1]	3	3	3	3	3	3	3	3	3	3	3(−1)	2[+1]	3
	3	2	2	2	4	4	4	4	4	4[+1]	5	5[+1]	6	6	6	6	6(−1)	4	4	4
	2	2	2	2[+1]	4	4[+1]	5	5[+1]	6	6	6	6	6[+1]	7	7[+1]	6	6(−1)	5	5	5
	1	2[+1]	3[+1]	4	4	4[+1]	5	5[+1]	6	6	6	6	6[+1]	7	7[+1]	8(−1)	7	7[+1]	8(−1)	7

74

以GCQC法配合第一种移动策略（一次移动一个）之阻尼器配置过程   表4

重排次数	0	1	2	3	4	5	6	7	8	9	10	11	12	13	14	PR(13)
最大层间变位(cm)	1.77	1.66	1.64	1.56	1.54	1.50	1.46	1.46	1.48	1.45	1.42	1.46	1.48	1.46	1.47	1.46(1.34)
各楼层对应之阻尼器个数																
12	2(−1)	1(−1)	0	0	0	0	0	0	0	0	0	0	0	0	0	0
11	2	2	2(−1)	1(−1)	0	0	0	0	0	0	0	0	0	0	0	0
10	2	2	2	2	2(−1)	1(−1)	0	0	0	0	0	0	0	0	0	0
9	2	2	2	2	2	2	2(−1)	1	1	1(−1)	0	0	0	0	0	0
8	2	2	2	2	2	2	2	2(−1)	1	0	0	0	0	0	0	0
7	2	2	2	2	2	2	2	2(−1)	1(−1)	0	2(−1)	1(−1)	0	0	0	0
6	2	2	2	2	2	2	2	2	2	2	2	2	2	2	2	2
5	2	2	2	2	2	2	2	2	2(+1)	3	3	3(+1)	4	4(−1)	3(+1)	4
4	2	2	2	2	2	2(+1)	3	2	2	2	2	2	2(+1)	3	3	3
3	2	2	2	3	3(+1)	4	4	3	3	3(+1)	4	4	4	4	4	4
2	2	2(+1)	4	4(+1)	5	5	5(+1)	4(+1)	5	5	5(+1)	6	6(−1)	5(+1)	6(−1)	5
1	2(+1)	3(+1)	4	4(+1)	5	5(+1)	6	6	6	6	6	6	6	6	6	6

75

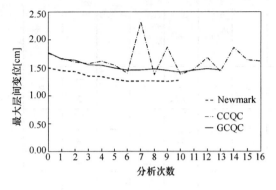 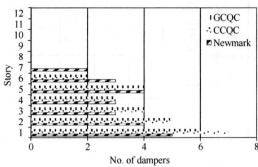

图 4　第一种移动策略最大层间变位变化图　　　图 5　第一种移动策略阻尼器最佳化配置图

接着讨论第二种移动策略搭配三种分析方法（直接积分法（Newmark's method）、CCQC 及 GCQC）。每一次移动楼层数一半（=6）之阻尼器，即将 6 个数层间变位较小位置的阻尼器移至同个数层间变位较大的位置。三种方法的配置过程详如表 5、表 6 及表 7 所示。由实际分析发现最后阻尼器配置的形态会重复出现时即达到停止条件。于表中三种分析方法所需分析次数依序为 4 次、4 次及 5 次；整个最佳化过程所需分析时间分别为 588s、196s 及 110s。三种分析方法之分析次数差不多，但是整体所花费的时间依分析方法的不同可节省三分之二至五分之四的时间，尤其以 GCQC 法所需分析时间最短。三种分析方法所得最佳配置下之最大层间变位分别为 1.27cm、1.36cm、1.38cm，折减 15.33%、9.33% 及 8.0%，其最大层间变位变化过程如图 6，三种分析方法收敛的速度很快，以直接积分法效果最好，但是另两种方法亦有 8%~9% 的抑制结构反应之效益。而最后三种分析方法之阻尼器最佳化配置如图 7，图中三种分析方法阻尼器配置皆集中于低楼层，且差异性不大。整体而言，虽然 GCQC 法之折减效果不是最好的，但是考虑分析中所花费的时间，以 GCQC 法是最具有效益性之分析方法。

以直接积分法配合第二种移动策略（每次移动个数 $n/2 \sim 6$）之阻尼器配置过程　　表 5

	重排次数	0	1	2	3	4	PR(3)
	最大层间变位(cm)	1.50	1.38	1.33	1.27	1.33	1.27
各楼层对应之阻尼器个数	12	2(−1)	1(−1)	0	0	0	0
	11	2(−1)	1(−1)	0	0	0	0
	10	2(−1)	1(−1)	0	0	0	0
	9	2(−1)	1(−1)	0	0	0	0
	8	2(−1)	1(−1)	0	0	0	0
	7	2(−1)	1[+1]	2[+1](−1)	2[+1]	3(−1)	2
	6	2[+1]	3(−1)	2[+1]	3[+1](−1)	3[+1](−1)	3
	5	2[+1]	3[+1]	4(−1)	3[+1]	4(−1)	3
	4	2[+1]	3[+1]	4(−1)	3(−1)	2[+1]	3
	3	2[+1]	3[+1]	4(−1)	3[+1]	4(−1)	3
	2	2[+1]	3[+1]	4[+1]	5(−1)	4[+1]	5
	1	2[+1]	3[+1]	4[+1]	5(−1)	4[+1]	5

以 CCQC 法配合第二种移动策略（每次移动个数 $n/2\sim6$）之阻尼器配置过程　　　　表 6

重排次数		0	1	2	3	4	PR(3)
最大层间变位(cm)		1.76	1.65	1.83	1.52	1.61	1.52(1.36)
各楼层对应之阻尼器个数	12	2(−1)	1(−1)	0	0	0	0
	11	2(−1)	1(−1)	0	0	0	0
	10	2(−1)	1(−1)	0	0	0	0
	9	2(−1)	1(−1)	0	0	0	0
	8	2(−1)	1(−1)	0	0	0	0
	7	2(−1)	1(−1)	0	0	0	0
	6	2[+1]	3[+1]	4(−1)	3(−1)	2[+1]	3
	5	2[+1]	3[+1]	4(−1)	3[+1]	4(−1)	3
	4	2[+1]	3[+1]	4(−1)	3(−1)	2[+1]	3
	3	2[+1]	3[+1]	4[+1]	5(−1)	4[+1]	5
	2	2[+1]	3[+1]	4[+1]	5[+1]	6(−1)	5
	1	2[+1]	3[+1]	4[+1]	5[+1]	6(−1)	5

以 GCQC 法配合第二种移动策略（每次移动个数 $n/2\sim6$）之阻尼器配置过程　　　　表 7

重排次数		0	1	2	3	4	5	PR(4)
最大层间变位(cm)		1.77	1.65	1.61	1.54	1.48	1.52	1.48(1.38)
各楼层对应之阻尼器个数	12	2(−1)	1(−1)	0	0	0	0	0
	11	2(−1)	1(−1)	0	0	0	0	0
	10	2(−1)	1(−1)	0	0	0	0	0
	9	2(−1)	1(−1)	0	0	0	0	0
	8	2(−1)	1(−1)	0	0	0	0	0
	7	2(−1)	1(−1)	0	0	0	0	0
	6	2[+1]	3[+1]	4(−1)	3(−1)	2(−1)	1[+1]	2
	5	2[+1]	3[+1]	4(−1)	3[+1]	4(−1)	3[+1]	4
	4	2[+1]	3[+1]	4(−1)	3(−1)	2[+1]	3(−1)	2
	3	2[+1]	3[+1]	4[+1]	5(−1)	4[+1]	5(−1)	4
	2	2[+1]	3[+1]	4[+1]	5[+1]	6(−1)	5[+1]	6
	1	2[+1]	3[+1]	4[+1]	5[+1]	6[+1]	7(−1)	6

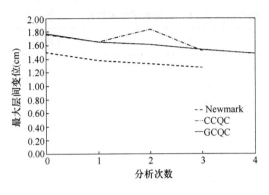 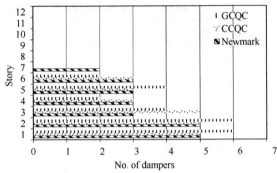

图 6 第二种移动策略最大层间变位变化图　　图 7 第二种移动策略阻尼器最佳化配置图

最后讨论第三种移动策略搭配三种分析方法（直接积分法（Newmark's method）、CCQC 及 GCQC）。第一次将半数楼层层间变位较小位置的阻尼器移至另一半楼层层间变位较大的位置，然后依次递减移动一半个数阻尼器。三种分析方法的配置过程如表 8、表 9 及表 10 所示。由实际分析发现最后阻尼器配置的形态会重复出现时即达到停止条件。表中三种分析方法所需之分析次数依序为 6 次、10 次及 7 次；整个分析过程所需时间分别为 882s、490s 及 154s，三种分析方法中之 CCQC 所需之分析次数最多，但是仍然可节省二分之一的时间，当中以 GCQC 法所需分析时间最短。三种分析方法所得最佳化配置下之最大层间变位分别为 1.28cm、1.34cm、1.34cm，折减 14.67%、10.67% 及 10.67%，最大层间变位变化过程如图 8 所示，GCQC 之折减效果较 CCQC 好，但是两者最后仍有近 11% 的抑制结构反应之效能。而三种分析方法阻尼器最佳化配置如图 9，同前面两种移动策略，在本移动策略中阻尼器之配置在三种分析方法中依然集中于低矮楼层。整体来说，考虑时间效益及结构反应折减，第三种移动策略以 GCQC 为分析方法最好。

以直接积分法配合第二种移动策略之阻尼器配置过程　　表 8

重排次数		0	1	2	3	4	5	6	PR(5)
最大层间变位(cm)		1.50	1.38	1.28	1.26	1.28	1.28	1.27	1.28
各楼层对应之阻尼器个数	12	2(−1)	1(−1)	0	0	0	0	0	0
	11	2(−1)	1(−1)	0	0	0	0	0	0
	10	2(−1)	1(−1)	0	0	0	0	0	0
	9	2(−1)	1	1(−1)	0	0	0	0	0
	8	2(−1)	1	1	1(−1)	0	0	0	0
	7	2(−1)	1	1	1	1[+1]	2	2	2
	6	2[+1]	3	3	3	3	3	3	3
	5	2[+1]	3	3	3[+1]	4	4(−1)	3[+1]	4
	4	2[+1]	3	3	3	3	3	3	3
	3	2[+1]	3[+1]	4	4	4(−1)	3[+1]	4(−1)	3
	2	2[+1]	3[+1]	4	4	4	4	4	4
	1	2[+1]	3[+1]	4[+1]	5	5	5	5	5

以CCQC配合第二种移动策略之阻尼器配置过程 表9

重排次数		0	1	2	3	4	5	6	7	8	9	10	PR(9)
最大层间变位(cm)		1.76	1.65	1.50	1.43	1.63	1.43	1.60	1.61	1.63	1.62	1.45	1.62(1.34)
各楼层对应之阻尼器个数	12	2(−1)	1(−1)	0	0	0	0	0	0	0	0	0	0
	11	2(−1)	1(−1)	0	0	0	0	0	0	0	0	0	0
	10	2(−1)	1(−1)	0	0	0	0	0	0	0	0	0	0
	9	2(−1)	1	1	1(−1)	0	0	0	0	0	0	0	0
	8	2(−1)	1	1(−1)	0	0	0	0	0	0	0	0	0
	7	2(−1)	1	1	1	1(−1)	0	0	0	0	0	0	0
	6	2[+1]	3	3	3	3	3(−1)	2	2	2	2	2	2
	5	2[+1]	3	3	3	3	3	3[+1]	4	4(−1)	3	3	3
	4	2[+1]	3	3	3	3	3	3(−1)	2[+1]	3	3(−1)	2[+1]	3
	3	2[+1]	3[+1]	4	4	4	4	4	4	4	4	4	4
	2	2[+1]	3[+1]	4	4	4[+1]	5[+1]	6	6(−1)	5	5	5	5
	1	2[+1]	3[+1]	4[+1]	5[+1]	6	6	6	6	6[+1]	7[+1]	8(−1)	7

以GCQC配合第二种移动策略之阻尼器配置过程 表10

重排次数		0	1	2	3	4	5	6	7	PR(6)
最大层间变位(cm)		1.77	1.65	1.55	1.47	1.50	1.42	1.47	1.46	1.47(1.34)
各楼层对应之阻尼器个数	12	2(−1)	1(−1)	0	0	0	0	0	0	0
	11	2(−1)	1(−1)	0	0	0	0	0	0	0
	10	2(−1)	1(−1)	0	0	0	0	0	0	0
	9	2(−1)	1	1	1(−1)	0	0	0	0	0
	8	2(−1)	1	1(−1)	0	0	0	0	0	0
	7	2(−1)	1	1	1	1	1(−1)	0	0	0
	6	2[+1]	3	3	3	3(−1)	2	2	2	2
	5	2[+1]	3	3	3	3	3	3[+1]	4(−1)	3
	4	2[+1]	3	3	3	3	3	3	3	3
	3	2[+1]	3[+1]	4	4	4	4	4	4	4
	2	2[+1]	3[+1]	4	4	4[+1]	5[+1]	6(−1)	5[+1]	6
	1	2[+1]	3[+1]	4[+1]	5[+1]	6	6	6	6	6

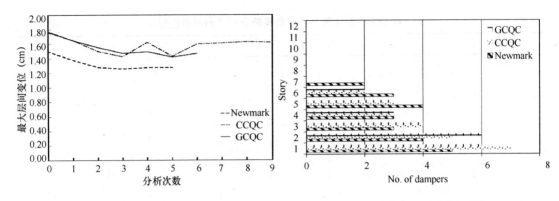

图8 第三种移动策略最大层间变位变化图　　图9 第三种移动策略阻尼器最佳化配置图

从本案例之探讨可总结从节省分析时间之角度上来看，以 GCQC 法最有效益，不论使用哪种移动策略皆为三种方法中分析时间最短者。虽然在上述共九种结果里皆以直接积分法之折减效益最好，但是 GCQC 及 CCQC 搭配第一或第三移动策略之折减效益也有近 11% 左右的成效，而第三移动策略之计算时间又比第一种策略少。因此，若在阻尼器最佳化配置中欲以反应谱分析法来作为动力分析之分析方法，透过本研究建议以第三种移动策略搭配 GCQC 法可以同时达到节省时间并且具有与直接积分法折减效益相近之成果。GCQC 及 CCQC 搭配第二移动策略虽然计算时间最少，但是精度略差。

7. 结论

在本研究当中，首先透过复数模态分析可得到非古典阻尼结构动力系统之模态频率与模态阻尼比，再分别以两种方法将非古典阻尼结构动力系统在实数域中解耦，求得其模态方程式；并且分别应用于反应谱分析法中。最后，在阻尼器最佳化配置中以反应谱分析法取代直接积分法。在本研究案例探讨中分别使用三种不同移动策略搭配三种分析方法（直接积分法、CCQC 及 GCQC 法）比较其分析所需时间及折减效益，最后得到以第三种移动策略（开始时移动楼层数一半之阻尼器，之后每次减半）搭配 GCQC 法不管是在分析时间上之减少或是结构系统反应之折减，为所有搭配方法中最具有效益性的一种。因此在阻尼器最佳化配置中，以反应谱分析法代替直接积分法，是一种可行且具有效益性的方法。

参考文献

[1] 吕良正，张仁德，张慈昕. 平面剪力屋架中黏性阻尼器的简易最佳配置法. 结构工程，2011，26(1)：26-39

[2] Lopez Garcia, D. and Soong, T. T. Efficiency of a simple approach to damper allocation in mdof structures. *Journal of Structural Control*, 2002, 9(1): 19-30

[3] Ma, F., Imam A. and Morzfeld, M. The decoupling of damped linear systems in oscillatory free vibration. *Journal of Sound and Vibration*, 2009, 324(1-2): 408-428

[4] Ma, F., Morzfeld, M. and Imam, A. The decoupling of damped linear systems in free or forced vibration. *Journal of Sound and Vibration*, 2010, 329(15): 3182-3202

[5] Zhou, X. Y., Yu. R. F. and Dong, D. Complex mode superposition algorithm for seismic responses of non-classically damped linear MDOF system. *Journal of Earthquake Engineering*, 2004, 8(4): 597-641

[6] Song, J., Chu, Y. L., Liang, Z., and Lee, G. C. Modal analysis of generally damped linear structures subjected to seismic excitations, *Technical Report MCEER*-08-0005, 2008

采用高性能材料和可变形剪切连接件的大跨组合梁有限元分析

钟国辉,陈松基

(香港理工大学土木及结构工程学系)

摘 要:本文采用有限元建模,综合分析了高性能材料与可变形剪切连接件所构成的大跨组合梁的结构性能。研究的目标是建立一个精确可靠的有限元模型,引入真实的几何尺寸与材料特性,有效地模拟大跨组合梁的结构特性。组合梁的有限元建模采用了 C3D8 实体单元,以及 S4 壳单元,并根据文献中 4 根高性能材料简支组合梁的数据,包括 C30/37 和 C70/85 的实体混凝土楼板和 S460 钢梁,进行了材料、几何与交界面的非线性分析。其分析结果与文献中简支组合梁的测试结果吻合。此有限元建模亦检验了 12m 长的简支梁,以不同的高性能材料组合,如 C30/37 或 C70/85 混凝土与 S355 或 S690 钢梁,进行了系统性的参数分析。其分析结果在本文中有充分的讨论。

关键词:高强度材料;组合梁;有限元建模;可变形的剪切连接件;端头滑移

An Investigation into Long Span Composite Beams with High Performance Materials and Deformable Shear Connections using Finite Element Modelling

K F Chung and C K Chan

(Department of Civil and Structural Engineering, The Hong Kong Polytechnic University,
Hung Hom, Kowloon, Hong Kong SAR, China)

Abstract: This paper presents a comprehensive investigation into the structural behaviour of long span composite beams with high performance materials and deformable shear connections using finite element modelling. The investigation aims to establish reliable finite element models for accurate quantification of the structural behaviour of composite beams with practical geometrical configurations and material specifications. Advanced finite element models of composite beams using solid elements C3D8 and shell elements S4 with material, geometrical and interfacial non-linearity are established, and they have been calibrated successfully against the test results of four beams of high performance materials reported in the literature, i. e. composite beams with C30/37 and C70/85 solid concrete slabs, and S460 steel sections. Based on the calibrated advanced finite element models, a parametric study is carried out to examine the structural behaviour of 12 m long simply supported composite beams with practical configurations and various combinations of

---

钟国辉(1960—),男,博士,教授,研究主要涉及钢结构,钢与混凝土组合结构,冷弯薄壁型钢结构的设计和建造;结构稳定性;有限元模拟分析;工程结构抗火。邮箱:cekchung@polyu.edu.hk。

陈松基(1986—),男,博士,研究员,研究主要涉及钢与混凝土组合结构的有限元模拟分析和设计。

high performance materials, i. e. C30/37 and C70/85 concrete with S355 and S690 steel sections. Various key findings of the investigation are fully discussed in this paper.

Keywords: composite beams; deformable shear connectors; high strength materials; finite element models; end slippage

Introduction

In the recent years, there is a steady trend to use high strength materials in building structures as they provide increased load carrying capacities without increasing the geometrical dimensions as well as the dead loads or the self-weights of a structure. In many quality steel mills around the world, high strength steel plates with yield strengths ranging from 460 to 890 N/mm^2 are readily available. Moreover, both the latest versions of the European material specification for high strength low alloy structural steel plates, BS EN 10025: Part 6: 2004 (BSI, 2004), and the Chinese material specification for high strength structural steel materials, GB 1591: 2008 (SAC, 2008) provide detailed information on both the chemical compositions and the mechanical properties, and many thousands of tonnes of high strength steel plates have been produced to these material specifications every year for several years. However, there is a lack in complementary design methods in modern structural design codes for practical design of high strength steel materials, and hence, effective construction using high strength steel is greatly prohibited.

In general, a composite beam is essentially a thin wide concrete slab connected with a steel section through a series of shear connectors, and a certain depth of the concrete slab is in compression while the steel section is largely in tension. The structural form is able to maximize the structural advantages of both the concrete slab and the steel section, and it is widely used in composite construction for many years owing to their high structural efficiency.

At present, the conventional plastic stress block method is recommended for composite beams with concrete up to grade C60/75 and steel sections up to S355 in BS EN 1994-1-1 (BSI, 2004). The fundamental assumption is that the strains in the steel sections are sufficiently high to allow the development of full moment resistances of the composite cross-sections, and hence, the stress blocks of the steel sections are rectangular with a constant stress level at its yield strength value, p_y. However, this method is not applicable to composite beams with S460 steel sections, and reduction to the moment resistances of the composite beams is found to be necessary (Johnson & Anderson, 2001).

According to the results of a pilot study reported by the authors (Chung & Chan, 2011) in examining the effectiveness of composite beams with high performance materials, i. e. C70/85 concrete, and S460 and S690 steel sections, the expected increases in the load carrying capacities of these composite beams are estimated to range typically from 30 to 75%. Consequently, it is highly desirable to establish the effective use of composite beams with high strength steel sections. Moreover, as the current design rules are developed and calibrated against composite beams with normal strength materials, their applications to composite beams with high strength concrete and steel sections have yet to be established.

1. Literature Review

The structural behaviour of composite beams and deformable shear connections has been a popular subject over the last three decades among the international research community of structural engineering, and many researches on finite element modelling using two and three dimensional finite elements have been reported from times to times. Among all, the following research works on finite element modelling of composite beams are considered to be of direct relevance to the present study:

1.1 Liang et al. (2004) proposed a three-dimensional finite element model to simulate the structural behaviour of a continuous composite beam in which shell elements were adopted to model both the steel section and the concrete slab. Failure of the continuous composite beam was defined either by a large deformation of the steel section or numerical divergence due to brittle behaviour of the concrete.

1.2 Queiroz et al. (2007) proposed a three-dimensional finite element model to simulate the structural behaviour of a simply supported composite beam with full and partial shear connection subjected to either concentrated or uniformly distributed loads. The steel sections were modelled with shell elements while the concrete slabs were modelled with three-dimensional solid elements. In general, comparisons between the experimental and the predicted results on a number of composite beams selected from the literature are found to be good.

1.3 In the past few years, there was a remarkable interest on the deformation characteristics of shear connections in push-out tests as well as in beam tests, and both numerical and experimental investigations (Ellobody & Young, 2006; Ranzi et al., 2009; Mirza & Uy, 2010; Qureshi et al., 2011) had been reported in the literature. Moreover, there was a growing concern on the differences in their deformation characteristics measured from standard push-out tests and beam tests.

2. Objectives and scope of work

This paper presents a comprehensive numerical investigation into the structural behaviour of simply supported composite beams with high performance materials and deformable shear connectors. The scope of the investigation includes:

2.1 To develop advanced finite element models with material, geometrical and interfacial non-linearity for composite beams with high strength concrete and steel sections, and deformable shear connectors, and to calibrate the proposed models against the test results of a total of four composite beams.

2.2 To carry out a parametric study of simply supported composite beams with practical configurations and various combinations of high strength concrete and steel materials, and to assess on their structural behaviour.

For full details of the numerical investigation into the structural behaviour of simply supported composite beams with a wide range of steel and concrete materials, refer to Chung & Chan (2011) and Chan (2012).

3. Tests of composite beams

In order to ensure the general applicability of the proposed finite element models, four simply supported composite beams with either C30/37 or C70/85 concrete and S460 steel sections (Hegger & Doinghau, 2004; Hegger & Goralski, 2006) have been selected as reference composite beams, and their test data are adopted for calibration of the proposed models. Table 1 presents the general information of these composite beam tests while Figure 1 illustrates the overall test set-ups of Beams B100, B300, B1 and B5 under one-point loads. Table 2 summarizes the measured data of the mechanical properties of the concrete and the steel materials as well as the measured failure loads of the beams. Moreover, the geometrical dimensions of various components of the composite cross-sections are also provided in Figure 1 together with the details of the shear connection arrangements. The measured load mid-span deflection curves of all these beams are also available for subsequent analysis. It should be noted that complementary push-out tests of the composite shear connections had also been carried out by the researchers (Hanswille & Hegger, 2002).

Summary of simply supported composite beam tests — Table 1

Beam test test	Nominal grade		Shear connection	Loading and boundary condition	Failure load, P_{Test} (kN)
	Concrete	Steel section			
B100	C70/85	S460	Full		725
B300	C70/85	S460	Partial		710
B1	C30/37	S460	Full		670
B5	C30/37	S460	Partial		570

Notes:
1) All beams have solid slabs.
2) All shear connectors are headed shear studs with a diameter of 19 mm.

Measured strengths of steel sections, steel reinforcements and concrete   Table 2

Beam test	Steel section				steel reinforcement	Cylinder strength of concrete (N/mm^2)
	Flange		Web			
	Yield strength, p_y (N/mm^2)	Tensile strength, p_u (N/mm^2)	Yield strength, p_y (N/mm^2)	Tensile strength, p_u (N/mm^2)	Yield Strength, p_y (N/mm^2)	
B100	458	581	488	584	591	79
B300	465	515	474	586	616	89
B1	508	581	508	581	579	31
B5	490	548	490	548	579	41

Botes:
1) All beams have solid slabs.
2) All shear connectors are headed shear studs with a diameter of 19mm.

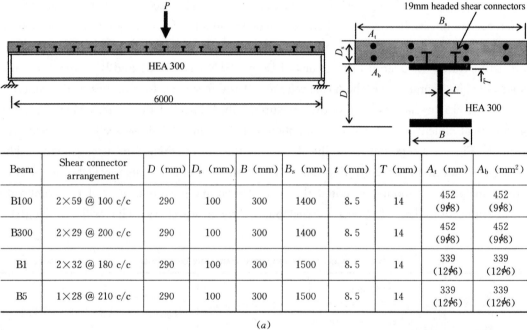

Beam	Shear connector arrangement	D (mm)	D_s (mm)	B (mm)	B_s (mm)	t (mm)	T (mm)	A_t (mm)	A_b (mm^2)
B100	2×59 @ 100 c/c	290	100	300	1400	8.5	14	452 (9ϕ8)	452 (9ϕ8)
B300	2×29 @ 200 c/c	290	100	300	1400	8.5	14	452 (9ϕ8)	452 (9ϕ8)
B1	2×32 @ 180 c/c	290	100	300	1500	8.5	14	339 (12ϕ6)	339 (12ϕ6)
B5	1×28 @ 210 c/c	290	100	300	1500	8.5	14	339 (12ϕ6)	339 (12ϕ6)

(a)

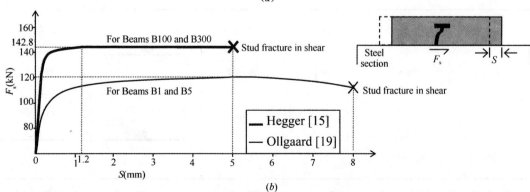

(b)

Figure 1   General layout of Beams B100, B300, B1 and B5
(a) Details of test specimens; (b) Load-slippage curves of composite shear connections

4. Finite element modelling

In order to simulate numerically the structural behaviour of composite beams with high strength materials and deformable shear connections, three dimensional finite element models are established using the general purpose finite element package ABAQUS (2010). In the proposed model of the composite beams, 4-noded shell elements S4, 8-noded solid elements C3D8 and 4-noded quadrilateral surface elements SFM3D4R are employed to model the steel sections, the concrete slabs and the steel reinforcements respectively. The proposed model is illustrated in Figure 2, and it should be noted that spring elements, SPRING2, are adopted to model the shear connections between the concrete slabs and the steel sections.

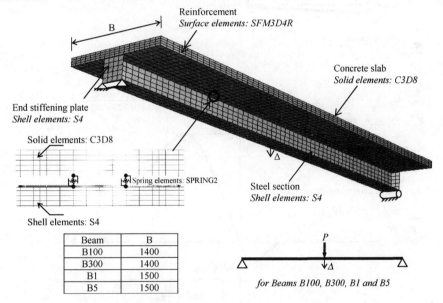

Figure 2   Finite element models of various composite beams

4.1 Material models

The material models of the concrete, the steel section and the steel reinforcement according to BS EN 1994-1-1 (BSI, 2004), as shown in Figure 3, are adopted into the three dimensional models, and hence, material and geometrical non-linearity is fully incorporated.

4.2 Initial imperfections

In general, residual stresses with simplified linear stress patterns in the hot rolled steel sections should be considered in finite element modelling, and these values are directly inputted as the initial stresses of the shell elements of the steel sections. Moreover, it is necessary to provide initial geometrical imperfections in the models to facilitate smooth

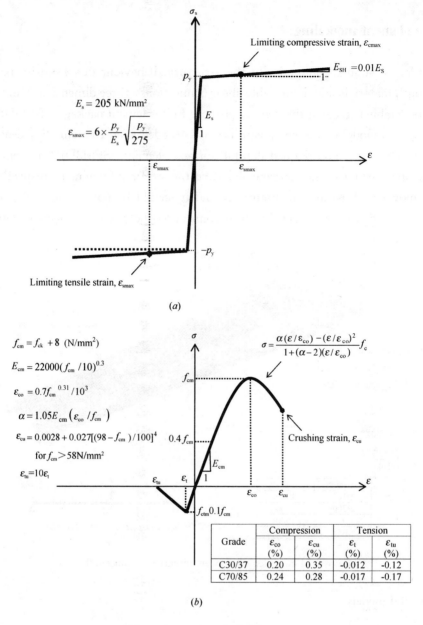

Figure 3 Material curves of steel and concrete materials
(a) Steel section and reinforcement; (b) Concrete

transition across bifurcation limits during equilibrium iterations to avoid numerical divergence. As part of a standard procedure of numerical modelling, elastic linear eigenvalue analyses on the models under the corresponding loading and boundary conditions are performed, and the eigenmodes corresponding to the lowest eigenvalues, are extracted. The eigenmodes are then superimposed to the perfect geometries of the composite beams as initial geometrical imperfections, as shown in Figure 4, and the magnitude of the maximum out-of-plane initial imperfections is taken to be 0.25 t, where t is the web thickness of the steel sections.

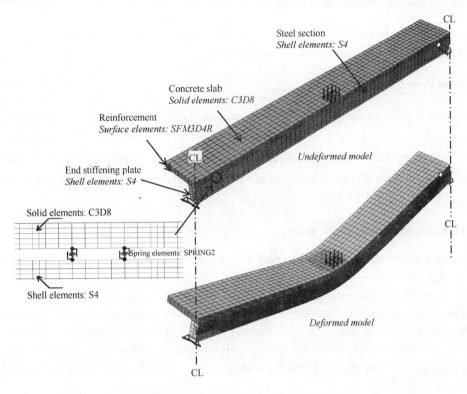

Figure 4　Deformed models of composite beams B100, B300, B1 and B5

4.3　Deformation characteristics of shear connections

In order to simulate various actions of the shear connections at the steel-concrete interface of the composite beam, spring elements SPRING2 are provided in the model to enable effective shear transfer along the steel-concrete interface (Wang & Chung, 2006 & 2008). It should be noted that each of the shear connectors is modelled with one longitudinal spring, one transverse spring and one vertical spring, all at the same time, in order to simulate the longitudinal shear force, the transverse shear force as well as the pull-out tensile force of the shear connector respectively during the deformation of the composite beam. Hence, the measured load-slippage curves of the shear connections shown in Figure 1 is incorporated into the material models of the spring elements.

4.4　Mesh intrusion

In order to avoid local intrusion between the finite elements of the concrete slabs and the steel sections during non-linear analyses, axial spring elements with extremely high compressive stiffness but zero tensile stiffness are provided along the interfaces between the concrete slabs and the steel sections. However, no friction between the underside of the concrete slab and the topside of the steel flange is incorporated into the model.

4.5 Numerical results and calibration

The proposed models have been developed for those composite beams, and successful runs of all the models are achieved. It is found that:

4.6 Load deflection curves

Figures 5 and 6 present the predicted load mid-span deflection curves obtained from the proposed models of the four composite beams together with the test data for direct comparison. It is shown that the predicted load mid-span deflection curves obtained from the three dimensional models follow closely to the test data not only in the elastic deformation ranges but also in the large deformation ranges.

4.7 Failure criteria and check points

In order to define failure in the composite beams, the following check points for failure are established:

Check point for concrete

When the maximum compressive strain of the concrete slab reaches the limiting compressive strain of the concrete, ε_{cmax}, crushing of the concrete is imminent, and thus, the composite beam is considered to be failed. Refer to Figure 3 for the limiting compressive strains in the concrete, ε_{cmax}, of various concrete grades adopted in the present study.

Check point for steel

Owing to the ductility of the steel material, it is rational to allow the steel material to work significantly beyond its first yield strain, and thus, the check point for the steel material is proposed to be equal to a large tensile or compressive strain, ε_{smax}, which is defined as follows:

$$\varepsilon_{smax} = c_o \times \frac{p_y}{E_s} \times \sqrt{\frac{p_y}{275}}$$

where c_o is the deformation coefficient which is taken to be 6; p_y is the yield strength of the steel material; and E_s is the Young's modulus of the steel material which is taken to be 205kN/mm².

Check point for slippage of shear connection

In order to register excessive slippage in the composite shear connections of the composite beam, two slippage limits, s_a and s_b, with their values equal to 5 and 10mm respectively, are proposed. It should be noted that after detailed data analyses on the numerical results of the proposed models, the values of various check points are identified, and they are plotted onto the predicted load deflection curves of the composite beams shown in Figures 5 and 6.

It is shown that in all these beams, crushing of the concrete is found to be critical, and thus, the corresponding check point is denoted as Point CP1 while excessive tensile strain in the bottom flange of the steel section is denoted as Point CP2. In general, the predicted failure loads of the proposed models are found to be very close to the measured failure loads according to the proposed failure criteria, and they are summarized in Table 3 for easy comparison. Moreover, a model factor, ψ, is established for easy comparison

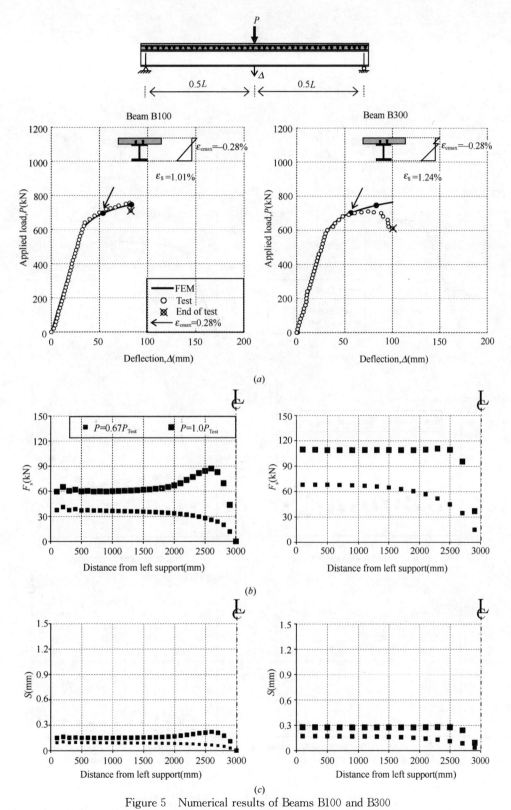

Figure 5 Numerical results of Beams B100 and B300
(a) Load deflection curves; (b) Longitudinal shear forces in shear connectors; (c) Slippage of shear connectors

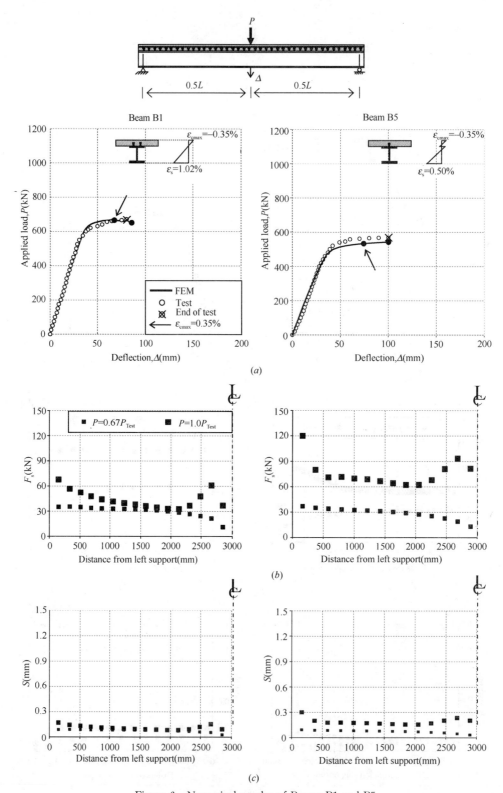

Figure 6　Numerical results of Beams B1 and B5
(a) Load deflection curves; (b) Longitudinal shear forces in shear connectors; (c) Slippage of shear connectors

which is defined as follows:

$$\Psi_{FEM} = \frac{P_{Test}}{P_{FEM}}$$

where

P_{Test} is the measured failure load of a beam obtained from test; and

P_{FEM} is the predicted failure load of the beam obtained from finite element analysis.

All the model factors of the proposed models are summarized in Table 3 for easy comparison, and it is shown that the model factors range from 0.98 to 1.03 with an average value of 1.01. Hence, the proposed models are demonstrated to be conservative and highly effective.

Table 3 Failure loads of composite beams

Beam test	P_{Test} (kN)	Numerical results							Plastic design method		
		Check point	P_{FEM} (kN)	Concrete strain (%)	Corresponding steel strain (%)	End slippage, S(mm)	Model factor, Ψ_{FEM}	k_{sc}	P_{PDM} (kN)	Model factor, Ψ_{PDM}
B100	725	CP1	695.2	−0.28	1.01	0.15	1.04	1.74	719.2	1.01
B300	710	CP1	701.7	−0.28	1.24	0.28	1.01	0.86	706.9	1.00
B1	670	CP1	664.5	−0.35	1.02	0.17	1.01	1.01	651.3	1.03
B5	570	CP1	529.9	−0.35	0.50	0.29	1.08	0.43	579.0	0.98

Check points

Beam test	CP1 for Concrete	CP2 for steel flange
B100	−0.28%	1.73%
B300	−0.28%	1.77%
B1	−0.35%	2.02%
B5	−0.35%	1.91%

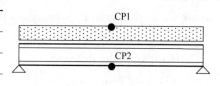

Notes:
P_{Test} is the measured failure load of the beam obtained from test.
P_{FEM} is the predicted failure load of the beam obtained from finite element analysis.
P_{PDM} is the predicted failure load of the beam obtained from plastic design method.
k_{sc} is the degree of shear connection in the composite beam.
Ψ_{FEM} is the model factor of finite element model.
Ψ_{PDM} is the model factor of plastic design method.
S is the maximum end-slippage of a composite beam predicted from finite element model.

5. Parametric Study

In order to examine the structural behaviour of simply supported composite beams with practical configurations, a parametric study on composite beams with high strength concrete and steel sections using the calibrated finite element models is carried out. As shown in Figure 7, the composite beam comprises of a UB $457 \times 152 \times 52$ kg/m, and a solid concrete slab of 125mm thick and 3m wide. The beam is 12m long, and it is subjected to uniformly distributed loads. During the study, various combinations of C30/37 and C70/85 concrete as well as S355 and S690 steel sections are considered. Moreover, the load slippage curves of the composite shear connections with two different concrete grades are also illustrated, with a partial safety factor of 1.25 for design propose. It should be noted that these curves are assumed to have a slippage ductility of 10mm. Figure 8a) illustrates the general arrangement of the finite element model while typical deformed shape of the finite element model at large deformation is shown in Figure 8b).

5.1 Numerical results

Figure 9 illustrates various deformation characteristics of the composite beams with C30/37 concrete and S355 steel sections under two different levels of shear connection, namely, i) $k_{sc}=1.0$, and ii) $k_{sc}=0.4$. It is shown that:

a) For the composite beam with $k_{sc}=1.0$, the load deflection curve is highly non-linear with a sharp change in slope after the elastic deformation range, as shown in Figure 9a). Moreover, the critical failure criterion is concrete crushing which occurs at the top of the concrete slab when the mid-span deflection of the beam reaches 375mm, or span/32. The corresponding strains in the concrete slab and the steel bottom flanges are 0.35% and 1.72% respectively. In general, owing to the presence of a large number of shear connectors provided along the beam span, large shear resistances are readily mobilized at small slippages, as shown in Figure 9b). The maximum slippage among all the shear connections is found to be merely 2.8mm, as shown in Figure 9c).

b) However, the deformation characteristics of the composite beam with $k_{sc}=0.4$ are found to be very different. As only a small number of shear connections are provided along the beam span, large slippages in the shear connections are often needed, especially towards the beam ends, in order to mobilize the load resistance of the beam. Hence, excessive slippage of shear connections will become critical while concrete crushing is unlikely to occur. Consequently, the beam is considered to be failed when the maximum slippage among all the shear connections reaches 10mm, and the corresponding mid-span deflection of the beam is 250mm, or span/48, while the corresponding strains in the concrete slab and the steel bottom flange are merely 0.12% and 0.75% respectively, as shown in Figure 9a). It is interesting to note that at a maximum slippage of 5mm among all the shear connections, the mid-span deflection of the beam reaches 120mm, or span/100; this corresponds to 93% of the load resistance of the beam.

Similarly, Figure 10 illustrates various deformation characteristics of the composite beams with C70/85 concrete and S690 steel sections under two different levels of shear connection, namely, i) $k_{sc}=1.0$, and ii) $k_{sc}=0.4$. Owing to the significant increases in the material strengths of both the concrete and the steel sections, more shear connections along the beam span are required to mobilize fully the moment resistances of the beams at

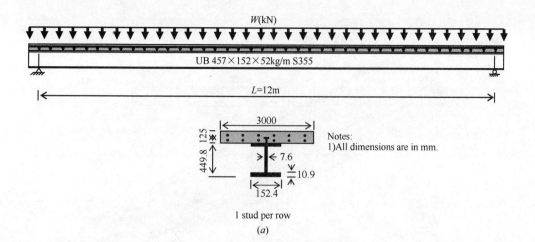

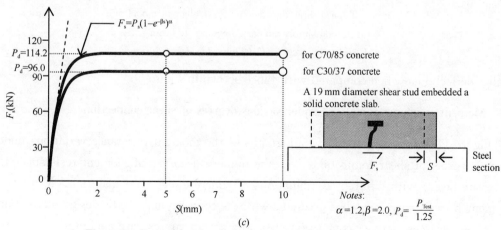

Figure 7 General layout of simply supported composite beams with different combinations of material strengths

(a) Geometrical dimensions and cross-sectional details; (b) Material strengths; (c) Assumed load-slippage curves of shear studs

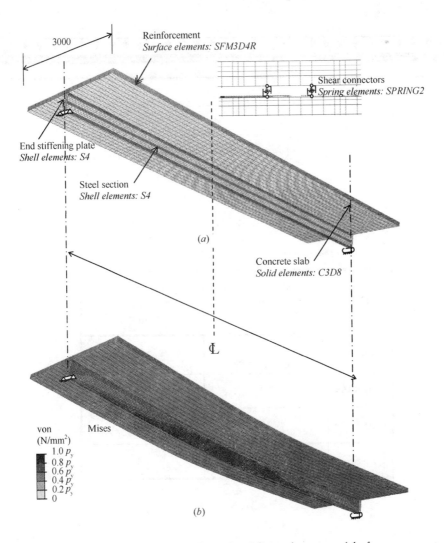

Figure 8 Typical three dimensional finite element model of a
12m long simply supported composite beam
(a) Undeformed model; (b) Deformed model at failure

smaller deflections, when compared with those shown in Figure 9.

5.2 Moment resistance variations under various degrees of shear connection

Figure 11 illustrates a total of four graphs of the moment resistance variations under various degrees of shear connection, i. e. the numerical curves of moment resistances for composite beams with different combinations of concrete and steel materials. Moreover, the moment resistance variations predicted with the conventional plastic design method under various degrees of shear connection, i. e. the design curves, are also plotted onto the graphs for direct comparison. Furthermore, the design resistances according to the reduction method given in BS EN 1994-1-1 for high strength concrete and steel sections are also provided for direct comparison. It is shown that

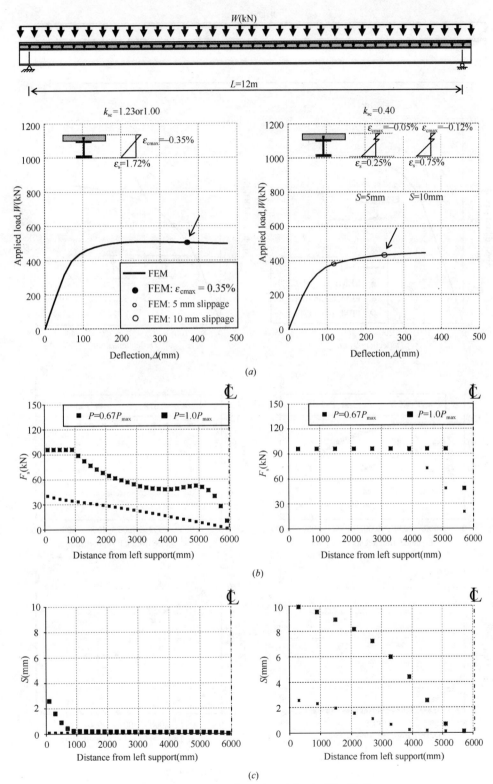

Figure 9 Numerical results of Study C30-S355

(a) Load deflection curves; (b) Longitudinal shear forces in shear connectors; (c) Slippage of shear connectors

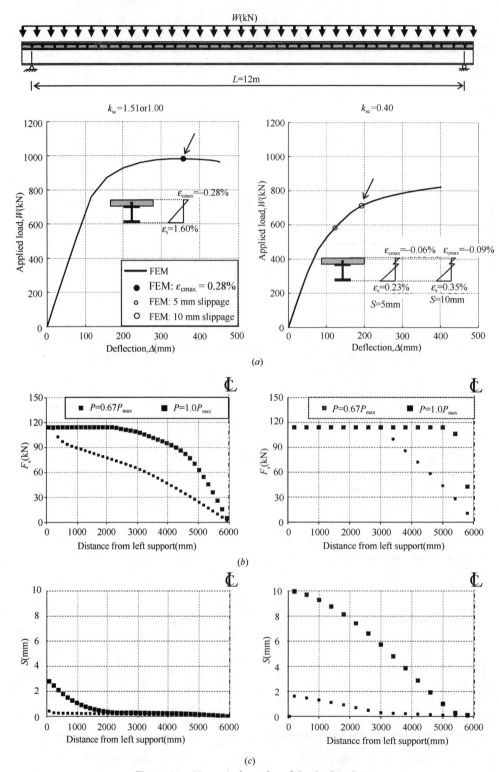

Figure 10 Numerical results of Study C70-S690
(a) Load deflection curves; (b) Longitudinal shear forces in shear connectors;
(c) Slippage of shear connectors

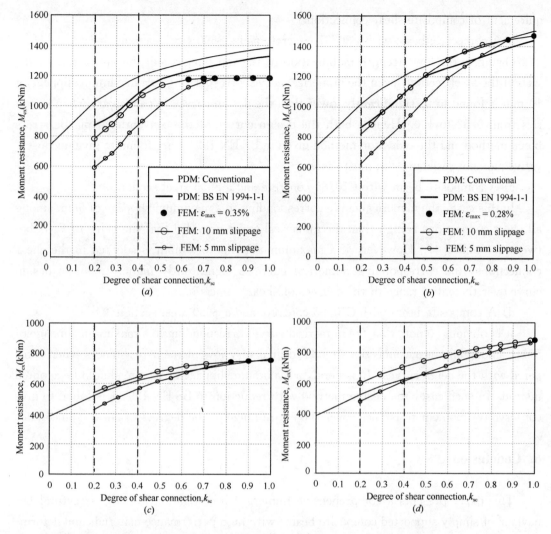

Figure 11 Moment resistances of composite beams with different degrees of shear connection
(a) Beam C30-S690; (b) Beam C70-S690; (c) Beam C30-S355; (d) Beam C70-S355

a) A composite beam with C30/37 concrete and a S355 steel section

In this case, the numerical curve corresponding to a maximum slippage of 5 mm among all shear connections follows closely to the design curve provided that the value of k_{sc} is between 1.0 and 0.62. For composite beams with k_{sc} smaller than 0.62, the design curve is significantly higher than the numerical curve, giving unconservative moment resistances. However, if a maximum slippage of 10mm is adopted in the shear connections, the corresponding numerical curve follows closely to the design curve over the entire range of the degrees of shear connection.

Consequently, it is concluded that a maximum slippage of the composite shear connection at 10mm, instead of 6mm, is assumed and implicitly embedded into the plastic design method for composite beams with C30/37 concrete and S355, especially with low degrees of shear connection. This is contradictory to what is commonly assumed in the ductility requirement of composite shear connection.

b) A composite beam with C30/37 concrete and a S690 steel section

Owing to the low strength mobilization of the top flanges and the upper part of the web of the steel section, even the numerical curve corresponding to a maximum slippage of 10 mm is found to be significantly lower than the design curve, in particular, when k_{sc} varies from 0.62 to 1.0. Hence, both the design curves of the conventional plastic stress block method and the reduction method given in BS EN 1994-1 are found to give unconservative moment resistances in most cases.

c) A composite beam with C70/85 concrete and a S355 steel section

In this case, the numerical curve corresponding to a maximum slippage of 5mm among all shear connections is high than the numerical curve provided that the value of k_{sc} is between 1.0 and 0.47. However, if a maximum slippage of 10mm is adopted in the shear connections, the corresponding numerical curve is significantly higher than the design curve over the entire range of the degrees of shear connection.

d) A composite beam with C70/85 concrete and a S690 steel section

It is obvious to note that the numerical curve corresponding to a maximum slippage of 5 mm among all shear connections tends to give very low moment resistances when compared with the deign curves. In general, a maximum slippage of 10 mm should be adopted instead. In such case, the design method with reduction to BS EN 1994-1 is found to give conservative moment resistances over the entire range of the degrees of shear connection.

6. Conclusion

This paper presents a comprehensive numerical investigation into the structural behaviour of simply supported composite beams with high performance materials and deformable shear connections. During the calibration exercise, it is shown that the predicted load mid-span deflection curves of the proposed models are shown to follow closely the measured curves of the tests along the entire deformation ranges.

Based on the calibrated advanced finite element models, a parametric study is carried out to examine the structural behaviour of 12m long simply supported composite beams with practical configurations and various combinations of high performance materials, i.e. C30/37 and C70/85 concrete with S355 and S690 steel sections. It is shown that the conventional plastic design method with rectangular stress blocks is able to give conservative moment resistances for composite beams with C30/37 concrete and S355, provided that a maximum slippage of 10mm in the shear connections is adopted. However, for composite beams with C30/37 and S690 steel sections, the design moment resistances of the composite beams according to the plastic design method with reduction given in BS EN 1994-1-1 are significantly larger than those numerical values obtained by the proposed models corresponding to a slippage ductility of 10 mm in the shear connections. Thus, the design method is shown to be inapplicable. Nevertheless, for composite beams with C70/85 concrete and S355 as well as S690 steel sections, the plastic design method with reduction is found to be conservative, and improvement for more efficient design is needed.

References

[1] ABAQUS (2010), User's Manual, Version 6.10, Hibbitt, Karlsson and Sorensen, Inc
[2] British Standards Institution, BS EN 1994-1-1 (2004). Eurocode 4: Design of composite steel and concrete structures. Part 1.1: General rules and rules for buildings. European Committee for Standardization
[3] British Standards Institution, BS EN 10025: Part 6 (2004). Hot rolled products of structural steels. Technical delivery conditions for flat products of high yield strength structural steels in the quenched and tempered condition. European Committee for Standardization
[4] Chan CK (2012). Structural performance of long span composite beams with high performance materials and practical constructional features. PhD thesis, The Hong Kong Polytechnic University
[5] Chung KF and Chan CK (2011). Numerical investigation of simply supported composite beams with high performance materials and deformable shear connectors. Proceeding of the 6th International Symposium on Steel Structures, pp278-285
[6] Ellobody E and Young B (2006). Performance of shear connection in composite beams with profiled steel sheeting. Journal of Constructional Steel Research, 62(7): 682-694
[7] Ernst S, Bridge R Q and Wheeler A (2010). Correlation of beam tests with push out tests in steel-concrete composite beams. Journal of Structural Engineering, ASCE, 183-192
[8] European Commission, EUR 20104 (2002). Use of high-strength steel S460, Technical Steel Research Series, ISBN 92-894-3108-3, pp1-339
[9] Hegger J and Doinghaus P (2004). High performance steel and high performance concrete in composite structures. Proceedings of the Composite Construction in Steel and Concrete IV Conference, Canada, pp891-902
[10] Hegger J and Goralski C (2006). Structural behavior of partially concrete encased composite sections with high strength concrete. Proceedings of the Composite Construction in Steel and Concrete V Conference, pp346-355
[11] Liang QQ, Uy B, Bradford MA and Ronagh HR (2004). Ultimate strength of continuous composite beams in combined bending and shear. Journal of Construction Steel Research, 60: 1109-1128
[12] Mirza O and Uy B (2010). Effects of the combination of axial and shear loading on the behaviour of headed stud steel anchors. Engineering Structures, 32(1): 93-105
[13] Queiroz FD, Vellasco PCGS and Nethercot DA (2007). Finite element modelling of composite beams with full and partial shear connection. Journal of Constructional Steel Research, 63: 505-521
[14] Qureshi J, Lam D and Yea JQ (2011). Effect of shear connector spacing and layout on the shear connector capacity in composite beams. Journal of Constructional Steel Research, 67: 706-719
[15] Ranzi G, Bradford MA, Ansourian P, Filonov A, Rasmussen KJR, Hogan TJ, and Uy B (2009). Full-scale tests on composite steel-concrete beams with steel trapezoidal decking. Journal of Construction Steel Research 65: 1490-1506
[16] Standardization Administration of the People's Republic of China (2008). GB1591-2008: High strength low alloy structural steels

[17] Wang AJ and Chung KF (2006). Integrated analysis and design of composite beams with flexible shear connectors under sagging and hogging moments. Steel and Composite Structures, 6(6): 459-478

[18] Wang AJ and Chung KF (2008). Advanced finite element modelling of perforated composite beams with flexible shear connectors. Engineering Structures 30: 2724-2738

高雄海洋文化及流行音乐中心结构设计概述

苏晴茂[1]，陈陆民[2]，陈焕炜[3]，王胜辉[4]

（1. 联邦工程顾问股份有限公司 负责人；
2. 科建联合结构技师事务所 董事长；
3. 联邦工程顾问股份有限公司 副总经理；
4. 联邦工程顾问股份有限公司 协理）

摘 要：高雄海洋文化及流行音乐中心基地坐落于台湾高雄市11至15号码头，基地面积约11.5hm^2，本工程为一国际竞图标案，希望借由建置国际艺术文化展演场所及海洋文化中心，带动高雄成为亚太流行音乐创作及表演中心与国际海洋文化交流平台。在众多参加的国际设计团队中，最后由西班牙Manuel Alvarez Monteserin Lahoz等建筑团队及台湾翁祖模建筑师共同投标获得第一名。本文目的在探讨本案最具地标性质的两幢外观有如海浪意象的高低塔楼、长跨度的大型室内展演空间、室外展演空间的大跨距雨庇以及本案其他各栋建筑物的结构分析与耐震设计考虑，以供工程界参考。

关键词：国际竞图标案；地标高低塔楼；结构分析；耐震设计考虑

Structural Design Overview of Kaohsiung Marine Culture and Pop Music Center

Ph. D. C. M. Su[1], L. M. Chen[2], H. W. Chen[3], S. H. Wang[4]
(1. CEO of Federal Engineering Consultant Inc. , Taiwan
2. CEO of Tec-Build Engineering Inc. , Taiwan
3. Vice President of Federal Engineering Consultant Inc. , Taiwan
4. Associate of Federal Engineering Consultant Inc. , Taiwan)

Abstract: The site of Kaohsiung Marine Culture and Pop Music Center is located in 11-15 Pier Kaohsiung, Taiwan. The planned area is about 11.5 hectares. This international competition was wined by the team of Spanish architect (Manuel Alvarez Monteserin Lahoz) and domestic architect (HOY). The main goal of this project is to create the center of Asia-Pacific pop-music industry and also provide high standard performance areas specified for pop-music shows. This article is focused on the features of structural and seismic design which includes the landmark twin towers imaged from sea waves, the long-span performance hall, the long-span canopy of stage, and other unique buildings imaged from marine creatures.

Key words: international competition; landmark twin towers; structural analysis; seismic design

1. 工程概述

高雄海洋文化及流行音乐中心为一项由高雄市政府出资的重大公共建设，希望在完工后能提供市民一个优质的休闲空间，打造高雄成为流行音乐重镇，并构建海洋文化与旅游观光中心。本案建筑设计包含 ZONE1 及 ZONE2 两个主要场区，ZONE1 区域主要由跨度约 95m 的大型室内展演空间，具地标性建筑的两栋高低塔楼以及可容纳约一万两千人的室外展演空间串联而成。ZONE2 区域则包含海豚造型的音乐艺术与文创产业空间，海洋文化展示中心以及小型室内表演空间等。本工程各栋建筑物合计总楼地板面积约 72,000m^2，各栋建筑物在两个区域的平面配置见图 1。

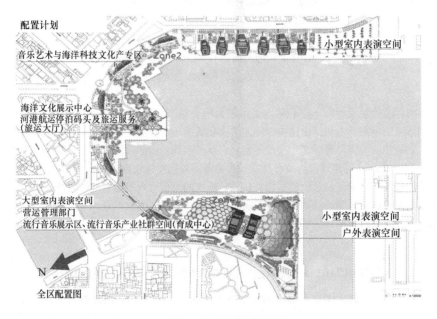

图 1　全区平面配置图

由于本工程各栋建筑物的外观设计都很有特色，结构系统的规划必须要能同时符合建筑设计与结构抗震的需求，本文拟针对本案结构系统规划以及设计过程较为特殊的考虑做一概要之说明与探讨。

本案原定于 2012 年 9 月完成所有设计工作并进行工程发包作业，设计过程因业主需求略有改变，截至投稿日前设计作业暂停中，将待业主需求确认之后再进行后续细部设计作业。

2. 结构系统说明

本工程结构系统规划主要考虑各栋建筑物其建筑外观设计、室内空间规划以及使用功能需求各有不同，必须选择能同时符合建筑设计与结构抗震需求的结构系统，兹就各栋建筑物的结构系统概略说明如下：

2.1　高低塔楼

位于 ZONE1 的高塔楼平面尺度长宽约 61m×33m，11 层楼总高度约 68m，低塔楼平面尺度长宽约 55m×28m，7 层楼总高度约 41m，高低两栋塔楼之间建筑规划以一宽度约

26m 高度约 40m 的中庭区相连接，高塔楼跟西南侧中庭区的结构以伸缩缝隔离，低塔楼与东北侧中庭区的结构则相连共构。此外，在 ZONE1 位于高塔楼东北侧的大型室内展演空间，以及位于低塔楼西南侧的室外展演空间大雨庇，两者与高低塔楼的连接处亦皆规划设置伸缩缝，使本区四个主要结构量体的结构系统均为各自独立。

高低塔楼的结构系统选用钢骨构造的二元系统，垂直力之传递路径经由楼板、梁、柱及斜撑等构件进入基础，抗侧力系统包含配置于两栋塔楼服务核外围的韧性斜撑构件、配合建筑物外观立面的斜撑构件以及内部的梁柱抗弯矩构架系统，基础形式选用基桩与筏式基础共构。

2.2 长跨度大型室内展演空间

长跨度大型室内展演空间平面尺度长宽约 95m×92m，其屋顶结构最大跨距约 70m，规划采用深度约 4m 的双向立体曲面桁架结构。配合建筑外观设计意象与室内使用空间需求，本工程之曲面桁架结构，其上下弦杆平面配置为六边形，立面配置为弧线。抗侧力系统选用承重墙系统，主要包含沿大型室内展演空间外围配置的斜柱以及位于看台区下方的 RC 承重墙，而在大型室内展演空间与高塔楼连接处，因建筑动线需求规划以巨型组合式集力梁柱框架取代该处的斜柱构件。垂直力的传递路径经由楼板、曲面桁架、承重墙及柱构件等进入基础。大型室内展演空间的基础与高低塔楼的基础相连共构，故其基础型式亦为桩筏共构。

2.3 室外展演空间大跨距雨庇

室外展演空间大跨距雨庇平面尺度长宽约 83m×31m，东西向跨度约 83m 的雨庇结构，规划采用深度约 4m 的桁架结构，抗侧力系统选用承重墙系统，主要为配置于室外展演空间东西侧的斜柱及斜撑构件。垂直力的传递路径经由楼板、桁架梁、斜柱及斜撑构件等进入基础。室外展演空间的基础亦与高低塔楼相连共构，故基础形式亦为桩筏共构。

2.4 音乐艺术与文创产业空间

建筑平面规划为海豚造型的五处音乐艺术与文创产业空间，分别坐落在 ZONE1 及 ZONE2 两个区域，每处平面尺度长度约 60m 宽度约 12～15m，结构系统短向规划采用桁架梁柱所组成的构架系统，长向则规划采用仅能承受张力的斜拉杆系统，垂直力之传递路径经由楼板、桁架梁、桁架柱进入基础。基础形式规划采用联合基脚。

2.5 海洋文化展示中心

海洋文化展示中心平面尺度长宽约 108m×82m，结构系统主要由三处核心筒状承重墙系统所构筑而成，筒状承重墙系统之内核心为厚度 50cm 的 RC 承重墙，外核心为具对角斜撑构件的钢骨构架，垂直力之传递路径经由楼板、钢梁、承重墙及柱构件进入基础，抗侧力系统主要借由各楼层的刚性楼版将侧力导入三处核心筒状承重墙系统。基础形式规划采用联合基脚。

2.6 小型室内表演空间

位于ZONE2的六个小型室内表演空间结构系统规划采用RC承重墙系统，屋顶结构跨距约24～28m则选用梁深约1400mm的钢梁结构，垂直力之传递路径经由楼板、钢梁、承重墙进入基础，侧力系统长向由40～50cm厚的剪力墙所承担，短向则借由屋顶层的刚性楼板将侧力导入前端厚度40cm的剪力墙。基础形式规划采用筏式基础。

前述各栋建筑物之结构系统整理如下表，至于各栋建筑物之外观透视图与结构分析模型示意图可详见图2～图13。

结构系统总表　　　　　　　　　　　　　　　　　　表1

分区	建物名称	构造类别	结构系统	基础形式
ZONE1	高塔楼	钢构造	二元系统	桩筏共构
	低塔楼	钢构造	二元系统	桩筏共构
	大型室内展演空间	钢构造	承重墙系统	桩筏共构
	室外展演空间雨庇	钢构造	承重墙系统	桩筏共构
ZONE2	音乐艺术与文创产业空间	钢构造	构架系统/拉杆系统	联合基脚
	海洋文化展示中心	钢构造	承重墙系统	联合基脚
	小型室内表演空间	RC构造	承重墙系统	筏式基础

图2　高低塔楼外观透视示意图

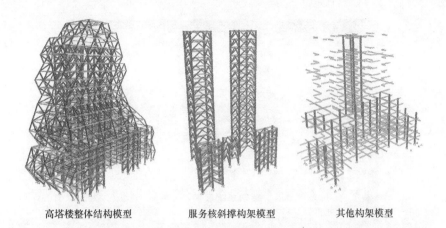

高塔楼整体结构模型　　服务核斜撑构架模型　　其他构架模型

图3　高塔楼结构分析模型示意图

图4　大型室内展演空间透视示意图

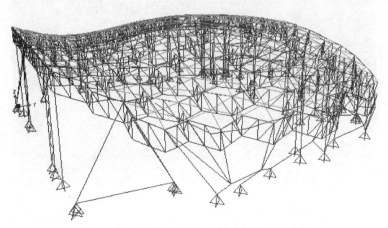

图5　大型室内展演空间结构分析模型示意图

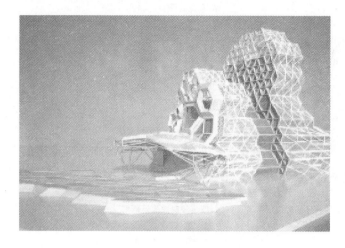

图 6　室外展演空间大跨距雨庇外观示意图

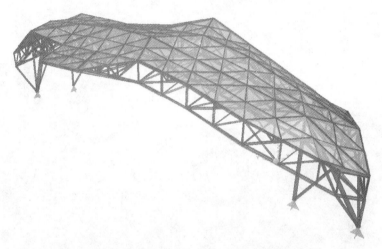

图 7　室外展演空间大跨距雨庇结构分析模型示意图

图 8　音乐艺术与文创产业空间外观示意图

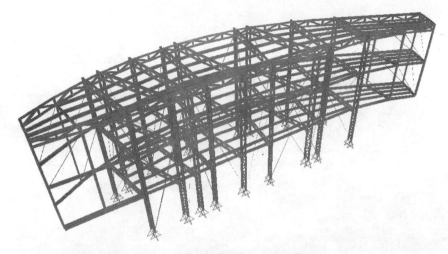

图 9 音乐艺术与文创产业空间结构分析模型示意图

图 10 海洋文化展示中心外观示意图

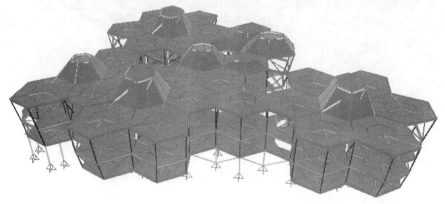

图 11 海洋文化展示中心结构分析模型示意图

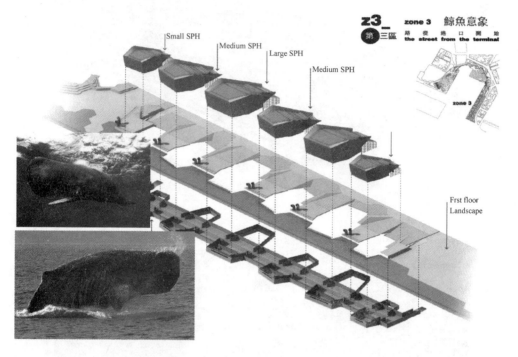

图 12 小型室内表演空间外观示意图

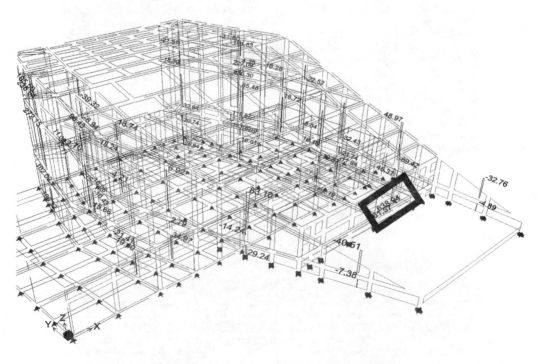

图 13 小型室内表演空间结构分析模型示意图

3. 结构材料

本工程抗震构件之主筋皆采用 SD420W 耐震钢筋，其 $f_y=4,200\mathrm{kgf/cm^2}$ 且降伏比小于 0.8。由于基地位于港口边考虑海风、盐害及建筑物之耐久性等问题，所有建筑物之基础及地下构造物，皆使用设计强度为 $350\mathrm{kgf/cm^2}$ 的混凝土，部分钢柱构件规划采用柱内灌浆方式提升其强度与劲度，为符合灌浆过程的施工性并确保混凝土能确实填充，柱内灌浆部分选用强度 $420\sim560\mathrm{kgf/cm^2}$ 具高流动性的高性能混凝土。主体结构耐震钢材的选择主要考虑强度与韧性的需求。组合 H 型大梁采用 SN490B 规格钢材，组合箱型柱采用 SN490B（板厚小于 40mm）或 SN490C（板厚大于或等于 40mm）规格钢材，挫屈束制斜撑之核心构材规划使用 LYS100 低降伏强度钢材，小梁及其他非抗震构件则采用 ASTM A572 Gr. 50 钢材，大跨距桁架结构之圆钢管构件采用 STK490 钢材，相关结构材料规格整理如下表。

结构材料规格　　　　　　　　表 2

钢　筋	
抗震构件主筋	SD420W
其他	SD280 / SD420
混凝土	
基础及地下室	fc' $350\mathrm{kgf/cm^2}$
柱内灌浆	fc' $420\sim560\mathrm{kgf/cm^2}$
其他	fc' $280\sim350\mathrm{kgf/cm^2}$
钢结构	
大梁	SN490B
组合箱型柱	SN490B 或 SN490C
挫屈束制斜撑核心构材	LYS100
小梁及其他非抗震构件	ASTM A572 Gr. 50
桁架结构之圆钢管构件	STK490

4. 结构分析模型

本工程为一跨国性的建筑设计合作案，由于各栋建筑物的外观设计都非常具有特色，在结构分析与设计过程中，如何建构与建筑设计可相应的结构分析模型为首要之课题，因此必须建立一个作业流程将建筑 3D 外观设计与结构构架系统及结构尺寸相互结合，本案建构结构分析模型的作业流程整理如下图。

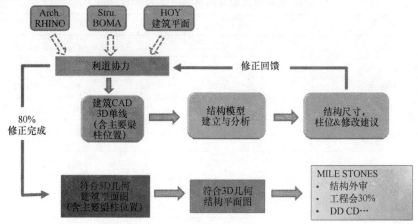

5. 设计成果

经初步规划设计本工程各栋建筑物的周期及层间相对位移角整理如下表3。

表3

建 物 名 称	周 期		层间相对位移角(‰)	
(1) 高塔楼(Tall Tower)	$T_x=1.157$	$T_y=0.820$	$D_x=2.10$	$D_y=1.09$
(2) 低塔楼(Short Tower)	$T_x=0.579$	$T_y=0.494$	$D_x=0.84$	$D_y=0.65$
(3) 长跨度大型室内展演空间(LPH)	$T_x=0.505$	$T_y=0.437$	$D_x=1.28$	$D_y=0.98$
(4) 室外展演空间大跨距雨庇(OPA)	$T_x=0.975$	$T_y=0.433$	$D_x=1.72$	$D_y=0.75$
(5) 音乐艺术与文创产业空间(大海豚)	$T_x=0.687$	$T_y=0.503$	$D_x=2.32$	$D_y=1.43$
(6) 音乐艺术与文创产业空间(小海豚)	$T_x=0.487$	$T_y=0.614$	$D_x=1.10$	$D_y=1.47$
(7) 海洋文化展示中心(Museum)	$T_x=0.214$	$T_y=0.649$	$D_x=0.58$	$D_y=0.62$
(8) 小型室内表演空间(SPH)	$T_x=0.311$	$T_y=0.178$	$D_x=0.12$	$D_y=0.05$

各栋建筑物主要的结构尺寸分别整理如下列尺寸表4。

A. 高低塔楼（单位：mm） 表4

主 分 类	分布楼层	尺 寸
外环立面斜撑 Façade Brace	1F～11MF	BOX 350～450
服务核柱 Core Column	1F～11MF	BOX 450～600
服务核梁 Core Beam	1F～11MF	H700×250
服务核斜撑 Core Brace	1F～11MF	BOX 450×450
柱	1F～R2F	BOX 450×600
B1F柱	B1F	RC600×600～1000×1000
梁	1F～PR	H700～800
地梁	基础层	FB 80×200(cm)
高楼区楼板	1F～PR	15cm DECK板
1F楼板	1F	S(20)～S(25)/20(cm) DECK板
地下室楼板	B1F	BS(20)
筏基板	基础层	FS(60)

B. 长跨度大型室内展演空间（LPH） 表5

主 分 类	分布楼层	尺寸(mm)
外围柱	1F～RF	Pipe 355.6×12 Pipe 457.2×12 Pipe 558.8×16 Pipe 609.6×22
看台柱(内环)	1F～2F	
看台柱(外环)	1F～3F	
电梯柱	1F～3F	RH400×400×13×21
南面集力桁架	1F～RF	Box700×28 Box700×32
看台后方斜撑	3F～RF	Pipe 355.6×12 Pipe 457.2×12 Pipe 558.8×16 Pipe 609.6×22

续表

主 分 类	细 分 类	尺寸(mm)
主桁架与次桁架	上弦杆 下弦杆	Pipe 267.4×8 Pipe 267.4×15.1 Pipe 457.2×12 Pipe 457.2×19 Pipe 558.8×22
	斜杆	Pipe 216.3×8 Pipe 216.3×10.3 Pipe 216.3×15.1 Pipe 267.4×15.1 Pipe 267.4×21.4
	垂直杆	Pipe 216.3×8 Pipe 216.3×10.3 Pipe 216.3×15.1
南面集力桁架	上弦杆 下弦杆	BH 500×500×16×25 BH 500×500×19×32
	斜杆 垂直杆	RH 400×400×13×21 RH 400×408×21×21 RH 414×408×18×28
小梁	屋面板小梁	RH 582×300×12×17 RH 700×300×13×24 BH900×300×16×25 BH900×300×19×36 BH900×400×22×40
	电梯小梁	RH 400×400×13×21

C. 室外展演空间大跨距雨庇（OPA） 表 6

主 分 类	细 分 类	尺寸(mm)
桁架	上弦杆 下弦杆	H300×300～H400×400
	斜杆 垂直杆	H200×200～H250×250
柱		Pipe 457.2 Pipe 508.0 Pipe 609.6 Pipe 711.2 Pipe 1016.0
小梁		RH400×200

D. 音乐艺术与文创产业空间（大海豚） 表 7

主 分 类	分布楼层	尺寸(mm)
组合钢柱 C1	1F～RF	直杆 PIPE 323.9 斜杆 PIPE 165.2
拉杆（BR2）	1F～RF	ROD 50～80

续表

主 分 类	分布楼层	尺寸(mm)
钢梁(G1)	2F~RF	H 450×200 ~ 400×200
钢小梁(sb)	2F~PR	H 400×200 ~ 250×125
钢柱(P1)	2F~RF	H 200×200
拉杆(BR1)	2F~RF	ROD 30~60
拉杆(T1)	2F	ROD 20
桁架(TR1)	2F~PR	上弦杆 RH 200×200~250×250 下弦杆 RH 200×200~250×250 斜杆 RH 150×150~175×175
桁架(TR2)	PR	上弦杆 PIPE 216.3 下弦杆 PIPE 216.3 斜杆 PIPE 165.2
钢承板(DS)	1F~RF	15 cm DECK 板

E. 音乐艺术与文创产业空间（小海豚） 表8

主 分 类	分布楼层	尺寸(mm)
组合钢柱 C2	1F~2F	直杆 PIPE 273.1 斜杆 PIPE 165.2
拉杆(BR2)	1F~2F	ROD 50~80
钢梁(G2)	2F	H 400×200
钢小梁(sb)	2F~RF	H400×200~250×125
钢柱(P1)	2F	H 200×200
拉杆(BR1)	2F	ROD 30~60
桁架(TR2)	2F~RF	上弦杆 RH 200×200~150×150 下弦杆 RH 200×200~150×150 斜杆 RH 150×150~100×100
桁架(TR4)	RF	上弦杆 PIPE 165.2 下弦杆 PIPE 165.2 斜杆 PIPE 139.8
钢承板(DS)	1F~RF	15 cm DECK 板

F. 海洋文化展示中心（Museum） 表9

主 分 类	分布楼层	尺寸(mm)
大梁	2F~RF	H600×200 H600×200~H700×300
小梁	2F~RF	H600×300 H600×300~H700×300 H300×150

续表

主 分 类	分布楼层	尺寸(mm)
斜柱和斜撑	1F	PIPE 355.6×12（TYP.） PIPE 457.2×19（局部）
斜撑和斜柱	2F～3F	PIPE 318.5×9（TYP.） PIPE 355.6×12（局部）
外环斜撑	2F～3F	PIPE 318.5×9
外 CORE 柱	1F～3F	PIPE 457.2×19 （1F需内灌 SCC fc'＝420kg/cm²）
CORE 大梁	1F～3F	H700×300
外 CORE 斜撑	1F～3F	PIPE 318.5×9
悬臂斜撑/背撑		PIPE 355.6
外 CORE 柱加劲支撑 （1F 半高以下）	1F	PIPE 318.5×9
内 CORE RC 墙	1F～3F	50cm
钢承板	1F～3F	3W DECK (15cm)
拉杆		ROD 20Φ

G. 小型室内表演空间（SPH） 表 10

主 分 类	分布楼层	尺寸(mm)
RC 承重墙/剪力墙	1F	45～50cm
长跨钢梁	RF	BH1400×450(mm)

6. 基础形式与开挖工法

本工程基地坐落于高雄市 11 至 15 号码头，基地下方之土层主要为砂性土层，位于浅层的砂土层较为疏松（$N=2\sim7$），位于地表 5m 以下的砂土层则属中等紧密（$N=8\sim20$）。基础型式之选择综合考虑结构安全性、施工工法、施工质量及经济性等因素，位于 ZONE1 区之塔楼、大型室内展演空间及室外展演空间大跨距雨庇等选用基桩与筏式基础共构，基桩型式规划采用桩径约 80～100cm 的预制植入式 PC 桩，有较大承载力需求之柱位下方采用群桩共同承载方式。ZONE 2 区的音乐艺术与文创产业空间及海洋文化展示中心基础型式选用联合基脚，同样位于 ZONE 2 的六个小型室内表演空间，基础型式选用筏式基础。考虑基地下方的砂性土层于大地震发生时会有液化之潜能，土壤容许承载力将因此折减，故于浅基础下方视结构实际载重需求配置地质改良桩提高土壤之容许承载力，本工程规划之地质改良桩改良率采用 12.5%，改良范围自地表下 1.5m～6m。预制植入式 PC 桩与地质改良桩在施工前与完工后，规划进行验证试桩以确保其施工质量可以符合结构设计承载力的需求。

本工程位于 ZONE 2 区无地下室之浅基础，其基础开挖深度约介于 1.5m～2m，开挖工法规划采用斜坡明开挖工法，必要时可配合打设钢板桩减少基础开挖范围。ZONE 1 区规划一层地下室之塔楼区基础范围，其开挖深度约 6m，考虑开挖深度不深但开挖范围较大，开挖工法规划采用岛区式开挖工法，首先沿地下室外墙范围打设钢板桩作为挡土结构，再以斜坡明开挖工法进行第一阶段中央区域的土方开挖，接着构筑中央区域之基础结构体并于其上设置反力墩座，利用反力墩座架设斜向钢支撑于钢板桩，进行第二阶段周边区域的土方开挖，最后构筑周边区域之结构体。

7. 后续工作

本案目前已经完成结构系统规划（SD）及结构设计发展（DD）阶段的工作，后续尚有细部设计及施工详图（CD）阶段的工作待完成，后续细部设计阶段的工作主要将着重于设计的验证以及构件接合细节的检讨，例如以非线性静力侧推分析方式（PUSH-OVER）确认本工程所有的设计成果其耐震能力皆可符合法规需求。

8. 结语

如何规划能与建筑物相应适合的结构系统，使建筑物能兼具结构安全性、成本合理性、空间使用性以及施工可行性，是结构设计团队必须尽力达成的主要目标。相信借由本会每年定期举办的工程技术交流，必定能使建筑结构设计技术不断的进步与提升。

福州市海峡奥体中心体育馆屋盖结构设计概述

傅学怡，周 颖

(中建国际（深圳）设计顾问有限公司，北京 100013)

摘 要：本工程为甲级大型体育馆，建筑面积 44426m²，座位 11172 席，综合比赛馆屋盖为椭圆形，跨度约 97m×116m，布置创新三环四边形环索弦支结构，边跨设置张弦梁结构，提高环索及撑杆的效率，改善屋盖网格梁结构的受力和变形性能。进行考虑拉索张拉全过程的施工过程模拟分析，指导施工全过程；对结构进行细致的分析，保证屋盖结构在正常使用荷载及偶然荷载作用下结构安全、经济合理。

关键词：四边形环索弦支结构；张弦梁结构；张拉全过程

ROOF STRUCTURE DESIGN OF THE GYMNASIUM IN FUZHOU STRAIT OLYMPIC SPORTS CENTER

X. Y. Fu, Y. Zhou

(China Construction Design International, Beijing 100013, China)

Abstract: This project is a large gymnasium with 11,172 seats and 44,426 square meters of building area. The competition hall has an elliptical roof with span of about 97m×116m. Three innovative quadrilateral ring cable chord-support structures and beam string structures in two sides are adopted, which lead to more effective ring cables and vertical bars, and the structural performance of grid beams is improved. Analysis of construction simulation including whole processes of prestressing is executed to guide the construction process, The detailed analysis has been carried out to gain better safety and economy of all roof structure under normal and accidental loads.

Keywords: quadrilateral rings cable chord-supported structure; beam string structure; whole processes of prestressing

1. 引言

福州海峡奥林匹克体育中心位于福州市南台岛仓山组团中部，包括体育场、体育馆、游泳馆、网球馆、室外观众平台、商业多个建筑子项，在体育竞赛功能上能满足全国运动会、城运会和世界单项比赛的要求，赛后满足群众体育健身、休闲娱乐、餐饮、商业购物等使用要求的城市综合体。其中体育馆建筑面积 44426m²，总坐席数 11172 席，建筑平面

第一作者：傅学怡（1945—），男，博导，中国工程勘察设计大师，E-mail：fu_xue_yi@yahoo.com.cn
通讯作者：周 颖（1980—），女，硕士，主要从事结构设计工作，E-mail：zhou.ying@ccdi.com.cn

是由多段圆弧组成的中轴对称形体，平面近似"水滴"形，体育馆单体南北长约 211m，东西宽约 153m，占地约 1.9hm²，建筑效果图见图 1。

体育馆混凝土结构采用框架剪力墙结构，比赛大厅混凝土结构看台最高点标高为 21.97m；体育馆覆盖整体钢结构屋盖及墙面，屋盖建筑完成面最高点为 40.67m，其中综合比赛馆上空屋盖为椭圆形，跨度约 97m×116m，采用四边形环索弦支-张弦组合结构体系，其余范围屋盖为平面主次桁架结构；比赛馆及训练馆周圈混凝土柱作为屋盖结构的内支座。体育馆外墙支承结构为单层方钢管交叉柱，北侧交叉柱直接连接于基础，东、西、南侧支撑于平台层下部混凝土柱上。交叉柱与屋面钢结构桁架之间设置周圈立体过渡桁架，整体结构模型见下图 2。

图 1 福州海峡体育中心体育馆鸟瞰效果图

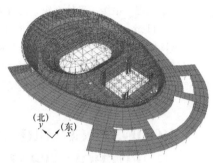
图 2 整体结构模型三维图

2. 结构构成

2.1 混凝土结构选型与布置

北侧的比赛大厅近似椭圆形，北侧为四层，局部五层，南侧三层，混凝土结构看台最高点标高为 21.97m；南侧的训练馆区部分为两层混凝土结构，首层为观众平台，局部有二层观众连廊；采用混凝土框架剪力墙结构体系。比赛区看台后排混凝土柱以及训练区周圈柱作为钢结构屋盖的竖向支承构件。混凝土结构整体布置以及剪力墙设置如下图 3 所示，典型剖面见图 4。

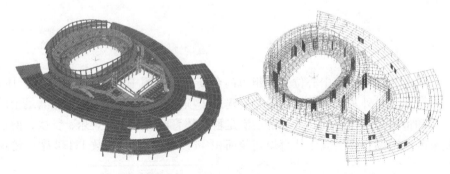

(a) 混凝土结构布置　　(b) 剪力墙布置

图 3 混凝土结构模型三维视图

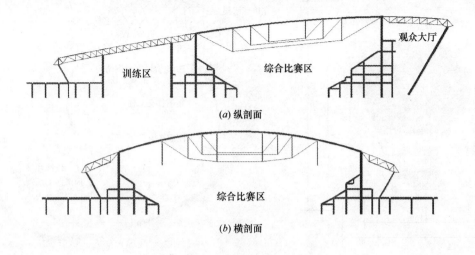

图4 结构典型剖面

混凝土结构首层平台长度南北约200m,不设永久变形缝,设计考虑地基的有限约束刚度,对混凝土结构进行整体温差分析。结合工程的施工过程、全年气候统计资料并考虑混凝土的收缩徐变效应,对温度变化效应进行分析。并进一步采取施工构造措施减小混凝土收缩及温度应力不利影响,留设后浇带,混凝土低温入模;加强混凝土养护、覆盖;降低水泥用量,减小水灰比等。

2.2 钢结构选型与布置

福州体育馆覆盖钢结构及墙面,屋盖平面似水滴形,南北长约298m,东西约152m。屋盖整体北高南低,北侧结构杆件中心线最高点标高为39.67m,正南侧端部结构杆件中心线标高为18.67m;沿东西方向为中间高,两端低,屋盖最低点标高约为16.2m。

整个屋盖利用下部混凝土结构的框架柱作为支座,分为三个区域(见图5):观众大厅、比赛区、训练区;其中比赛区屋面为近椭圆形,长边116m,短边97.5m,北侧最高点标高39.68m(结构中心线标高),向南及东西两侧逐渐降低,南侧最低点标高31.42m,东西最低点标高21.65m;训练区平面为矩形,长边50.4m,短边36m;其余部分为观众大厅。墙面采用单层矩形钢管交叉网格结构,交叉柱柱底连接于混凝土框架柱顶或基础。在墙面及屋面结构交界处设置立体过渡环桁架,则屋盖结构支承于内部混凝土柱以及周圈立体环桁架上。钢结构屋盖的支座布置示意如图6所示。

(1) 比赛区屋盖结构

体育馆比赛区屋盖采用四边形环索弦支-张弦组合结构。四边形环索弦支结构是一种结构新型体系,由网格梁、斜索、下撑钢管及四边形环向索构成,通过斜索的预应力张拉,使下撑钢管受压力,提高改善网格梁结构的受力和变形性能。同时结合本工程特点,边跨设置张弦构件进一步提高结构受力性能。多重四边形环索弦支结构在本工程此规模的椭圆形屋盖中应用属首创,具有线条简单明快,用钢节约,传力明确的特点。

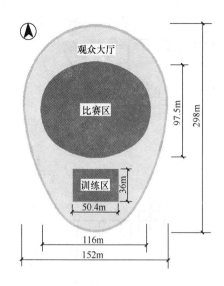

图5 福州体育馆屋面平面尺寸示意

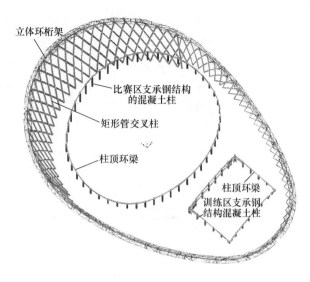

图6 钢屋盖支座布置示意

比赛区四边形环索弦支-张弦组合结构体系结构布置三维视图及立面视图如下图7所示。

比赛区看台后排Φ1200混凝土作为比赛区屋盖的支承柱，柱顶设置1200（宽）mm×900（高）mm环梁。比赛区网格梁布置见图8，屋盖中部正交网格尺寸为7.2～9.5m，东西两侧采用放射状布置，网格尺寸5.5～8.5m。大部分网格梁采用400mm×700mm矩形钢管截面，材料为Q345B。

布置三环四边形环索，如下图9所示。第1环（内环）撑杆高度9.0m，四边形环索长21.6m、19.0m，斜索长13.9m、15.0m；第2环（中环）撑杆9.0m，四边形环索长37.9m、36.0m，斜索长13.4m、

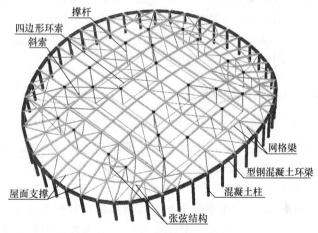

图7 比赛区屋盖结构布置三维视图
（■点标记为竖向撑杆支撑位置）

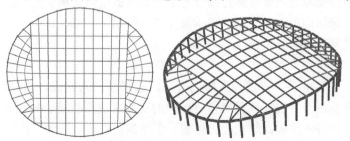

图8 比赛区网格梁布置

14.85m；第3环（外环）撑杆9.5m，四边形环索长56.9m、51.1m，斜索长度15.8m、18.2m。由于屋盖四边负荷面积较大，在四边形环索第三环增设中间撑杆（撑杆高13.0m），在东西两侧设置独立张弦梁结构（撑杆高度8.5m、11.0m），增加结构竖向刚度，见图10。撑杆截面采用$\phi325$及$\phi351$的圆钢管截面，材料为Q345B。

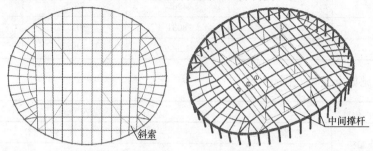

图9 四边形环索的布置（三环）

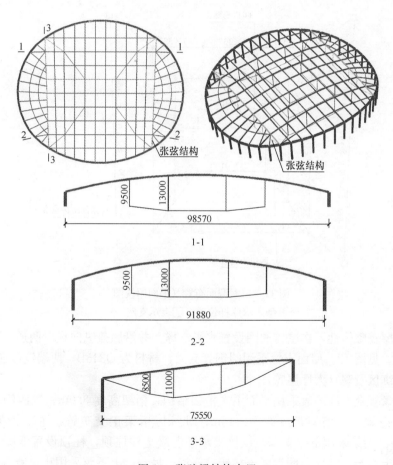

图10 张弦梁结构布置

拉索均采用高强钢丝组成的成品索，高强钢丝抗拔强度不小于1670MPa，弹性模量不小于$1.9×10$MPa，选用的拉索截面参数主要如下：

规格	钢丝束直径（mm）	钢丝束截面积（mm²）	折算直径（mm）	破断荷载（kN）
φ5×91	55	1787	47.7	2984
φ5×139	66	2729	58.9	4557
φ5×253	87	4968	79.5	8297
φ5×409	110	8031	101.1	13412
φ7×337	141	12969	128.5	21658

拉索布置见如下表格及图11。

构件		拉索规格（mm）
四边形环索弦支结构	斜索	φ5×139　φ5×253　φ7×337
	环索	φ5×91　φ5×139　φ5×409　φ7×337
张弦梁结构		φ5×409

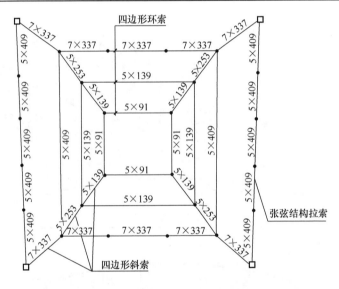

图11 拉索截面布置（平面俯视图，
●表示撑杆位置，□表示支座）

为增强屋盖整体性，在屋盖周圈设置水平支撑，并沿屋盖纵向设置两道、横向设置一道水平支撑，见图12。屋面支撑采用圆钢管截面，材料为Q345B，两端铰接连接。

（2）训练区及观众大厅屋盖

训练区及观众大厅屋盖采用平面主次桁架结构，桁架高度3.0m，东西侧观众大厅桁架呈放射状布置，次桁架跨度约7～10m；训练区桁架正交布置，主次桁架节间长度7.2m。训练区桁架采用下弦支承，铰接支承于混凝土柱柱顶，柱顶设置混凝土环梁。观众大厅桁架上弦与比赛厅周圈混凝土柱铰接连接，另一侧上下弦采用相贯节点连接于立体环桁架上。

（3）墙面交叉网格

体育馆墙面采用单层交叉曲面网格结构，采用矩形钢管截面，并与水平面成60度倾

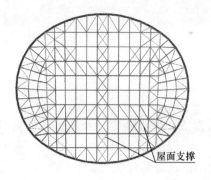
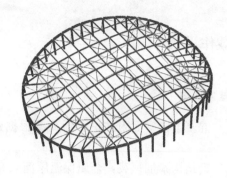

图 12 屋面水平支撑布置

角。北侧墙面结构直接连接于基础，最大高度约 34.5m，网格平面内节间长度约 5.6～8.9m，东西及南两侧网格连接于 6.0m 标高观众平台框架柱顶，墙面高度约 6.5～14.0m，网格平面内节间长度约 4.2～7.5m。墙面交叉柱上下端均采用铰接计算假定，上下端截面经计算及分析均采用杆件缩径与支座以及立体环桁架弦杆相连。墙面交叉网格模型见图 13。

(4) 交叉支撑及系杆

为增强结构整体稳定性，在比赛区、训练区内支座附近周圈布置支撑，同时沿图中屋盖竖向中轴线布置一道连续的交叉支撑。为防止桁架弦杆受压屈曲，在受压弦杆之间布置了系杆。钢结构平面布置如图 14 所示。

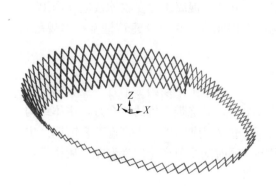
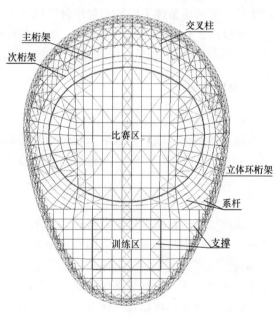

图 13 墙面交叉网格三维视图　　　　图 14 钢结构屋盖平面视图

(5) 比赛区屋盖经济指标

比赛区屋盖型钢理论用钢量 890.4t，按展开面积 9120m² 计 97.6kg/m²，索理论用量 41.3t，即 4.52kg/m²。考虑节点加劲肋以及夹节点重量，实际用量较理论用量约放大 10%。

3. 荷载作用

3.1 重力荷载

(1) 混凝土楼屋面附加恒、活荷载根据建筑作法、降板高度以及建筑功能荷载规范确定。

(2) 钢结构附加恒载：金属保温屋面 $0.85kN/m^2$，玻璃自动天窗 $1.5kN/m^2$，金属幕墙 $0.6kN/m^2$，太阳能板不小于 $0.2kN/m^2$，比赛馆训练馆屋盖内表面吸声层 $0.2kN/m^2$；马道上悬挂荷载按照设备的实际情况施加，并适当预留大型演出灯光、音响等荷载。考虑最不利上吸风荷载时，该预留悬挂荷载不与风荷载组合。

钢结构屋盖不上人屋面活荷载：$0.5kN/m^2$。

3.2 风荷载

基本风压钢结构按 100 年重现期取值 $\omega_0 = 0.85kN/m^2$，混凝土结构按 50 年重现期取值 $\omega_0 = 0.70kN/m^2$，地面粗糙度 B 类。

本工程为形状复杂、柔性大、阻尼小的大跨度风敏感柔性屋盖结构，且位于台风频发的福州市，福州市公共建设项目管理处委托同济大学土木工程防灾国家重点实验室进行了该项目常规风场条件下以及非常规台风风场下表面风压分布及风振分析的研究，风洞模型为刚体模型，几何缩尺比为 1/200。考虑结构的对称性，设计时取 $\beta=0°$、$30°$、$60°$、$90°$、$270°$、$300°$、$330°$ 七种风向输入角度下的试验结果进行验算。

3.3 温度作用

(1) 计算模型

随主体结构生成，逐层输入随时间变化的温差，并考虑混凝土徐变收缩效应（CEB-FIP MODEL CODE 1990）及桩基础对上部结构的有限刚度约束（《建筑桩基技术规范》JGJ94—2008）。

(2) 施工顺序

假设体育馆首层平台及看台施工工期约 1 个月，二层施工工期约 1 个月，三、四层施工约 1 个月，钢结构屋盖、墙面安装施工工期约 3 个月，后浇带合拢约 2 个月，主体结构全部施工完成约 8 个月，主体结构全部施工完成后，进入为期 12 个月的施工装饰期。主体结构逐层生成，后浇带逐层生成，逐层施加温差，同时随季节变化改变温差，后浇带选择在低温月合拢，可以在主体混凝土完工后一次浇筑。

(3) 温差

由于局部温差可通过施工覆盖措施予以降低，且影响较小。温差计算仅考虑结构所经历的整体温差的影响。结构施工阶段混凝土合拢温度取为施工结构组相对应时段内的平均气温，温差取为施工期阶段内最低（高）气温与相应合拢温度的差值。

装饰期内结构外围护已形成，室内温差变化较小，装饰期内温度变化不起控制作用。且由于降温时混凝土结构的变形趋势与混凝土收缩变形基本一致，二者的叠加较不利于混凝土缩裂变形及受拉应力水平的控制，从最热月 8 月份开始施工时结构将经历最不利负温差，此时混凝土结构已经经历了施工中较为不利的负温差过程，使用过程中的温度变化不起主要控制作用。钢结构合拢后将经历使用过程中的最不利温差，因此钢结构合拢应选择在低温月的较高温时段，高温月的较低温时段合拢，若钢结构在 2 月合拢，则升温温差：42－8＝36℃，降温温差：－2－8＝－10℃，若钢结构在 8 月合拢，则升温温差：42－25＝17℃，降温温差：－2－25＝－27℃。

3.4 地震作用

地震作用重现期 50 年，抗震设防烈度 7 度，设计基本地震加速度值 0.10g，设计地震分组为第三组，Ⅲ类场地。总装分析中，下部混凝土结构阻尼比按照 0.04 取值，上部钢结构按照 0.02 取值，结构构件承载力计算考虑三向地震作用。

根据建筑抗震设计规范（GB 50011—2001）50 年超越概率 63%（小震）场地特征周期 T_g 为 0.65s，水平地震影响系数最大值 0.08，水平峰值加速度 35g；50 年超越概率 10%（中震）水平地震影响系数最大值 0.24。

根据福建地震地质工程勘察院提供《福州海峡奥林匹克体育中心工程场地地震安全性评价报告》，50 年超越概率 63%（小震）场地特征周期 T_g 为 0.60s，水平地震影响系数最大值 0.101，水平峰值加速度 43g；50 年超越概率 10%（中震）场地特征周期 T_g 为 0.70s，水平地震影响系数最大值 0.249。

4. 抗震性能化标准

（1）总装结构（钢结构、混凝土结构）弹性分析下的结构位移控制以及构件截面承载力弹性设计满足安评小震弹性要求。

（2）上部钢结构杆件、节点、混凝土支承柱、混凝土支承柱柱顶环梁截面承载力满足规范中震弹性要求。

（3）其他下部混凝土框架柱、剪力墙（不包含连梁）、斜梁、转换梁截面承载力满足规范中震不屈服要求。

5. 比赛区屋盖施工全过程模拟分析

5.1 施工模拟计算步骤

本工程结构体系决定其施工方案的特殊性。合理选取初始张拉力，制订合理的张拉工艺，控制支撑构件及屋面的侧向变形及竖向挠度，尽量减小重复张拉次数，确保施工顺序顺利进行。施工全过程模拟采用 SAP2000 静力非线性分析的方法。

（1）施工下部混凝土结构及钢结构的支承柱。添加结构时，结构自重同时生成，并施加施工荷载 0.5kN/m²。

（2）搭设临时支撑，施工比赛区网格梁以及其他范围内屋盖。设计初定临时支撑点位如图 15，待施工方案确定后按照施工方案进行复核计算。临时支撑点采用竖向只压不拉单元模拟，预应力逐步施加时，克服自重作用屋盖逐渐产生向上变形，临时支撑单元随之逐步退出工作。

(3) 安装竖向撑杆及预应力拉索，此时拉索均松弛未张拉，撑杆零应力状态。

(4) 张拉四边形环索结构外环斜索（张拉步骤一）。

(5) 张拉四边形环索结构中环斜索（张拉步骤二）。

(6) 张拉四边形环索结构内环斜索（张拉步骤三）。

(7) 张拉东西两侧弦支结构拉索（张拉步骤四）。

(8) 施加金属屋面、马道等附加恒荷载。

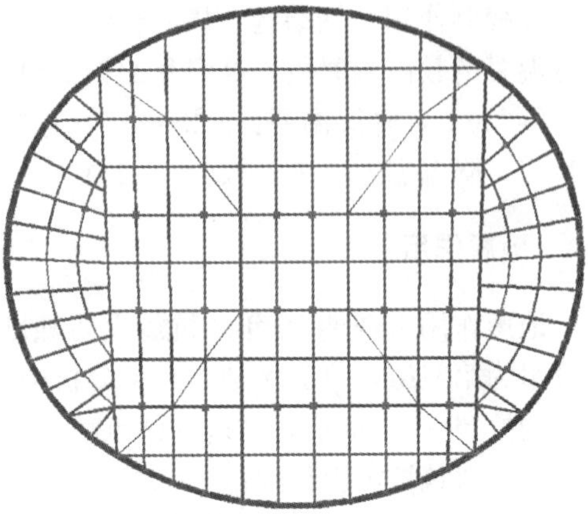

图 15 设计初定临时支撑点位
（■表示临时支撑位置）

5.2 初始预应力控制原则

拉索初始张拉应力控制在 0.1~0.2fyk。

最不利荷载组合下最小拉应力>0.05fyk，最不利荷载组合下最大拉应力<0.5fyk。

5.3 张拉过程中拉索预应力

由于各环四边形环索结构并非完全对称结构，内力平衡状态下各环的四根斜索以及环索内力每根并不完全相同，但差别不大。每环各选取一根斜索及一根环索，用图 16 表示其在张拉过程中内力及应力变化情况，其中表格横坐标表示张拉步骤，纵坐标表示拉索内力，曲线上数值表示应力水平。

由上图可见，拉索初始张拉以及在张拉过程结束后，预应力水平都在 0.1~0.2 fyk。

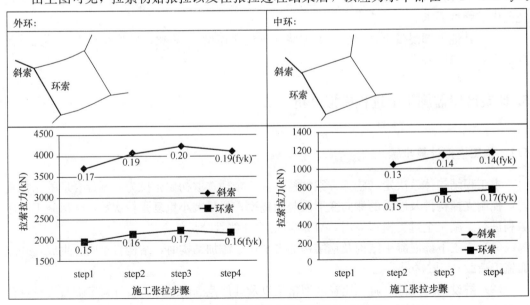

图 16 施工张拉过程中拉索预应力（一）

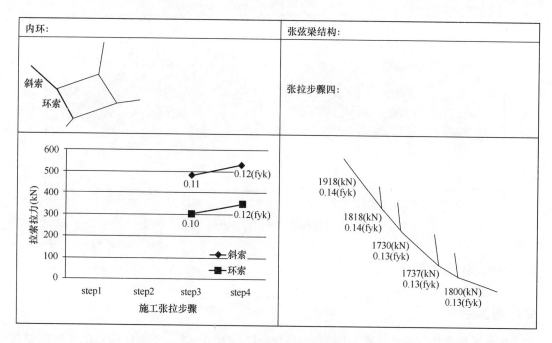

图16 施工张拉过程中拉索预应力（二）

张拉下一步时，会对已张拉的构件内力结果有所影响，但拉索在张拉各步内力变化较为均匀。

5.4 张拉过程中屋盖竖向位移

张拉过程中，随着预应力的施加临时支撑逐渐退出工作，屋盖逐渐克服自重产生上拱位移。外环斜索张拉后，外环角部撑杆上方屋盖最大向上位移71mm，继中环、内环斜索张拉后，外圈撑杆内轴力逐渐卸荷，角部屋盖向上位移逐渐减小，张拉结束后屋面最大竖向位移46mm，如图17所示。

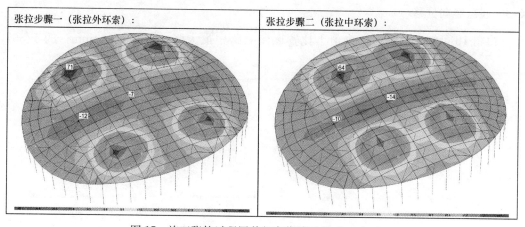

图17 施工张拉过程屋盖竖向位移云图（mm）（一）

127

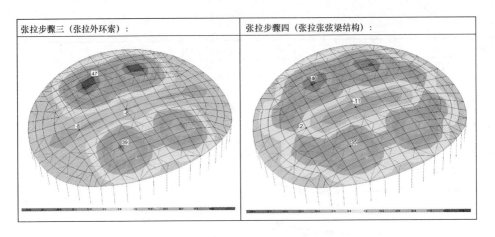

图 17　施工张拉过程屋盖竖向位移云图（mm）（二）

6. 结构整体受力性能

6.1　模态分析

比赛区屋盖竖向振动频率 $1/0.888=1.126 Hz>1.0 Hz$，上部钢结构整体扭转周期与第一平动周期的比值 $0.467/0.680=0.686<0.9$，满足要求；整体结构（包含混凝土结构）的扭转周期与平动周期的比值 $0.266/0.313=0.85<0.9$，满足要求。

6.2　安评反应谱基底剪力分析

结构总重力荷载代表值879913kN，基底反力输出方向沿结构整体坐标，阻尼比0.02。

荷载工况	FX	FY	剪重比	
	kN	kN	X向	Y向
X向小震	57864.1	1218.5	6.58%	
Y向小震	1294.6	62308.3		7.08%
以X向为主双向小震	57874.6	53026.9	6.58%	6.03%
以Y向为主双向小震	49248.7	62316.9	5.60%	7.08%

6.3　屋盖竖向位移

由图18可知，恒+活标准值下比赛区屋面跨中最大竖向位移-144mm，约1/667，

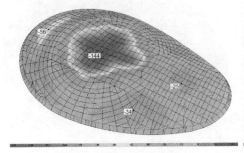
恒+活标准值下屋盖竖向位移

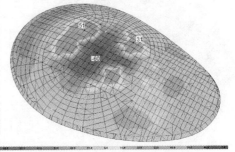
恒+风洞试验各风向角包络风荷载下屋盖竖向位移

图 18　屋盖竖向位移云图（mm）

与风荷载（上吸风）组合后，比赛区屋盖跨中最大竖向位移-60mm，边跨附近产生向上位移，最大向上位移51mm。

6.4 比赛区构件内力及应力水平

（1）矩形网格梁

矩形网格梁在张拉步骤一局部杆件短时应力最高 $0.52 \sim 0.62 f_y$，在下步张拉后此局部应力显著降低，在随后的张拉中网格梁应力最高不超过 $0.42 f_y$，普遍在 $0.25 f_y$ 以内。

恒、活荷载设计组合下网格梁应力水平最高 $0.74 f_y$，大部分应力水平 $0.2 \sim 0.6 f_y$ 之间；在小震、风、温度设计组合下应力增长不多，大部分杆件应力由竖向荷载设计组合控制；中震复核下网格梁应力水平有一定增长，尤其是东西两侧与混凝土柱连接的边缘构件以及屋面跨中构件，但应力水平均低于 $0.9 f_y$，满足中震弹性要求。

（2）屋面支撑

在外环、中环斜索附近屋面水平支撑在张拉过程中应力水平最高 $0.34 f_y$；恒、活荷载设计组合下，四边形环索结构斜索附近屋面支撑轴压力较高，应力水平最高 $0.87 f_y$；小震、风、温度设计组合下斜索附近屋面支撑在应力变化不大，屋盖跨中及边跨支撑应力有所增加。中震作用下应力普遍增长，但均低于 $0.90 f_y$，满足要求。

（3）竖向撑杆

四环形环索结构外环撑杆在张拉过程中受力最大，约 1100kN，应力水平最高 $0.28 f_y$；恒、活荷载设计组合下四环形环索结构外环撑杆受力最大约 2100kN，应力水平最高 $0.58 f_y$，张弦梁结构撑杆内最大压力约 760kN，应力水平 $0.38 f_y$；撑杆受地震作用影响不大，主要由竖向荷载及预应力的组合控制。

（4）拉索

拉索一般在竖向荷载设计组合及中震组合下产生最大拉力，应力水平最高为四边形环索结构外环斜索，拉力 7600kN，$0.35 f_{yk}$；在上吸风组合下产生最小拉力，应力水平最低为内环环索，拉力 165kN，$0.05 f_{yk}$。

6.5 结构稳定及极限承载力分析

（1）比赛区屋盖构件线性屈曲稳定分析

在自重及预应力基础上，加载模式：1.0附加恒载（不包含预留后期悬挂荷载及悬挂大设备荷载）+1.0活载，对屋盖进行线性屈曲分析，构件局部屈曲模态见图19。

由上图19可知，局部压应力水平较高的屋面水平支撑在竖向荷载下将最先产生杆件屈曲，调整支撑布置方式使得高应力区支撑的内力分布尽量均匀，将局部高应力单斜撑改为交叉撑，提高支撑的局部稳定性。

由欧拉临界荷载公式反算构件的计算长度系数，图19中屈曲支撑截面为圆钢管 $\phi 325 \times 12$，其临界荷载 $N_{cr} = K \times N_{D+L} + N_{self+press} = 7.97 \times 231 + 288 = 2129$kN（式中 N_{D+L} 为附加恒+活初始荷载下杆件的轴向力，$N_{self+press}$ 为预应力+自重下杆件的轴向力，此处为压力 288kN），该杆件长度 11742mm，$I = 1.447E8$ mm^4，可得该支撑构件计算长度系数：

$$\mu = \frac{\pi}{l}\sqrt{\frac{EI}{N_{cr}}} = \frac{3.14}{11742}\sqrt{\frac{206e3 \times 1.4476e8}{2129e3}} = 1.03$$

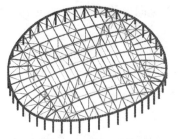
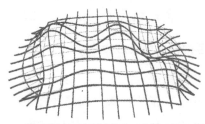

(a) 支撑构件屈曲模态（荷载系数7.97）　　(b) 网格梁局部屈曲模态（荷载系数18.57）

(c) 竖向撑杆构件屈曲模态（荷载系数35.15）

图19　屋盖构件局部屈曲稳定模态（浅灰色为未变形形状）

屋盖支撑计算长度系数按可按1.05统一取值。

同样计算得到屋盖网格梁计算长度系数可按2.3取值，竖向撑杆计算长度系数按可按1.0统一取值。

（2）比赛区屋盖整体线性屈曲稳定分析

结构整体前两阶线性屈曲振型如图20，第一阶线性屈曲荷载临界系数为9.81。

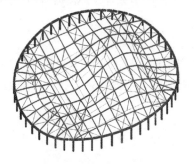
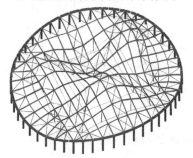

第一阶整体线性屈曲模态（荷载系数9.81）　　第二阶整体线性屈曲模态（荷载系数13.16）

图20　比赛区屋盖线性屈曲分析模态

（3）比赛区屋盖弹塑性极限承载力分析（pushdown）

在自重及预应力基础上，加载模式：1.0附加恒载（不包含预留后期悬挂荷载及悬挂大设备荷载）+1.0活载，考虑几何非线性及材料非线性。初始缺陷按照第一阶整体线性屈曲模态分布方式（图20），缺陷最大值按照屋盖跨度1/300选取。

在网格梁两端设置轴力和双向弯矩共同

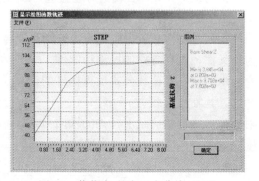

图21　荷载步-基底反力曲线（kN）

作用的塑性铰（PMM 铰），屋面支撑和竖向撑杆设置轴力铰。塑性铰采用 SAP2000 默认的 PMM 和 M 铰，其具体的力学性能参考 FEMA356 和 ATC-40。

结构稳定极限承载力安全系数：

结构达到极限承载力时，基底反力 97080kN（仅屋盖及支承柱部分），自重及预应力下结构基底反力 38454kN，1.0 附加恒+1.0 活结构基底反力 13914kN，故弹塑性全过程分析结构稳定性极限承载力安全系数 $K=$（97080-38454）/13914＝4.21＞2，满足要求。

塑性铰的开展情况：

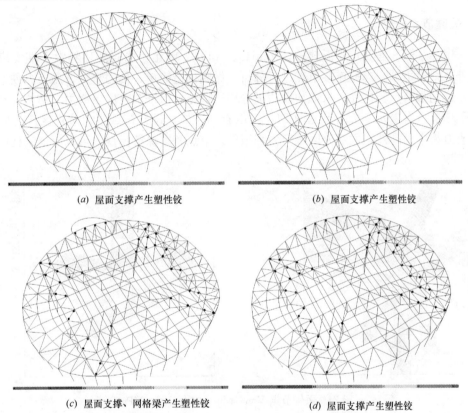

图 22　竖向荷载下比赛区屋盖塑性铰的开展情况

在竖向荷载逐渐加载过程中，屋面支撑最先出现塑性铰，继续加载，更多的屋面支撑以及网格梁支座附近出现塑性铰，随后更多屋面支撑产生塑性铰，塑性铰均在轻微破坏阶段。继续加载，结构承载不再继续升高，达到稳定极限承载力。弹塑性分析的极限承载力安全系数高于规范要求，结构具有较好的承受竖向荷载能力，有较强的安全储备。

6.6　拉索加工、安装误差影响分析

拉索在加工、安装过程中，由于工厂加工存在下料长度误差，施工存在埋件定位误差等，当拉索张拉至设计所需的预应力值时，拉索以及撑杆的几何位置可能与设计位置存在一定的偏差，如图 23 所示。

假定由于拉索加工、埋件偏差，拉索张拉后四边形环索结构外环撑杆偏离理论位置 Δ50mm，内环以及中环撑杆偏离理论位置 Δ20mm。经计算，考虑索长误差后对屋盖竖向位移基本无影响；在同样的预应力值作用下，恒、活荷载设计组合下结构典型构件内力基本变化不大，主要控制力变幅在5%以内，个别网格梁内剪力增幅较多，但因绝对数值较小，对截面应力水平影响很小。

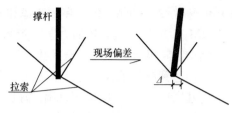

图23 拉索现场张拉后的几何位置偏差示意

6.7 关键节点分析

采用ANSYS有限元分析软件，solid45、solid92单元，选取外环斜索上端支座节点以及外环斜索下端与撑杆连接节点进行分析，做到构造简单、传力直接、施工方便、安全可靠，达到强节点弱构件。采取加设肋板等构造措施，保证节点区承载力。结构外环斜索上、下端节点有限元分析结果见图24，可以看到在最大设计荷载下，节点区应力除个别点因应力集中稍高外，大部分区域应力都很低，节点基本保持弹性，可以保证结构的正常安全工作。

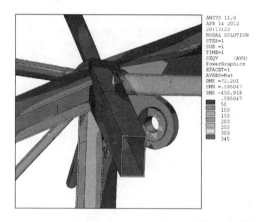
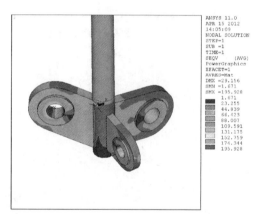

图24 关键节点 Von—Mises 应力云图（MPa）

7. 结论

四边形环索弦支结构的多重应用，在国内此规模的大跨度屋盖的应用尚属首次，多重弦支的合理设置，同时结合在边跨设置的张弦梁结构，使得屋面网格梁整体受力分布均匀，屋盖刚度合理。假定屋盖的施工支承条件，分析施工分步张拉的全过程，确定主动索的初始预拉力，并分析预应力张拉过程中屋盖受力及位移情况，有效的指导施工全过程。

进行竖向荷载分析、风荷载分析、地震反应谱分析、地震波时程分析、线性稳定分析、极限承载力分析、考虑下部支承构件刚度退化、拉索加工安装误差的影响分析，以及在偶然荷载作用下的断索分析，对结构薄弱部位进行加强，使屋盖结构刚度及构件内力始终在安全范围。进行节点有限元分析，通过在节点区域设置加劲肋等构造措施，保证强节点弱构件，提高结构延性。整体结构形式简洁，传力明确，经济指标较好。

参考文献

[1] 徐培福，傅学怡，王翠坤，肖从真．复杂高层建筑结构设计．北京：中国建筑工业出版社，2005
[2] 傅学怡．实用高层建筑结构设计．北京：中国建筑工业出版社，2010

大偏心单柱双层高架桥之设计与施工——以台湾国道1号五股杨梅段拓宽工程泰山至林口段为例

王泓文[1]，蔡益成[2]，陈光辉[3]，林曜沧[4]，王照烈[5]，张荻薇[6]

(1. 台湾世曦工程顾问股份有限公司，第一结构部计划工程师；
2. 台湾世曦工程顾问股份有限公司 五杨北段监造项目经理；
3. 台湾世曦工程顾问股份有限公司，第一结构部副理；
4. 台湾世曦工程顾问股份有限公司，第一结构部协理；
5. 台湾世曦工程顾问股份有限公司，资深协理；
6. 台湾世曦工程顾问股份有限公司，总经理)

摘　要：高架道路常因交通量大须有较宽之桥面，惟受限于路权宽度，部分路段必须采双层高架配置因应。就桥梁整体结构配置及减少道路用地等因素考虑，以往双层桥梁上下层均尽量重叠布设，也因此大多采框架式桥墩配置，例如台北市水源快速道路、台北市环东快速道路…等。惟基于下层采光、视野感受及桥墩外观造型上之考虑，加上受用地及施工条件限制下，必须采双层仅少部分重叠方式布设，并以双层不同方向外悬帽梁之树状式单柱及结构大偏心方式配置桥墩。此种结构配置为台湾双层高架桥首见，桥墩柱之断面力变化与传递及整体桥梁结构动力行为较复杂难掌握，必须深入研析了解其结构力学行为，始能确实掌握结构之安全度。

台湾国道1号五股杨梅段拓宽工程泰山收费站至林口段即为此种配置之典型案例，该路段位于林口台地爬坡路段，并沿国道1号高速公路及大窠坑溪间布设，因受限于高速公路及大窠坑溪之严苛用地限制，上下双层桥采部分桥面重叠之大偏心单柱式桥墩配置，导致常时静载重及活重作用下，即对柱底产生较大的偏心弯矩，此一弯矩对墩柱在常时情况下即可能产生挠曲裂缝，且此一不对称双层桥之桥墩系有别于以往一般高桥及规则性双层桥构架或桥墩之结构行为。本文将以此为例，首先针对大偏心单柱双层高架桥作一整体性介绍，并进行动力分析以了解其在地震力作用下之动力行为。除此外，也针对大偏心单柱桥墩配合柱内 H 型钢和预力施拉作进一步之分析与探讨，以提供作为未来工程界相关工程之参考。

关键词：双层桥梁；大偏心桥墩；桥柱内置 H 型钢；常时情况

第一作者：王泓文（1970—），男，硕士，硕士，结构技师，主要从事桥梁及结构工程规划设计，E-mail：wanghw@ceci.com.tw

Design and construction of the double deck verdict with eccentric single pier-Take National Freeway No. 1 wilding works form Wuku to Yangmei as an example

Wang Hung-Wen [1], Tsai I-Cheng [2], Chen Kuang-Hui [3],
Lin Yew-Tsang [4], Wang Jaw-Lieh [5], Dyi-Wei Chang [6]

(1. CECI Engineering Consultants, Inc. Taiwan, Structural Engineering Department (1) Project Engineer
2. CECI Engineering Consultants, Inc. Taiwan, WuYang Project Manager
3. CECI Engineering Consultants, Inc. Taiwan, Structural Engineering Department (1) Deputy Manager
4. CECI Engineering Consultants, Inc. Taiwan, Structural Engineering Department
(1) Assistant Vice President
5. CECI Engineering Consultants, Inc. Taiwan, Associate Vice President
6. CECI Engineering Consultants, Inc. Taiwan, President)

Abstract: Freeway verdict often has the wider deck because of the great traffic volume. Some particular section adopt double deck configuration due to limited right of way. Consider the entirety bridge structure configuration and to decrease the land usage, double deck use to place one on top of another and use frame type pier, shuiyuan express road and huantung express road in Taipei are typical examples for this kind of bridge. Consider the natural light of the lower deck, driving vision, appearance of pier, and the limited construction site, the double deck verdict in this project can only allow little overlap and should overhang in different direction which leads to a tree-like eccentric single pier. This kind of structure configuration is first seen in the double deck bridge in Taiwan. The change and transfer of section force as long as the dynamic behavior turns out to be very complicate, which needs detail research in order to understand the structure behavior and its safety and reliability.

Bridges between Taishan toll station and Linkou section of National Freeway No. 1 wilding works from Wuku to Yangmei is a typical example for double deck verdict with eccentric single pier. Located in the uphill section of Linkou tableland, along National Freeway No. 1 and Dakekeng River, the double deck verdict only allows little overlap and adopts an eccentric single pier. The bottom of the piers in this configuration has major eccentric moment under dead load and live load, and causes flexural crack during normal situation. The structural behavior of the asymmetric pier is different from the general double deck verdict. This article takes this project as an example, carry out the overall introduction of the double deck verdict with eccentric pier. Dynamic analysis is proceeded in order to understand the behavior of the pier under earthquake. In addition, analysis and research for the H-beam and the prestressed tendon placed in the eccentric pier is done and provide as a reference for the engineering related projects.

Keywords: double deck verdict; eccentric pier; H-beam placed inside pier; normal situation

1. 前言

随着经济快速成长及人口都市化现象，使得用地取得成为交通建设的难题之一，因此，现今都市建设规划常以高架道路取代以往平面道路以减少用地征收的问题。此一类型桥梁却常因路权宽度限制及高交通量需求须采双层高架配置，例如台北市水源快速道路、台北市环东快速道路、台 64 线新北市中和区路段等，皆属此种框架式双层高架桥墩配置，然而考虑下层桥面之采光、视野感受及桥墩外观造型以及施工条件限制下，改采上下层桥面部份重叠方式布设，并以双层不同方向外悬帽梁之树状式单柱及大偏心方式配置桥墩。

目前台湾的重大工程建设"国道 1 号五股杨梅段拓宽工程计划"之泰山收费站至林口段桥墩配置即属此大偏心单柱双层高架桥。其工程位置图如图 1 所示；此一工程范围跨越泰山收费站后，南行至北上线里程 37K＋500～40K＋150 区间将接近岩屑崩滑高敏感区（37K＋400～39K＋100），以及直接危害可能性之土石流高危险潜势区（37K＋600、37k＋800 等两处），依据该计划环境影响评估之承诺，计划路线须回避，故北上线高架桥于里程 37K＋000 附近跨越国道 1 号至南下线侧，并与南下线采双层共线设置（北上线置于上层），后于里程 40K＋000 附近再度跨越国道 1 号回归原北上线侧，对于泰山收费站至林口段，大寨坑溪紧邻国道 1 号并与路线平行，高架桥系沿国道 1 号与大寨坑溪间狭窄廊带布设，如图 2 及图 3 所示。

图 1　国道 1 号五股杨梅段拓宽工程位置图

图 2　双层高架桥工程位置图

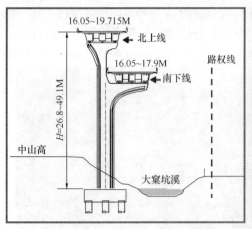 

图3 双层高架桥断面图及计算机仿真图

2. 双层高架桥工程概要

2.1 工址环境

本路段双层高架桥属五杨段拓宽工程计划C903标泰山林口段,路线长约2,855m,计有53座桥墩,南下线与北上线共计26个单元104跨桥梁,上部结构为多跨连续钢箱型梁＋RC桥面板(如图3所示)。因于大窠坑溪河中不可落墩及大窠坑溪对岸私有地征收困难且耗时,又考虑大窠坑溪部分已整治且仍为多样生态之自然溪流,桥梁布设空间受限等设计条件,如图4所示,致仅能于国道1号与大窠坑溪间狭窄廊带立墩布设南下线与北上线共线并行之双层桥梁。为使共构桥柱可立墩于国道1号既有边坡,且尽量避免桥梁侵入大窠坑溪与遮蔽日照,路线系采上下层桥面仅少部分重叠方式布设,并采双层不同方向外悬伸帽梁之树状式单柱及结构大偏心方式配置桥墩,如此大窠坑溪河中不须立墩,河道不须改道,故亦无新增用地及建物拆迁之问题。

图4 工址施工前鸟瞰图

2.2 桥梁结构形式研选

本路段桥梁工程配合道路线形,系以双层桥面部分重叠之共构方式布设,下层桥面距地面高约 18～33m，上下层桥面间高约 12m，柱高将高达约 30～45m，其结构行为较一般框架式双层共构桥梁复杂，国外虽有类似案例，但柱高与桥面宽等规模均远小于本计划。

本路段上层桥梁因高度较高，为减轻地震作用力，采用自重轻的钢箱型梁，下层桥梁则可选择预力箱型梁或钢箱型梁。考虑本路段之施工条件，下层桥梁下方为大窠坑溪，如选择预力箱型梁无法采用传统逐跨架设工法进行施工，如采用支撑先进工法施工，则上层桥墩将与工作车位置相冲突，且工址处附近均属保护区，无法设置预筑场，因此，基于上述理由，下层桥梁亦选择钢箱型梁。

此外，本路段桥墩高度最大达 45m，且须配合河川行水区域采偏心设置，柱底弯矩相当大，因此，桥墩断面尺寸亦须配

图 5　单柱式结构大偏心外总帽梁桥墩
（施工中）

合加大，故桥墩形式之研选较为受限，预铸桥墩较不适宜，评估可选择场筑 RC 桥墩或钢桥墩。由于本标工程主要于国道 1 号与大窠坑溪间之下边坡施工，全线均于便桥及构台上施工，工区狭窄、施工空间小、出入口少、施工动线不佳。因为桥墩柱及帽梁量体大，如采钢桥墩现场施工量大，质量不易控制。另于便桥及构台上施工，钢桥墩地组空间不足、活动式吊车难以布设吊装。故选择单柱式 R.C 桥墩内配置 H 型钢柱及内施预力（如图 5 所示）。

3. 结构大偏心单柱双层高架桥设计概要

3.1　结构大偏心单柱双层高架桥特性

此一规划桥梁上层高度约 30～45m，下层高约为 18～33m，上层桥宽为 16.05～19.715m，下层桥宽为 16.05～17.9m，结构动力特性将有二个明显的主要振态，且上下层质量中心与纵向的劲度中心不一致，将有明显之偏心扭转行为，其力学行为复杂，因此设计时亦须进行动力分析以了解其在地震力作用下之动力行为。另由于上层桥高、劲度小、位移量大，风力作用及防落措施须特别考虑。

3.2　上部结构配置

全在线下层共构长 2.857km，上部结构采自重轻之中跨径钢箱型梁桥配置，上、下层各 13 单元 52 跨（52.5+3@55+12－4@55），并因应环评之环境保护对策："林口北上与南下共线路段，其北上线两度跨越中山高平面道路不落墩，以维护道路景观质量及减少施工干扰"之环评承诺，为维持较短跨径，以降低桥梁量体，及工期与经费之考虑，二处跨越段曲线半径采 390m 布设，并以加宽路肩（6.75～7.15m）之方式，满足最短停车视距之规范要求。

3.3 桥墩 3H 工法之特性

以往高桥墩（high bridge piers）设计大都采用钢筋混凝土式桥墩，其所需钢筋数量庞大，因此，施工时将导致钢筋绑扎与混凝土浇置的困难，且此路段紧邻高速公路，行车交通量大。基于以上考虑，参考日本发展所谓 3H（Hybrid、Hollow、High Pier）工法，其最大的特色在于中空桥墩内置钢柱或钢管，由钢柱与钢管取代大部分桥墩的主筋，如图 6 所示，大幅减少钢筋绑扎与混凝土浇置的困难，进而增加施工便利性与缩短施工工期，并具有较佳的耐震性能。

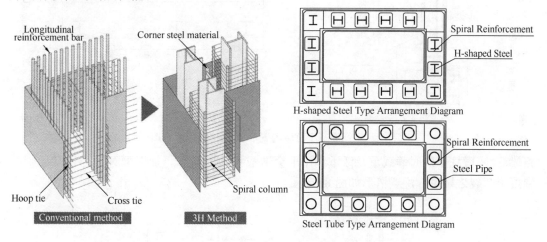

图 6　桥墩 3H 工法设计断面示意图

3.4 柱内设置 H 型钢

本路段之桥墩平均高度达 40m，属于高桥墩，经评估，桥墩施工应为本标工程之要径工程项目，因此，为提升本工程桥墩施工效率及安全性，故参考日本 3H 工法的构想，于每座桥墩内设置 H 型钢柱，一方面 H 型钢柱可取代部分钢筋承受荷载，可改善传统高桥墩的钢筋间距过于紧密，导致钢筋绑扎困难与混凝土浇置问题，另一方面 H 型钢柱可作为模板设置与钢筋组立时的支撑构架，降低施工意外之风险。本标工程全线共配置 53 座桥墩，各座桥墩所承担的载重大小不尽相同，若全面采用同一断面尺寸，或依据各座桥墩的载重大小设计不同的断面尺寸，均不符合经济成本，故为同时兼顾工程成本与统合桥墩断面外观尺寸，将全线桥墩断面尺寸划分成六种断面尺寸，其中垂直行车向宽度有 400cm 与 450cm 尺寸，行车向尺寸则有 500cm、540cm 与 580cm 三种变化。

每座桥墩配置 6 根 H 型钢柱，其中四根 H 型钢柱从桥墩基座延伸至上层帽梁，另外 2 根 H 型钢柱则从桥墩基座延伸至下层帽梁，如图 7 所示。

3.5 柱内设置预力钢腱

本标工程之上下双层采部分桥面重叠之结构大偏心单柱式桥墩配置，导致常时静载重及活重作用下，即对柱底及大悬臂帽梁产生较大的偏心弯矩，此一弯矩对墩柱及帽梁在常时情况下即可能产生挠曲裂缝，影响用路人对结构安全性和景观视觉上的感受，因此就本

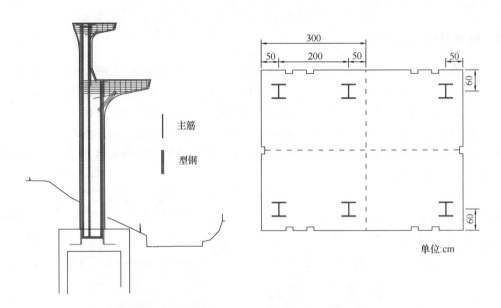

图 7 钢筋混凝土桥墩内置 H 型钢柱断面示意图

路段 53 座墩柱于常时静载重及活重作用下会造成挠曲裂缝者，加以配置预力钢腱以消除混凝土开裂之疑虑，配置情形如图 8 所示。

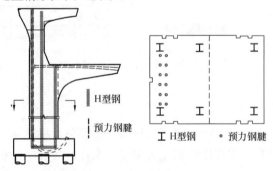

图 8 钢筋混凝土桥墩柱内置预力钢腱断面示意图

4. 桥梁之结构动力分析与检核

4.1 桥梁分析实例

以四跨连续双层之钢箱型梁桥为例，由大梁、帽梁与桥墩三大部分所组成。大梁由 RC 桥面板与钢箱梁组成，跨径为 4@55m＝220m，上层桥面宽度 16.6m、下层宽度 17.9m。桥面分为上层桥面之 GA、GB 与 GC 三根主梁；以及下层桥面之 GD、GE 与 GF 三根主梁。GB 内梁位于左桥面板中央，与 GA 及 GC 外梁间距皆为 4.15m。GE 内梁位于右桥面板中央，与 GD 及 GF 外梁间距皆为 4.8m。钢箱梁间设有横梁，横梁间距约 5.5m。

桥墩采用单柱式桥墩，高度各为 46.5m、47.2m、47.9m、48.2m、48.9m。下层桥面坐落于单柱悬臂侧，高度各为 34.5m、37.9m、38.3m、40.27m、40.7m。

分析模型坐标原点接近 GB 内梁轴线与 P1 墩交点。所有构件均以梁-柱元素（beam-column element）仿真之，定义各杆件第 1 轴为沿杆件轴向、第 2 轴为断面深度方向（铅直轴）、第 3 轴为断面宽度方向（水平轴）。

支承条件，沿行车方向于 P1 至 P5 桥墩分别为无变位束制（Roller）、有变位束制（Hinge）、有变位束制（Hinge）、有变位束制（Hinge）与无变位束制（Roller）。沿垂直行车方向皆为有变位束制（Hinge）。

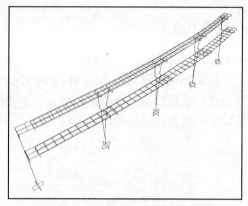

图 9　第一模态，周期 1.82s（一）

4.2　动力分析检核

模态分析结果如表 1 与图 9 所示。列出前 10 个周期与模态参与质量比。第 1 模态为行车向振动，周期为 1.82s、第 2 模态为垂直行车向振动，周期为 1.44s、第 3 模态为扭转振动，周期为 1.34s。

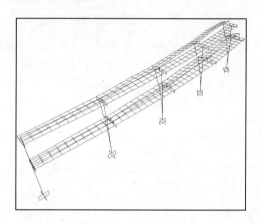

图 9　第二模态，周期 1.44s（二）

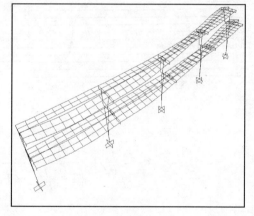

图 9　第三模态，周期 1.34s（三）

振态分析结果　　　　表 1

OutputCase Text	StepType Text	StepNum Unitless	Period Sec
MODAL	Mode	1	1.820
MODAL	Mode	2	1.439
MODAL	Mode	3	1.340
MODAL	Mode	4	1.212
MODAL	Mode	5	1.159
MODAL	Mode	6	1.068
MODAL	Mode	7	0.969
MODAL	Mode	8	0.828
MODAL	Mode	9	0.822
MODAL	Mode	10	0.772

4.3 应力检核

4.3.1 主梁断面应力比检核

桥梁钢板材质采用 ASTM A709 Gr.50，载重组合依"公路桥梁设计规范"之规定加以组合，根据载重组合分析结果进行主梁断面应力比检核，如图10所示，结果尚属安全。主梁断面应力检核公式如下所示：

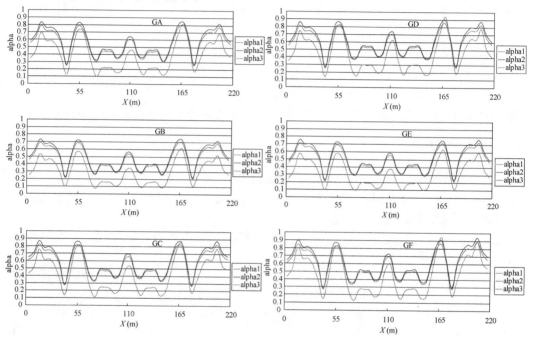

图 10 主梁应力比检核

$$\alpha_1 = \frac{f_a}{F_a} + \frac{C_{mx} \times f_{bx}}{\left(1 - \frac{f_a}{F'_{ex}}\right) \times F_{bx}} + \frac{C_{my} \times f_{by}}{\left(1 - \frac{f_a}{F'_{ey}}\right) \times F_{by}} \leqslant 1.0$$

$$\alpha_2 = \frac{f_a}{0.472 F_y} + \frac{f_{bx}}{F_{bx}} + \frac{f_{by}}{F_{by}} \leqslant 1.0$$

$$\alpha_3 = \left(\frac{f_a + f_{bx} + f_{by}}{0.55 F_y}\right)^2 + \left(\frac{f_V}{F_V}\right)^2 \leqslant 1.2$$

其中

$$C_{my} = C_{mz} = 0.85$$
$$F_y = 3500 \text{kgf/cm}^2$$

$\left(1 - \frac{f_a}{F'_{ex}}\right)\left(1 - \frac{f_a}{F'_{ey}}\right)$ 采用 0.9

F_a 采用 $0.4 F_y = F_a = 1400 \text{kgf/cm}^2$

$F_{bx} F_{by}$ 采用 $0.5 F_y = F_b = 1925 \text{kgf/cm}^2$

F_v 采用 $0.3 F_y = F_v = 1155 \text{kgf/cm}^2$

4.3.2 桥柱帽梁断面应力检核

分析显示桥柱断面配筋量主筋最大为 2.5%。墩柱内力与交互影响图比较结果如图 11 所示，分析结果均属安全。

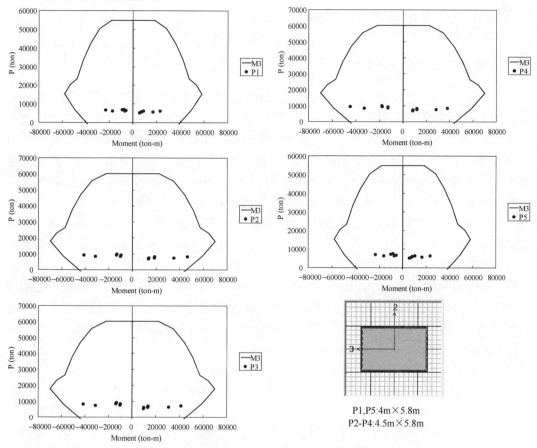

图 11 桥柱 P1～P5 交互影响图内力检核

4.3.3 梁端间隙检核

在地震发生时，为避免相邻两主梁间发生碰撞，依据交通部"公路桥梁耐震设计规范"规定，需检讨行车向地震力 EQX 作用下，同一桥墩上之相邻两主梁间之梁端间隙 S_B。

$$S_B \geqslant u_{Se} c_\phi + L_A, \quad u_s = 1.2 \alpha_y R_a u_{se}$$

其中：

S_B：所需梁端间隙，（cm）。

u_S：设计地震引致之主梁与桥台间相对位移，或同一桥墩上相邻两主梁对地相对位移之大值（cm）。

u_{Se}：长周期上部结构与下部结构间最大相对变位弹性变位。

c_ϕ：与相邻两振动单元的基本振动周期有关之调整系数。

L_A：上部结构施工误差所需之梁端间距余裕量，其值可取 1.5，（cm）。

以本标桥墩最高之单元进行检核，其分析结果为：

$$u_{Se} = 28.93 \mathrm{cm}, u_S = 57.28 \mathrm{cm}, c_\phi = 1.4$$

求得所需梁端间隙 $S_B=42.0$cm

实际配置梁端间隙为 48cm$>S_B$

5. 桥墩内置 H 型钢及预力钢腱设计

5.1 桥墩内置 H 型钢设计要领

依据内政部"钢骨钢筋混凝土构造设计规范与解说"之规定，在进行耐震设计时，一般 SRC 构造较少采用降伏强度 F_{ys} 大于 3,520kgf/cm² （50ksi）的钢材，其主要原因系为钢材之 F_{ys} 大于 3,520kgf/cm² 时，其极限应变较小，韧性相对降低。故本标工程所采用之 H 型钢柱为 A572（Grade 50）钢材，其目的在于提供施工方便性及安全性，而非完全取代主筋，因此，H 型钢柱所采用的尺寸为 H400×400×13×21，所对应的钢筋比约 0.5%（即 6 根 H 型钢柱之总断面积除以桥墩断面积）。关于 H 型钢柱配置于桥墩内的位置，依据内政部"钢骨钢筋混凝土构造设计规范与解说"之规定：当钢骨钢筋混凝土构材之主筋为 D22（♯7）以上时，钢骨之混凝土保护层须为 12.5 cm 以上；主筋与钢骨板面平行时，其净间距应保持 2.5 cm 以上，且不得小于粗骨材最大粒径之 1.25 倍。并考虑主钢筋配置所需空间，因此设置 H 型钢柱距面层 50cm 与 60cm。一般钢构件在设计时需检核整体挫屈与局部挫屈，本标桥墩的 H 型钢柱设置于钢筋混凝土内，钢筋混凝土可以提供 H 型钢柱良好的围束，防止 H 型钢柱发生整体与局部挫屈的情形，针对此问题与日商大日本工程顾问株式会社专家讨论也得到相同答复。

计算桥墩断面主筋量可依据内政部"钢骨钢筋混凝土构造设计规范与解说"，设计强度则可依据中国土木水利工程学会"混凝土工程设计规范与解说"中的设计，该法主要是将 SRC 构材中之钢骨视为等量的钢筋来设计（依 ACI-318 规范订定），并假设钢骨与 RC 之界面无相对滑动发生，亦即属于完全合成作用（Fully Composite）之状况。其中关于钢骨视为等量的钢筋的配置方式，依据日商大日本工程顾问株式会社的建议，将等量的钢筋均布在 H 型钢柱位置上，如图 12 所示。此外，为符合钢筋混凝土桥柱之韧性要求，依据交通部"公路桥梁耐震设计规范"之规定，桥墩主钢筋比不得小于 1%，亦不得大于 4%。

由于 H 型钢柱与混凝土的黏结力不如钢筋与混凝土的黏结力，若无配置剪力连接物，将产生剪力破坏而非弯矩破坏，进而降低桥墩断面之韧性，因此，在 H 型钢柱的特定部位（如塑性区或梁柱接头范围内）加设剪力钉。剪力钉应符合的规定：栓钉直径一般为 19mm，长度不宜小于 4 倍栓钉直径，间距不宜小于 6 倍栓钉直径，焊接质量应满足内政部"钢构造建筑物钢结构设计技术规范"要求。H 型钢柱埋入桥墩基座深度不应小于 3 倍型钢柱截面高度，在柱脚部位和柱脚至上一层的范围内，钢骨柱翼缘外侧设置栓钉，栓钉直径不小于 19mm，间距不大于

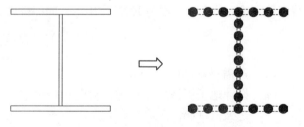

图 12 H 型钢柱转换等量钢筋示意图

20cm，且栓钉至翼缘板边缘的距离大于5cm。

5.2 桥墩内置预力钢腱设计要领

考虑本桥墩形式之墩柱与悬臂帽梁在常时载重作用下承受较大的静重（Dead Load）和活重（Live Load），因此，需检核墩柱在常时外力作用下是否造成开裂，以下式计算开裂弯矩，并求得需求之预力量，本文采用19T-15.2mm 预力钢绞线，预力钢材极限应力为 $F_{pu}=19000 kgf/cm^2$，并考虑所有预力损失取预力钢材服务阶段之应力为：$0.6F_{pu}$，取有效预力量为 $1000kgf/cm^2$，计算需求之钢腱数。

$$\frac{P}{A} - \frac{M_{cr} y}{I_g} = -2\sqrt{F'_c}$$

其中：P 为柱底承受之垂直荷重；A 为柱底之断面积。

M_{cr} 为墩柱所能承受之开裂弯矩。

y 为受拉钢筋之有效深。

I_g 为全断面惯性矩。

$F'_c = 420 kgf/cm^2$

6. 结构大偏心单柱双层高架桥之施工

6.1 钢桥自动化吊装工法

本标工程须考虑于此严苛用地限制及狭窄空间施工，因此，桥梁结构方案应符合"标准制式化跨度"、"施工不影响交通"、"不宜采现地支撑"、"保存大窠坑溪生态"等原则。桥梁上部结构宜采制式化及营建自动化之桥梁工法，经评估国内外各先进桥梁自动化工法，如预力桥梁之"支撑先进工法"、"预制节块吊装"、"场制悬臂工法"、"节块推进工法"、"预制全跨吊装"或钢桥梁之"推进工法"、"悬臂吊装工法"等均有其困难性，在综整上述工法后，汲取"预制节块吊装工法"及"全跨吊装工法"之精髓，以自走式桁架工作车进行钢梁吊装，统称为"钢梁逐跨吊装工法"，如图13所示，将成为台湾最具规模之钢桥自动化吊装工程。

此种工法具有吊梁速度快、起吊点

图13 钢梁逐跨吊装工法

少、钢梁免落地、工区免节块暂置场、免暂撑架、吊梁用地小及对地表扰动小等优点。此外，钢梁生产制造、试拼装钢梁运输及工区内之动线、钢梁在便桥上地组、吊装时程等作业，均需缜密规划控制，即可达到经济、快速、安全、环保的目标，堪称为最有效率之钢桥自动化吊装。

钢梁逐跨吊装工法主要分两大作业工项，一为钢梁吊装作业，二为桁架工作车推进作业，分别详见图 14 及图 15。

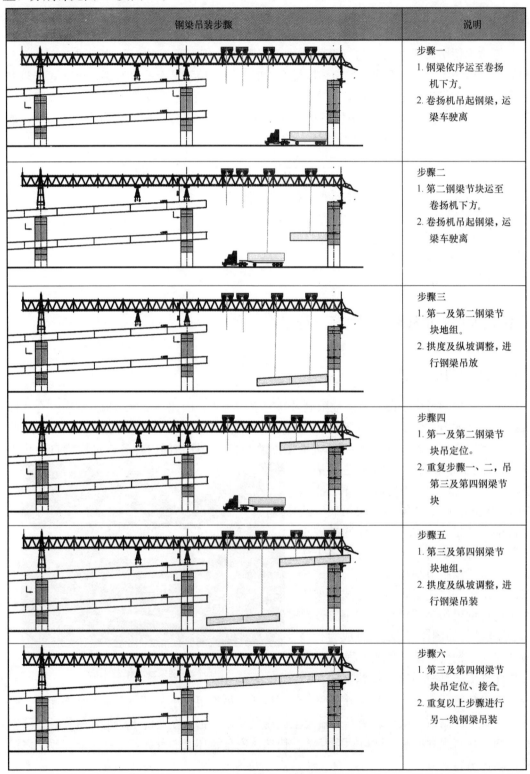

图 14 钢梁吊装步骤示意图

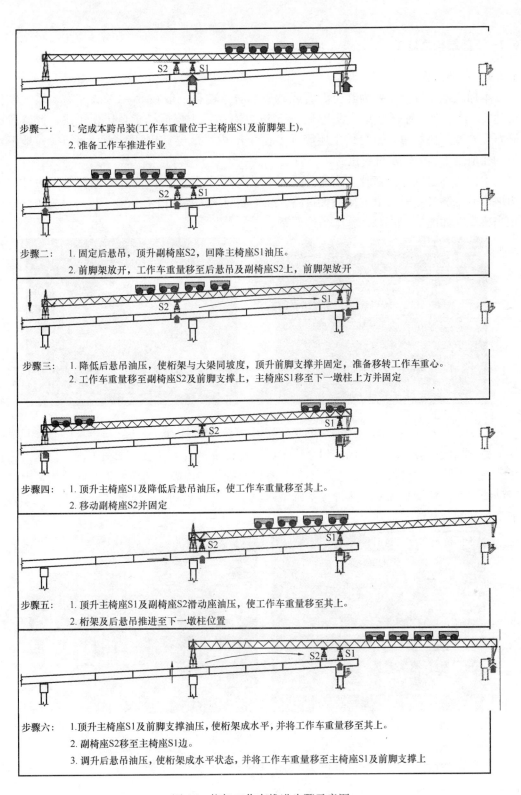

图 15 桁架工作车推进步骤示意图

6.2 下部结构之施工

6.2.1 基础工程

本标之地质特性主要为红土砾石或砂质砾石，粒径一般在5～10cm左右，最大粒径可达50cm以上，砾石上方一般上覆1～5cm左右之红土层或棕黄色风化土层，此路段卵砾石上方覆土层甚薄，一般覆土层厚度多在5cm内，地层N值几乎都大于100。地下水位约在地表下5～13m间变化。

本标基础之位置紧临既有高速公路（中山高），且位于中山高速公路旁之下边坡或上边坡，另一侧又受限大窠坑溪或大科路，施工期间须维持其既有功能，故可供施工之空间相当狭窄（如图16所示），施工条件严苛。

图16　桥梁可供施工之空间相当狭窄

另为减低植被扰动，本标工程全线于设置便桥及构台上施工（如图17所示），工区狭窄、施工空间小、出入口少、动线不佳。考虑基础与中山高距离、施工空间及构台上之施工动线，本标桥梁基础采用两种深基础型式，一为直径2.0m全套管基桩之基础；二为竹削型护基配合井筒式基础。

图17　便桥及构台

全套管基桩施作时，因基桩位于中山高陡峭下边坡，全套管基组于构台上施工，构台高达16m，基桩之空打长度最长超过20m，增加不少施工难度。另桥墩基础临中山高侧之开挖深也达16m，基础开挖挡土措施则采土钉逐阶降挖喷拟土工法，如图18、图19所示。

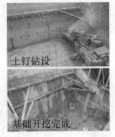

图18 施工构台

图19 基础开挖挡土措施则采土钉逐阶降挖喷拟土工法

竹削型护基配合井筒式基础，施工所需面积较小，主要于中山高与大窠坑溪间之空间无法以桩基础施工时采用，本工程于中山高陡峭下边坡施作井式基础，并配合竹削式开挖挡土措施（如图21所示），其施工效率、安全性高。另因本基础临近大窠坑溪，又地质属透水性高之砂质砾石，施工开挖时须适度抽水祛水。

图20 竹削型护基配合井筒式基础施工

6.2.2 墩柱工程

本标桥墩平均高度达40m，墩柱纵向宽度4.0及4.5m，横向宽度5.0、5.4及5.8m，混凝土采用 $f'_c = 420\,\text{kgf/cm}^2$ 之高强度自充填混凝土（SCC，Self-Compacting Concrete），以缩小量体。施工时，墩柱采用系统性模板，每一升层为4m，因桥墩内设置H型钢柱及预力钢腱须预埋预力套管，施工界面繁琐，另下层悬臂帽梁长度最大达13m，因地形条件，下层悬臂帽梁之支撑架需以钢质托架锁固于已完成之墩柱（如图21所示），悬臂帽梁之量体大，施工难度高。

图21 悬臂帽梁之支撑架及钢质托架

桥墩柱之预力钢腱需采先预埋预力套管，待桥墩柱整体完成后，才可穿预力钢腱，端锚设于墩柱顶部及墩柱底部进行预力施拉；另下层悬臂帽梁之预力钢腱需与预力套管一并埋设，预力钢腱为一固定端锚预埋于墩柱内，施力端锚设于悬臂帽梁之外侧进行预力施拉（如图 22 所示）。

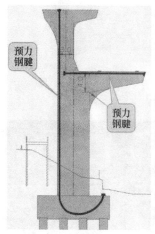

图 22　桥墩内置预力钢腱柱断面示意图及施工照片

7. 结论与展望

台湾地幅狭小，人口稠密，工程建设用地受限，本案考虑下层桥采光及视野感受采用一不对称双层不同方向外悬帽梁之树状形单柱及结构大偏心之桥墩，可大幅减少用地，减少用地范围征收，并考虑施工安全以避免影响高速公路行车及避免常时载重偏心桥柱及大外悬桥墩帽梁产生张力裂缝，故配置柱内 H 型钢和预力钢腱。期望此一配置型式得以满足用地及施工上之需求，提供作为未来工程界相关工程规划设计与施工之参考。

致谢

本桥在分析与设计过程中，台北科技大学宋裕祺教授、国家地震中心洪晓慧博士及大日本工程顾问株式会社对本设计提供相关参考数据及许多宝贵意见，谨此一并致谢。

参考文献

[1] Michio, O.；Jiro, F.；Takuya, A.；Moriyuki, O.；Yasuyuki, K. Development of New High Bridge Piers Containing Spiral Reinforcement，NIST SP 931；August 1998
[2] T KATO And Y TAKAHASHI. Earthquake Design And Construction of Tal Composite Bridge Pires
[3] 交通部，"公路桥梁设计规范"，台北：2009 年 12 月
[4] 交通部，"公路桥梁耐震设计规范"，台北：2008 年 12 月
[5] 交通部，"公路桥梁耐震设计规范"修订条文，台北：2009 年 6 月
[6] 内政部，"钢骨钢筋混凝土构造设计规范与解说"，台北：2004 年 4 月
[7] 中国土木水利工程学会，"混凝土工程设计规范与解说"，台北：2004 年 12 月

[8]   内政部,"钢构造建筑物钢结构设计技术规范",台北:2007年6月
[9]   林曜沧,陈光辉,王泓文,杨景华,翁新钧."大偏心单柱双层高架桥之设计-国道1号五股杨梅段拓宽工程C903标",中华技术第89期,2010年10月
[10] 林曜沧,陈光辉,王泓文,翁新钧."大偏心单柱双层高架桥之分析与设计-以国道1号五股杨梅段拓宽工程泰山收费站至林口段为例",第十届结构工程研讨会,2010年12月

空间钢结构无线传感监测技术研究与实践*

罗尧治

(浙江大学空间结构研究中心,浙江省空间结构重点实验室,杭州 310058)

摘 要：空间钢结构在大型建筑和公共设施中得到广泛应用,由于外形独特、体系复杂,建造难度大,通常也是人员聚集、大型活动的场所,其安全性尤为重要,而健康监测技术是保证空间钢结构施工与服役期安全的有效手段之一。本文从各类传感器研发、无线组网技术、安全测评技术等方面阐述了基于无线传感的空间钢结构健康监测系统,并在国家体育场、世博会英国馆、印象大红袍、杭州火车东站等进行工程实践,表明空间钢结构监测技术已逐渐成熟并对工程建设安全发挥重要作用。

关键词：空间钢结构；健康监测；无线传感；远程监控

中图分类号：TU393.3

STUDY AND APPLICATION ON WIRELESS MONITORING OF SPATIAL STEEL STRUCTURES

Yaozhi Luo

(Space Structures Research Center of Zhejiang University, Zhejiang Provincial Key Laboratory of Space Structures, Hangzhou 310058)

Abstract: Spatial steel structures are widely used in large-scale buildings and public infrastructures which are sites for large-scale activities with many people. They usually have complex structural systems, extraordinaire shapes, and are difficult to construction. As a result, the safety of this kind of structures is very important. Health monitoring is one of the effective means to guarantee the safety of spatial structures during construction and service. This paper presents a wireless structural health monitoring system for spatial steel structures in aspects of sensor development, wireless networking, safety evaluation and so on. The system has been applied in the national stadium, UK Pavilion of 2010 Shanghai World Expo, Impression Dahongpao, Hangzhou East Railway Station and so on. It shows that the health monitoring technology for spatial steel structures is going to mature and playing an important role in safety guarantee of structures.

Key words: spatial steel structures; health monitoring; wireless sensing; remote monitoring

1. 引言

空间钢结构工程广泛应用于各种大型体育场馆、火车站、会展中心、机场航站楼等重

基金项目：国家科技支撑计划资助项目（2012BAJ07B03）.
作者：罗尧治（1966—）,男,博士,教授,博导,副院长,主要从事空间结构方面的研究,E-mail: luoyz@zju.edu.cn

要的标志性建筑,由于结构体系新颖、体形巨大、受力性能复杂,在建设过程中大量采用新技术、新材料和新工艺,很多方面超出了现行建筑结构相关规范的限制范畴,在设计阶段完全掌握和预测结构的力学特性和行为是难以实现的,在施工过程中和建造完成后,结构的荷载传递及变形性能,乃至使用状态在很大程度上都存在着不可预知性和不可控制性[1]。因而,伴随着大型结构的发展与应用,重大结构安全事故开始增多,给社会经济和人民安全带来不可估量的损失,随着人们安全意识的提升,如何保障大型结构的施工过程安全有效及建成后运营的平稳有序,是当今迫切需要解决的重大课题。因此,建立一个相对完备的结构健康监测系统,对大型结构在长期运行和不利荷载作用下的结构状态进行监测和安全性评估正成为国内外研究的热点。

相对于大型桥梁、大坝、海洋平台、石油管道等国家大型土木工程设施,对于大跨度空间钢结构的健康监测技术研究起步较晚,目前也未形成成熟的技术手段与相关的行业标准。大跨度空间钢结构由于自身大跨度、大面域分布以及钢材料为主的建筑特征,明显区别于桥梁、大坝、石油管道等线性分布的混凝土结构类为主的建筑。本文针对大跨度空间钢结构特性,结合国家 863 课题和"十二五"国家科技支撑计划,对空间钢结构的破坏机理、安全性评价标准以及适用于空间钢结构类型的监测技术手段进行了系统的研究。

2. 空间钢结构监测的软硬件研发

无线传感系统作为一种新兴的传感技术,有着传统有线传感产品无法比拟的优点。虽然有线传感技术因其抗干扰性、稳定性、精确性等优点普遍用于各类重大土木工程设施的健康监测系统中,并将在今后相当长一段时间内仍然作为监测硬件的主流设备,但其有线采集、布线困难的弊病同样明显。桥梁、管道、大坝等结构由于线性特征,相对影响较小,但大型空间钢结构的特点是面积大、空间宽,若使用有线监测设备,尤其是在测点较多的情况下,用于布线的成本有可能大大超过监测设备本身,同时复杂大量的布线对于结构的美观、日后的维护,故障的检修都存在着无法回避的问题。目前国际上日益流行的无线传感技术,以其安装方便,成本低廉,维护简单,性能可靠等优点,将逐渐取代传统的监测手段而成为主流。

2.1 各类传感器研发

从设计的小型化、低功耗、低成本、高可靠性角度出发,根据模块化设计思路,即供能模块、中央处理模块、传感模块及无线收发模块,开发适合大型空间钢结构特点的无线传感设备系列。供能模块通过太阳能板、干电池或现场电源实现电能的收集、电量的存储、智能开关与稳压输出功能,从而保证整个系统运行的电能需求;中央处理模块对系统各个模块进行管理,并实现简单的数据处理、存储、系统休眠等功能;传感模块由中央处理模块控制,通过采用不同的传感元件实现对钢结构振动、应力、位移、温度等各种参数的检测,将检测到的模拟电压信号进行模数转换,转变为数字信号输出至中央处理模块;无线收发模块由中央处理模块控制,将无线信号进行调制解调,并采用灵活的组网,实现对检测数据、检测命令的无线通讯传输。将以上四大模块集成安装于保护盒中,组成一个检测单元,采用防水接头与密封设计,实现设备在晴雨天气下的全天候工作。由若干个检测单元、计算机工作站、信号接收基站即可组成一个完整的钢结构建筑无线检测系统。单

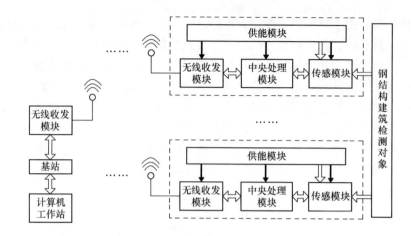

图 1 无线传感设备模块化设计框架

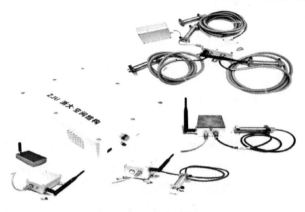

图 2 无线传感系列产品

个基本检测单元可根据测点数量、布置方案在钢结构建筑检测对象上进行灵活布置，各个基本检测单元之间可进行数据交互或传输接力，以实现灵活组网，并最终将数据传输至基站。整个模块化设计思路及检测系统架构如图1所示。目前已开发的无线传感产品有：无线加速度传感器（WAS）、无线应力应变传感器（WSS）、无线振弦应力应变传感器（WCS）、无线温度传感器（WTS）、无线位移传感器（WDS）、无线风速风压传感器（WWS）等，如图2。

2.2 组网与远程监控技术研究

系统所开发的是一种先进的现场无线组网及远程实时监控模式，有效避免了大量布线可能造成的现场维护难度及系统不稳定性。整个无线监控网络系统的整体运行过程简单描述为：各类监测测点相互之间形成智能网络，将采集到的数据通过路由节点，最终传输到基站节点，然后由基站节点经USB或串口传输给现场服务器；现场服务器通过3G无线网络接入Internet，任何终端设备只要连接到Internet就可以实现与现场服务器的数据交换，对采集到的数据进行显示、分析和管理，从而实现远程监控。图3为几种网络拓扑结

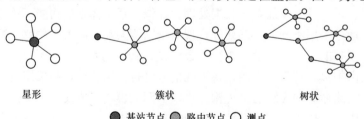

星形　　　　　簇状　　　　　树状

● 基站节点　● 路由节点　○ 测点

图 3 几种网络拓扑结构

构，图 4 为远程信息传输示意。

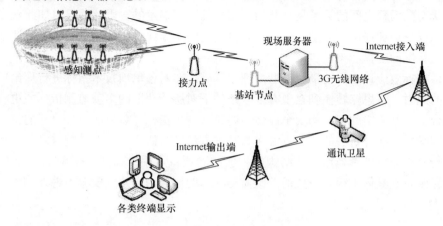

图 4 远程信息传输示意

2.3 监测软件与硬件系统

空间钢结构故障预警及安全评估系统采用模块化设计思路，以总系统与子系统的架构进行规划，主要包括空间钢结构故障预警及安全评估总系统 HMST 和若干子系统。总系统涵盖了现役结构在施工及服役期间所有常规参数如应变、加速度、位移等的测量、分析与预警。子系统则主要针对某项特定功能进行专业增强与分析，如结构振动模态分析、核心拉索监控、风速风压测控及风谱分析等等。

空间钢结构故障预警及安全评估系统采集软件界面如图 5 所示。软件系统主要包含：工程管理模块、结构建模与显示模块、结构分析与计算模块、测点布置与组网模块、数据收发模块、数据处理模块、结构评估与预警模块、文件存储模块。

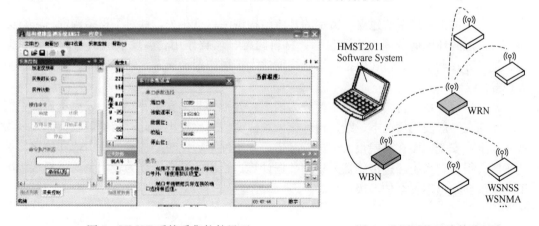

图 5 HMST 系统采集软件界面 　　　　图 6 监测系统无线传感组成

3. 结构健康状况测评技术

空间钢结构为一典型的开放式复杂巨系统，由于其自身大规模、大面积空间布局、静态应力和变形为主的特性，以及所处复杂环境中众多因素的影响，致使实际营运过程中影响安全状态的因素之间的关系是复杂的，相关的，不确定的。因此通过对如此众多而杂乱的因素加以挖掘和提炼，合理构建出空间钢结构的安全评估指标体系，是正确评估和把握结构安全状况的前提和关键。

以各指标效应量之间的耦合模型为理论基础，在对各单项技术指标进行理论推导和实测研究的同时，将思路转换到在现有技术条件下对所能获取的多项监测指标的更有效的挖掘和综合上，基于空间钢结构安全评估问题显著的多指标、多层次、多目标的特点，引入系统工程学中层次分析法的系统方法作为基本方法进行研究。在对影响空间钢结构安全的因素进行全面充分的考虑后，从结构的一般监测项目入手（如图7），结合评估指标的科学拟定原则，在具体工程所设置的健康监测项目的范畴内，建立安全预警评估的层次指标体系模型。

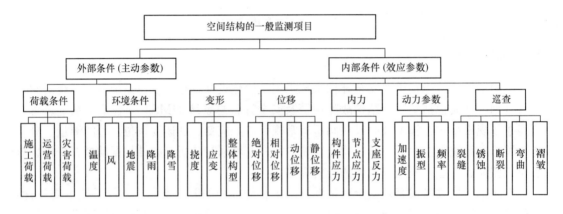

图7 普遍意义下空间钢结构的一般监测项目

安全评估指标体系的建立与特定的具体结构特性息息相关，针对工程结构资料进行分析，应考虑的环境荷载因素主要包括：材料问题、风载作用、温度效应、超载现象、振动、施工等；典型失事模式包括：结构挠度过大、结构振动异常、约束破坏、构件损伤与失稳等。

根据指标体系，将结构健康指标分为四级，分别为非常安全、安全、较大负荷、不安全。比如在一般使用情况下，结构构件应力和应变值均比较小，此时结构健康情况等级为非常安全或安全，而在温差很大或极端恶劣天气条件下，此时结构构件应力值和位移必然比平时大很多，系统经过传感器采集的数据计算，给出的健康等级为较大负荷，极端情况下，系统会给出不安全的等级评估。

4. 评估与预警分析

评估与预警系统具体包括监测系统、预警指标系统、预警警级判断系统、预警分析系统，报警系统。通过对警源的分析和确定，建立起预警系统的指标系统，然后通过制定监测方案，采用先进的监测仪器来获取这些指标在结构实际工作过程中各种指标的变化情况，通过预警机制的逻辑判断分析和预警境界的比较，来判断结构在当前情况下结构所处的安全警级，在某些特定条件下一旦结构处于超出安全工作状态的范围，发出警报，并且指导采取一定的应急修复措施来保护结构的安全。同时，监测数据都会记录在结构的健康历史档案中，可以随时调出查看先前的变化情况，可以从长期预警的角度掌握整个结构的变化情况。

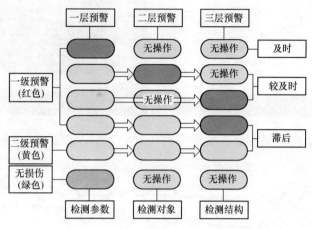

图8　预警机制

5. 工程实践

5.1　国家体育场鸟巢钢结构监测

国家体育场鸟巢主体钢结构平面尺寸为333m×294m，监测的主要内容包括对国家体育场的受力、变形、振动以及现场气候环境进行长期监测，综合利用多项结构性能指标，对结构的安全性与功能性进行评价。整个网络分为8个区，采用单线接力的簇状组网方式（图9）。

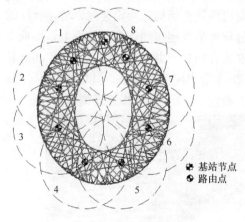

图9　鸟巢结构分区组网图

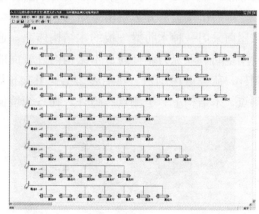

图10　无线网络测点拓扑关系

图 11　自动采集程序界面　　　　　　图 12　供游客参观的现场展示屏

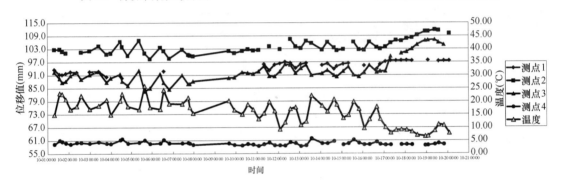

图 13　部分测点监测数据

5.2　上海世博会英国馆

英国馆主展馆的总尺寸为 25m（宽）×25m（长）×20m（高），内结构盒尺寸为 15m（宽）×15m（长）×10.4m（高），结构采用钢木结合体系，8 根箱形斜钢柱作为结构的竖向支承构件，主展馆楼板部分为钢桁架结构，主展馆与周边混凝土结构之间设连接钢桥，展厅的墙和屋面采用一种木肋梁—板的木结构体系，通过钢板节点连接到钢桁架楼板上，形成盒状封闭空间。亚克力刺杆通过木结构内外表面的安装孔固定，共安装 6.4 万余根 7.6m 长的刺杆，荷载重达 512t。在主展馆结构上共布设 20 个应力应变监测点，其中钢结构上 12 个测点，木结构上 8 个测点。另外，在各个墙面上布设了 13 个振动加速度测点。

图 14　英国馆建筑与入口处亚克力杆

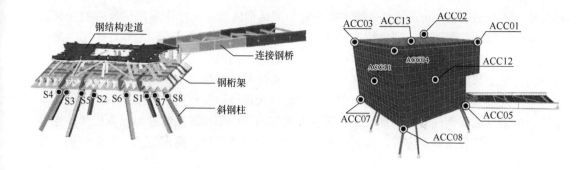

图15 英国馆测点布设情况

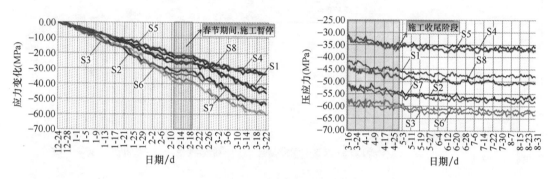

图16 英国馆监测数据

5.3 杭州火车东站

新建杭州火车东站是客运专线、城际铁路、普速铁路、磁浮轨道、城市轨道和道路交通等多种交通方式衔接的现代化大型综合交通枢纽，为全国最大的火车站之一。主站房平

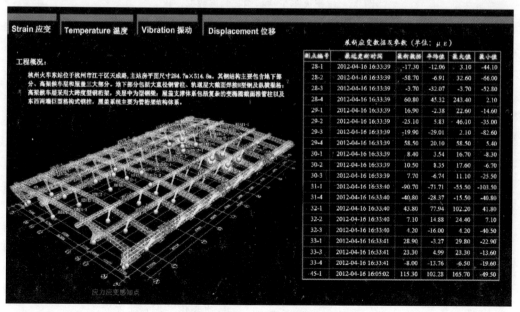

图17 杭州火车东站监控画面

159

面尺寸 284.7m×514.8m，屋盖系统主要为管桁架结构体系，支撑体系包括复杂的变椭圆截面椎管柱以及东西两端巨型格构式钢柱，高架候车层采用大跨度型钢桁架，夹层中采用大型钢梁。监测工作的范围主要集中在结构几个较为关键的受力部位，将其简要划分为核心部位与重要部位两个层次，核心部位主要包括跨度较大的屋盖主次桁架跨中部位、施工过程中出现较大应力的桁架、变截面钢柱的受力最大部位以及格构柱的主要承力构件；重要部位包括变截面钢柱的其余部位、一般主次桁架的跨中部位、支撑柱与屋盖桁架的连接部位以及"鱼眼"天窗等部位等。监测的参数包括应力应变监测、温度监测、振动监测、变形监测等。

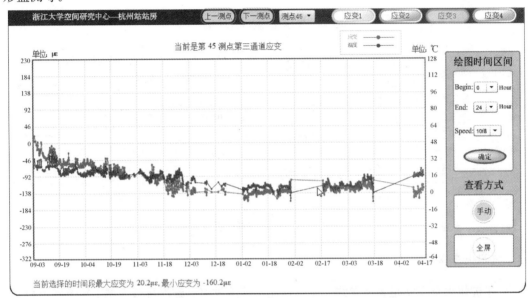

图 18 杭州火车东站结构某测点数据变化

5.4 印象大红袍旋转看台

旋转看台直径为 46.6m，设计承载观众 2000 人。看台自身可 360°自旋，观众位于看台上并与其同步运动，可实现全角度观看周围的表演。看台主体为钢结构，主承力骨架由多榀竖向平面桁架构成，底部基座设有环向轨道，上部结构通过刚性导轮传力于环向轨道。结构运行时，由液压马达驱动齿轮，通过链条带动整个上部结构运动。在看台结构上布置了应力应变传感器和振动加速度传感器，从而能够很好地监测到旋转看台结构在运动过程中的内力和振动变化情况(图 20)。

图 19 印象大红袍旋转看台

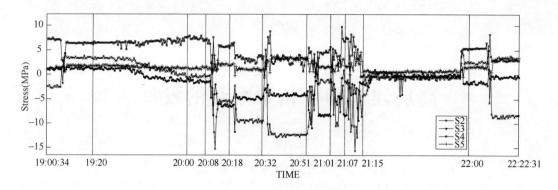

图 20 看台结构监测数据

6. 结语

无线传感技术以其灵活的拓扑布局，不受空间限制、不影响结构使用等特性，正在成为当前监测领域的技术主流。本文阐述了基于无线传感设备的空间钢结构安全监测系统，主要从硬件产品的开发、组网技术及专业软件系统的研发等方面进行了探索性研究，并在若干项大型工程结构中得到了应用，为大跨度空间钢结构的健康监测和安全评估技术积累了重要的经验。

参考文献

[1] 罗尧治，沈雁彬，童若飞，王小波. 空间结构健康监测与预警技术. 施工技术，2009，38(3)：4-8

[2] 罗尧治，王小波，杨鹏程，童若飞，沈雁彬. 某钢结构廊桥施工吊装过程监测. 施工技术，2010，39(2)：10-13

[3] 苑佳谦. 大跨度空间结构灾害预警评估理论研究与系统开发，浙江大学，2011

[4] 罗尧治，王治亲，童若飞，程华强，沈雁彬，毛德灿. 上海世博会英国馆结构健康监测. 施工技术，2011，40(2)：24-27，55

[5] LUO Yaozhi, SHEN Yanbin, et al. Development of a wireless sensor system potentially applied to large-span spatial structures; Proceeding of The 4th China-Japan-US Symposium on Structural Control and Monitoring; Hangzhou, Oct. 16-17, 2006

[6] 国家体育场结构监测报告，浙江大学空间结构研究中心，2011

[7] 印象大红袍旋转看台结构监测报告，浙江大学空间结构研究中心，2011

[8] 杭州火车东站站房钢结构施工监测报告，浙江大学空间结构研究中心，2012

高层建筑伸臂桁架系统的发展

何伟明

(奥雅纳,香港九龙塘达之路 80 号又一城 5 楼)

摘　要:高层建筑的结构性能极大程度上取决于其横向刚度和抵抗能力,在高层建筑的结构体系中,伸臂桁架系统是最常见的体系之一,特别是在相对称的平面布置。在建筑结构中使用伸臂桁架可追溯到 20 世纪 50 年代初的深梁概念,随着建筑物的高度上升,深梁变为混凝土墙,或成为至少一层高的钢桁架伸臂系统。由于扩大了材料的选择,伸臂桁架的形式、甚至使用目的也发生了改变。除了结构分析以外,伸臂桁架的设计和施工方案是不可分割的,核心筒和外框架之间轴向压缩的影响也是不可避免。本文提出高层建筑的伸臂桁架体系,当中包括伸臂桁架的发展史和应用,同时也会讨论伸臂桁架体系的概念、几何结构的优化、设计和施工方面的考虑。

关键词:高层建筑;伸臂桁架系统;结构设计;施工

THE EVOLUTION OF OUTRIGGERS SYSTEM IN TALL BUILDINGS[1]

Goman W. M. Ho

(Arup, Level 5, 80 Tat Chee Avenue, Kowloon Tong, Hong Kong SAR, China)

Abstract: The structural efficiency of tall buildings heavily depends on the lateral stiffness and resistance capacity. Among those structural systems for tall buildings, outriggers system is one of the most common systems especially for those with relatively regular floor plan. The use of outriggers in building structures can be traced back from early 50's from the concept of deep beams. With the rise of building height, deep beams become concrete walls or now in a form of at least one story high steel truss type of outriggers. Because of the widened choice in material to be adopted in outriggers, the form and even the objective of using outriggers is also changed. Besides analysis, the design and construction issue of outriggers is somehow cannot be separated. Axial shortening effect between core and perimeter structures is unavoidable. This paper presents a state of the art review on the outriggers system in tall buildings including development history and applications of outriggers in tall buildings. The concept of outriggers, optimum topology, and design and construction consideration will also be discussed and presented.

Keywords: tall Buildings; outriggers system; structural design; construction

---

作者:何伟明(1963—),男,博士,董事,主要从事结构工程设计

1. 引言

香港本是一个平地有限的小渔村，二次大战结束后，香港人口由 1945 年的 60 万人快速增长至 1952 年的 230 万人（陈 2006）。随着 1953 年 12 月 25 日石硖尾村的一场大火，数千间房屋被烧毁，5 万 8 千人无家可归，被迫住在两层高的"宝灵平房"（美荷楼资源包 2012）。后来，六层高、只有简单剪力墙结构的马克Ⅰ型房屋于 1955 年取代了"宝灵平房"。到了 20 世纪 70 年代中期，随着二次大战后经济复苏及人口增长，香港与其他大城市一样，开始发展高层建筑。

香港地处强风地带，平均每年有 6 至 8 个月面临热带台风的威胁，随着建筑越来越高，设计将受位移及横向稳定的控制。由于人口和有限的钢材资源，20 世纪 80 年代前香港大部分的高层建筑都是钢筋混凝土建成的住宅建筑。

高层办公楼的其中一个里程碑是拥有 64 层、共 216m 高的合和中心。利用当时有限的计算机运算能力和科技，合和中心采用筒中筒体系，半径为 22.9m，共有 48 根边柱，边柱间距只略小于 3m（图 1）。筒中筒结构的其中一个主要缺点就是外围的柱间距比较小，这种体系与在 20 世纪 70 年代末广为使用的核心筒剪力墙结构大不相同。

 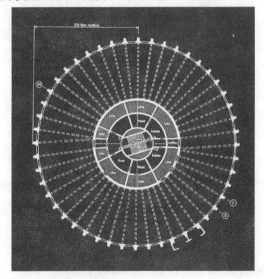

图 1　合和中心及其标准平面图

从 20 世纪 80 年代开始，香港上海汇丰银行总部犹如剪开钢结构的彩带的剪刀一样，钢和组合结构在香港变得更加常见。虽然钢材具有非常高的强度质量比，但钢结构的刚度还是不如钢筋混凝土结构，随着大间距的柱和窗的发展趋势，伸臂桁架系统便成为了解决高层建筑问题的最好方法。

2. 伸臂桁架结构系统的概念

建筑结构中伸臂桁架系统的概念是把外围和内部结构连接成一个整体来抵抗横向荷载，考虑图 2 所示的结构，核心筒和外框架的共同作用很小，因此，核心筒和外框架都只是以纯悬臂的方式抵抗横向荷载。理论上，如果核心筒和外框架之间的梁可以加深和变得更坚固，核心筒和外框架便可以共同合作，抵抗横向荷载。但是，核心筒和外框架间的标准跨度在9～15m，导致很难提供有足够刚度或深度的梁以连接核心筒和外框架。

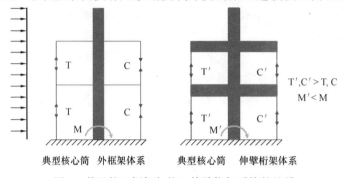

图2　普通核心框架与核心伸臂桁架系统的差异

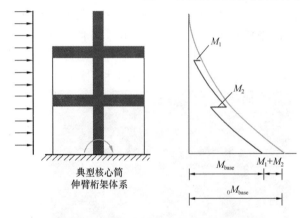

图3　普通框架与伸臂桁架弯矩图的分别

所有高层建筑都设有避难层、一些设有机械楼层，工程师一般会充分利用这些空间来加强结构刚度（可达二或三层楼的高度）。为了便于表述，伸臂桁架将如图3般绘画成深梁，假设伸臂桁架有足够刚度，可以产生抑制弯矩 M_1 和 M_2，基底弯矩 $_oM_{base}$ 将减少（M_1+M_2），即：

$$M_{base} = {_oM_{base}} + M_1 + M_2 \quad (1)$$

方程式（1）可以改写成以下形式：

$$_oM_{base} = M_{base} - \Sigma M_i \quad (2)$$

其中 M_i 是第 i 个伸臂桁架的抑制弯矩。

根据方程式（2），不论增加 M_i 的数值和/或伸臂桁架的数量（即 i），基底弯矩都会变小。但是，如果 M_i 的数值有限或很少，尽管有很多伸臂桁架，$_oM_{base}$ 都会接近 M_{base}。换言之，建筑中利用大刚度的伸臂桁架比增加伸臂桁架数量更为有效。

3. 优化伸臂桁架的位置和几何形状

尽管使用伸臂桁架已经越来越普遍，但是关于伸臂桁架的研究还是很有限。Taranath（1988），Stafford Smith（1991），Gerasimidis et al（2009）和 Fawiza et al（2011）的研究都关注在高层建筑中伸臂桁架在控制位移及优化其位置的整体效果上。尽管如此，设计师安装伸臂桁架系统的位置仍受到大厦功能的限制，实际上，伸臂桁架只能安装在机械层或者避难层，而不一定是"最佳"位置。受到安装位置所限及根据方程式（2），伸臂桁架的刚度成为优化结构的基本因素。但是，关于伸臂桁架的几何形状和刚度的公开资料有限，Ho（2012）研究伸臂桁架的几何结构、刚度和强度关系是其中一个主要资料。在相同的环境限制下，Ho（2012）研究了多种伸臂桁架的几何结构，发现伸臂桁架的强度并非与刚度成正比，该论文的结论归纳在表1。

关于伸臂桁架的几何结构总结　　　　　　　　　　　　　　　　　　　　　表 1

几何结构	A	B	C	D
材料用量	1.00	1.04	1.80	1.49
刚度	∨∨	∨	∨∨∨∨	∨∨∨
强度	∨∨∨∨	∨∨∨	∨	∨∨

4. 伸臂桁架的设计问题

　　伸臂桁架构件与梁柱的设计是相同的，由于核心筒和外框架之间压缩所造成的差别，伸臂桁架设计的主要考虑之一就是内应力。核心筒与外框架的缩短主要是由于弹性变形，压缩和徐变，由于伸臂桁架的刚度普遍都很高，一个很小的垂直变形都会导致伸臂桁架构件产生很大轴向力。虽然工程师可以预测弹性缩短量，但压缩和徐变量都是随着时间变化，这也意味着在建筑物完成或正常使用期间，压缩和徐变将不会完全发生。为了消除弹性压缩效果，工程师可以提供延迟接合点，当结构和主要荷载加到全部垂直构件上，伸臂桁架就可以被连接。但是，工程师必须找到方法，以保证施工阶段或伸臂桁架系统不在工作的情况下的横向稳定性。

5. 伸臂桁架系统的调整

　　液压平衡系统

　　Kwok 和 Vesey（1997）提出一种关于使用液压来平衡伸臂桁架/柱连接位置的荷载的想法，这个概念允许伸臂桁架在施工时已连接，通过调整伸臂桁架顶端和底端的液压系

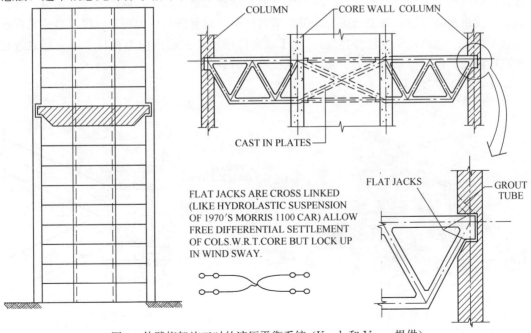

图 4　伸臂桁架施工时的液压平衡系统（Kwok 和 Vesey 提供）

统，便可调整伸臂桁架的高度。在施工阶段完结后，液压系统的空间将被灌浆，提供永久连接。经 Kwok 和 Vesey 的同意，他们的草图复制于图 4。

尽管液压平衡系统解决了弹性压缩问题，并同时保持了横向稳定性，但是它无法应付大厦使用期间的压缩、徐变和可变活荷载而产生的竖向变形。

6. 垫板法

本文的作者（Ho 等人（1999））提出液压平衡系统的修改版本，并于 1995 年应用于香港长江中心的设计和施工中。

长江中心是一座 300m 高的高层建筑，由钢筋混凝土核心墙和外框架组成。外框架是由混凝土填满的柱以 6m 间距所构成，由于核心筒是长方形，伸臂桁架无法与柱直接连接，因此采用腰桁架系统以分散伸臂桁架的垂直力到外围的柱上。

核心筒的最大高宽比是 15，而且在施工阶段极有可能遭遇台风，因此，伸臂桁架安装后，便须立即工作，以提高结构在施工阶段的横向刚度。所有伸臂桁架/外围连接，其顶部和底部都有垂直间隙，随着腰桁架和伸臂桁架的建成，伸臂桁架的顶端和底端都会安装钢板，以传递垂直荷载，这就是"垫板校正法"。这种想法类似于上述的"液压平衡系统"，只是在暂时和永久阶段，以更坚固的垫板取代液压系统。此外，伸臂桁架构件安装了应变计，间隙的大小可以在建筑物的使用阶段调整，以适应如 10 年或 30 年徐变。

外框架的轴向压缩可能令伸臂桁架的内应力非常大，根据笔者的经验，可能与横向力度的数值相约。换言之，如果压缩引致的轴向力不能释放，伸臂桁架的大小将会加倍。利用垫板法，当顶部垫板被移走，内应力将会 100% 被释放，由于伸臂桁架仍然在弹性范围内，所以会返回未变形前的状态。压力释放后，垫板会重新装上，以传递垂直荷载，并在施工阶段重复此过程。施工结束后，将会进行监察并重复进行压力释放程序，直到轴向压缩趋于稳定（压缩或徐变已经完成）。由于轴向压缩所产生的内应力在每个压力释放过程中都被释放，所以伸臂桁架将主要用于抵抗横向荷载。

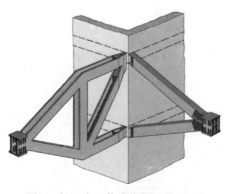

图 5　长江中心伸臂桁架系统的 3D 效果图（奥雅纳提供）

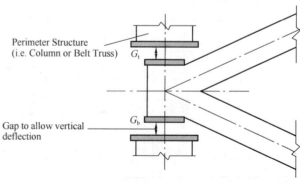

图 6　长江中心伸臂桁架尖端的标准细节

7. 伸臂桁架施工中的后安装技术

随着1995年开始设计、1997年落成的长江中心,人们注意到建筑中使用伸臂桁架的关键是伸臂桁架在建筑物中的高度。在20世纪90年代末,奥雅纳开始设计国际金融中心二期时,使用了一种新的伸臂桁架施工方法,令超高层结构的核心筒和伸臂桁架系统的发展迈进了一大步。为了提供弹性的办公室楼层配置,让租户可以在这座88层、412m高、24m柱间距的大楼与金融行业连接,所以选用了巨型柱和伸臂桁架系统。

三个分别三层楼高的伸臂桁架安装在与核心墙边缘直线对齐的位置,腰桁架安装于伸臂桁架的相应位置,用于将荷载从次角柱传送到大型柱及作加强处理。

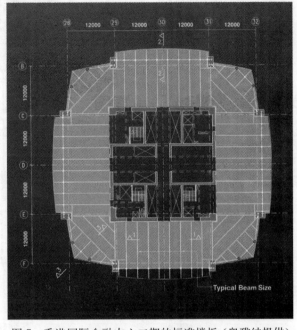

图7 香港国际金融中心二期的标准楼板(奥雅纳提供)

在传统的施工方法中,兴建一层核心墙需时3～4d,但在伸臂桁架楼层,由于构件需要吊运、焊接和安装,需时将近1个月。为了快速施工并不耽误核心墙施工,伸臂桁架楼层的核心筒在施工阶段将部分被阻隔。如图8所示:

使用后安装技术,核心墙施工将与伸臂桁架分开,伸臂桁架安装后,核心筒会用混凝土填满,成为一个整体的构件。按照后安装过程和详细的施工计划,施工速度可以达到没有伸臂桁架时的一般核心墙的建造速度。

8. 阻尼伸臂桁架系统

由于伸臂桁架的发展概念是在增强横向刚度,所以多数工程师认为伸臂桁架刚度越大越好,某种程度上,这种概念是正确的,但是未必适合于地震区的建筑物。

对于地震区的建筑物,设计目标是考虑风舒适度和抗震性能,而处理风和地震的共同因素就是结构的阻尼。

奥雅纳发展了阻尼伸臂桁架系统,并运用于菲律宾马尼拉的圣弗朗西斯香格里拉大酒店的设计中,该项目不但距离活跃断层只有2km,还会面临强风的吹袭。该混凝土建筑高217m,有一个核心筒和一层混凝土伸臂桁架,伸臂桁架的每个构件都通过黏性阻尼器与柱连接,由于阻尼器允许相对移动,该系统明显增加了阻尼,但是相对于传统的伸臂桁架系统,刚度增幅则较少。

利用最新的科技,黏性阻尼可设计用于非线性反应,并调整至符合多层性能的要求。

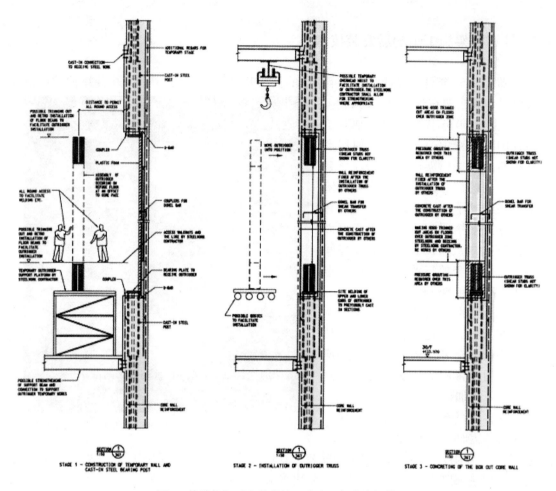

图 8　伸臂桁架后安装的施工次序（奥雅纳提供）

一旦阻尼器失效，伸臂桁架设计成以延性的方式屈服，但仍保持其完整性。

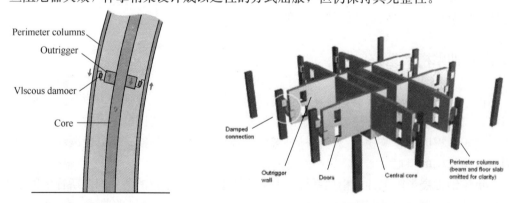

图 9　阻尼伸臂桁架概念
（奥雅纳提供）

图 10　伸臂桁架楼层的 3D 图
（奥雅纳提供）

由于增加了阻尼，风引致的反应亦降低了结构所需的横向刚度，从而减少材料成本。节省的材料成本抵消了附加阻尼、测试和安装该系统的费用，以及增加了使用面积和整体结构的结构性能。

9. 总结

历史的光辉和里程碑都证明了高层建筑中伸臂桁架系统的发展和应用,为了使工程师可以快速和容易地作出参考,本文对伸臂桁架的理论、概念和最佳几何形状作出简要介绍。轴向压缩是影响伸臂桁架设计的其中一个主要因素,利用现今计算机和软件的运算能力,复杂精细的分析,如轴向压缩的影响便能够轻易地解答。但是,如果将伸臂桁架设计为可以抵抗内应力,这将不是一个最好的设计,本文解释了调节伸臂桁架的方法,在施工和大厦使用期间,释放伸臂桁架的压缩压力可以使轴向压缩的影响减到最低。在高层建筑施工中,核心墙的施工都是在关键路径,后安装方法就是允许核心墙在不受因安装伸臂桁架而产生的延误下,可以维持原本的建造周期。高刚度并不是伸臂桁架系统的唯一目标,阻尼伸臂桁架系统可以调节刚度,向建筑物提供最佳的多目标性能。本文以历史回顾的方式陈述,并希望本文可以成为美好历史的一部分,伸臂桁架系统的发展是永无止境的。

10. 致谢

笔者非常感谢奥雅纳公司对他研究期间休假的支持,同时感谢香港理工大学在2011年9月至2012年2月期间提供的良好科研环境。

参考文献

Chan T. (2006),"*Burning Down the House*:*Hong Kong Public Housing*",Proceeding of Pacific Crossings Conference,AIA HK,KS-HK6

Heritage Office (2012),"*Mei Ho House Resource Kit*",published by Heritage's Office,Hong Kong SAR Government

Kwok M. K. Y. and Vesey D. G. (1997),'*Reaching for the moon— A view on the future of tall buildings*',Structures in the New Millennium,Lee (ed.),pp. 199-205,Balkema

Ho W. M. G.,Scott D. M. and Nuttall H. (1999),"*Design and Construction of 62—storey Cheung Kong Center*",Symposium on Tall Building Design and Construction Technology,II,162-170,June 1999

Taranath,B. S. (1988),"*Structural Analysis & Design of Tall Buildings*",McGraw-Hill

Stafford Smith,B and Coull (1991),A,"*Tall Building Structures— Analysis and Design*",John Wiley & Sons,INC

Gerasimidis S.,Efthymiou E. & Baniotopoulos C. C. (2009),"*Optimum outrigger locations of high-rise steel buildings for wind loading*",5th European & African Conference on Wind Engineering (PROCEEDINGS)

Fawzia S.,Nasir A. and Fatima T. (2011),"*Study of the Effectiveness of Outrigger System for High-Rise Composite Buildings for Cyclonic Region*",World Academy of Science,Engineering and Technology,60

Ho W. M. G. (2012),"*Outrigger Topology and Behavior*",accepted for publication,International Journal of Advanced Steel Construction

陈T. (2006),"烧毁的房屋:香港公共房屋",Proceeding of Pacific Crossings Conference,AIA HK,

KS-HK6

文物保育办事处(2012),"美荷楼资源包",文物保育办事处出版,香港特别行政区政府

郭 M. K. Y. 和 Vesey D. G. (1997),'到达月球——对未来高楼大厦的看法',新时代的结构,李(等人),pp. 199-205,Balkema

Ho W. M. G. ,Scott D. M. 和 Nuttall H. (1999),"设计和建造62层高的长江中心",高层建筑的设计和施工科技的评论,II,162-170,1999年6月

Taranath,B. S. (1988),"高层建筑的结构分析和设计",McGraw-Hill

Stafford Smith,B 和 Coull (1991),A,"高层建筑结构—分析和设计",John Wiley & Sons,INC

Gerasimidis S. ,Efthymiou E. & Baniotopoulos C. C. (2009),"高层钢结构中伸臂桁架受风荷载时的最佳位置",5th European & African Conference on Wind Engineering (PROCEEDINGS)

Fawzia S. ,Nasir A. 和 Fatima T. (2012),"伸臂桁架在飓风地区的高层组合结构中的效用研究",World Academy of Science,Engineering and Technology,60

Ho W. M. G. (2012),"伸臂桁架的几何结构及性能",accepted for publication,Advanced Steel Construction

CONSTRUCTION OF STEEL STRUCTURE IN HYBRID STRUCTURE FOR HIGH-RISE BUILDING

H. Wang

(China Construction Steel Structure Company, ShenZhen, 518040, China)

Abstract: In recent years, high-rise buildings of more than 400m have sprung up in China. The high-rise building construction technology has also developed rapidly. Combined with the author's engineering practice of many years, on the sub-projects of steel structure, three key control technology in the past high-rise steel structure construction was summarized in the paper, that was "the structural deformation control + structural internal force control + welding quality control". In addition, according to the concept of coordination of steel structure and civil engineering, some new problems and countermeasures such as the "construction period and security" faces were elaborated.

Keywords: high-rise; steel structure; construction control; vertical deformation difference; dissimilar materials welding; hydraulic integrated jack-up framework system; tower crane attachment

1. Introduction

Since the Jinmao Tower in Shanghai, high-rise buildings of more than 400m have sprung up in China. At the same time, the high-rise building construction technology has also developed rapidly. Combined with the author's engineering practice of many years, on the sub-projects of steel structure, the key control technology in the past high-rise steel structure build was summarized in the paper, and some new problems and countermeasures we were facing currently were elaborated.

Fig 1   World Financial Center in Shanghai   Fig 2   Pingan Financial Center in Shenzhen

---

Author: Wang Hong (1960—), Male, Professor of Engineering, mainly engaged in steel structure construction, E-mail: Wangh@cscec.com.

2. Steel construction control technology

Almost all the structural system of high-rise buildings were using an "outer frame and core-wall" giant hybrid structures. In the construction process of giant steel structure, the "deformation control + internal force control + welding control" was considered to be the key to the quality control of the steel structure engineering. Then the high-rise projects, such as World Financial Center in Shanghai, the Pingan Financial Center in Shenzhen, and the Zhujiang New Town West Tower in Guangzhou, etc. were taken for examples to outline the specific implementation of the three control technology.

2.1 Structural deformation control

In the construction process, the vertical deformation of the structure was inevitable, and it will cause three key problems:

1. the top of the structure elevation was inconsistent with the designed elevation.
2. the vertical deformation difference between outer frame and core-wall will cause the floor tilt and cracking, affecting the usage function of the building.
3. the vertical deformation difference of outer frame and core-wall will cause additional internal forces of structural members.

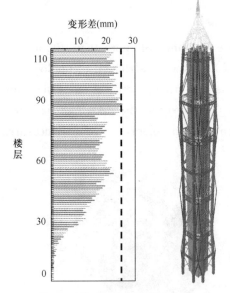

Fig 3 The vertical deformation difference distribution between outer frame and core-wall

The analysis shows that, in general, the vertical deformation of frame SRC columns exceeds 100mm after considering the shrinkage and creep contraction of the concrete when the structure height is higher than 400m. In order to solve the problem of inconsistent of actual elevation and designed elevation caused by vertical deformation, the frame SRC columns between the belt trusses can be taken as a segment unit to actively compensate the length loss of columns. Moreover, the compensation can also be carried out late in one or several layers at the top of the structure. The methods elaborated above, namely active compensation control method, then make the SRC columns reached the designed elevation after the building has been put into use for one year.

About the understanding of the vertical deformation difference between the core-wall and outer frames, there were some changes in recent years. Because in the actual construction process, the inner cylinder concrete was "leveling layer by layer" according to designed elevation, it exactly compensates the vertical deformation of a portion of the lower part (60% to 70%) caused by concrete shrinkage and creep.

Based on this important time varying characteristics, the theoretical analysis and the monitoring data show that the vertical deformation difference between the core-wall and outer frame was far less than that of previous forecasts, and this conclusion has changed the previous understanding in the industry. Nevertheless, the vertical deformation difference have an adverse impact on the use function, such as the floor tilted, non-structural members deformation and so on, and in the construction process, how to control or weaken the core-wall and outer frame deformation difference was always a challenge to high-rise construction quality.

Take the financial center of Shenzhen Pingan for example, the main structure was 554m height excluding the tower cap and the antenna, the vertical deformation difference distribution between the outer frames and core-wall was shown in Figure 3. In order to control the vertical deformation difference between the core-wall and outer frame, an idea of compensation in the construction treatment was adopted instead of previous way. Only the floor steel beam construction elevation was adjusted without considering each SRC columns and core-wall elevation compensation, that was to say, the location of the "corbel" and "embedded parts in core-wall" were elevated in different degrees in the stage of construction blueprint.

Fig 4   Floor steel beams in high-rise building

To sum up: (1) the deviation of the overall height of the structure and design elevation can be compensate centrally at the top of several layers, not necessary to compensate at each layer; (2) in order to control the vertical deformation difference between outer frame and core-wall, the elevation of "corbel" and "embedded parts in core-wall", which connects with the floor steel beams, can be elevated more or less in the stage of construction blueprint instead of length compensation in the way of layer by layer.

2.2 Structural internal force control

As mentioned above, the vertical deformation difference between the core-wall and outer frames also cause additional internal forces in the structure, especially the additional internal forces in outrigger trusses, used to decrease the overall lateral deflection of highrise building, can not be ignored.

In order to weaken the additional forces in outrigger trusses, some construction treatments were employed. At first, the top and bottom chord in outrigger trusses can be con-

Fig 5　Outrigger trusses installation

nected permanently, while the diagonal web member was temporary fixed by high-strength bolts, and then connected by welding when the vertical deformation of outer frames and core-wall was stabilized. The bolt holes temporary fixing diagonal web member were designed as bidirectional long hole, used to adapt the deformation of the outer frame and core-wall. Otherwise, when encountering execrable climatic conditions, high-strength bolts can be fixed by temporary final screw to ensure the structural safety. The outrigger truss installation method Above was considered to be "passive adaptation" to accommodate the inevitable vertical deformation difference between the core-wall and outer frames, and has been widely introduced in high-rise building construction.

In addition to the outriggers truss connecting the outer fames and core-wall, another structural member damaged due to the additional internal forces was the reinforced concrete floors between the outer frame and the core-wall. Being with the develop of the vertical deformation, the floors poured between the outer frames and core-wall will crack, thus, in construction practice of some high-rise building, especially in the position shown in Figure 6, there left a late poured band. To sum up, this construction method of outrigger trusses and floors can be unified called "Generalized late poured band processing".

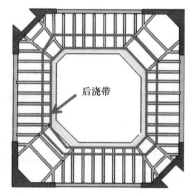

Fig 6　The late poured band position

2.3　Welding quality control

There is no doubt about the importance of the welding quality. Welding quality is another key of the steel structure engineering quality. In recent years, cast steel joint has been widely used in the site of the complex joint, thus, in ultra-high-altitude field operating conditions, the quality of the dissimilar materials, steel components and steel castings, welding control is the challenge of the industry. In Shanghai World Financial Center project, Shanghai World Financial Center project, the dissimilar material welding of a giant belt truss and outrigger trusses

Fig 7　Welding quality control of cast steel joint

steel castings, adopt the program of dissimilar material transition section processing (see Figure 5) in particular.

The past traditional practices of simply seeking to improve the operation environment was completely changed, and the dissimilar material welding whose quality is difficult to control in site is transferred to factory in better conditions for industrial production, and only the same kind of materials were welded in ultra-high-altitude construction site. Changes in welding technology broke through the quality bottleneck of the field welding of ultra-high-altitude dissimilar material, and to boost the primary welding qualification rate in Shanghai World 1.7 million linear meters welding reaches 98.60 percent, and accumulated valuable dissimilar material welding technology for structural steel engineering.

3. Steel construction facing new problems and countermeasures

Practice has found that the construction of steel structure and civil were indivisible in hybrid structure for high-rise building, the construction process. Their cross coordination not only throughout the whole construction process, and penetrated into every step of the process. It believe that the cross and coordination of the steel structure and civil engineering, is probably one of the most prominent characteristics in the construction process of the hybrid structure for high-rise building. Based on the idea of "steel structure" and "civil" coordination, the steel structure construction is facing some new problems such as duration and security, the following brief exposition.

3.1 Issue of duration

Concerned with duration, some digression is to be discussed. Seen from the technical bid, put aside of the business, the advantages of duration is the most important goal which ultra-high-level technical bid pursuits recent years, and it can also be considered to be one of the core competitiveness of "impressing the owners". Since the quality and safety are not easy to quantify, then how to make the technology bid impresses owners? The answer is only the duration.

What been just about is overall speed of the entire structure need to be higher. Then about the frame steel structure and the core-wall, the experience is that, since the core-wall construction employed a new framework system, the speed of steel structure construction cannot catch that of the civil. So, this leads to the speed coordination problems of steel structure and civil, and the steel structure should not hold duration back. This "construction speed upside down" phenomenon occurs because many core-wall constructions of hybrid structures, such as Zhujiang New Town West Tower in Guangzhou and international Financial Center in Shenzhen, have adopted a new so-called "integrated hydraulic jack-up framework system" (seen from Figure 8).

Compared with the traditional climbing formwork system, the supporting position of

Fig 8　Integrated hydraulic jack-up framework system employed in West Tower

the "hydraulic jack-up framework system" is lower, unlike general "climbing formwork" that relies on the strength of the fresh concrete; in addition, the heap load capacity of hydraulic jack-up framework system is higher, and lifting speed faster. Moreover, the practice shows that the higher the floor in the construction is, the greater pressure the steel structures suffers, and that is because when floor is higher, the core-wall becomes thinning, the amount of steel banding becomes less, and concreting becomes less, but steel structures lifting and welding workload don't reduce such much. Seen from practical effect of Zhujiang New Town West Tower in Guangzhou and international Financial Center in Shenzhen, on average speed of each layer, the hydraulic jack-up framework system is saving about two days than climbing formwork system. If there are 100 layers, the top mold system can save six months of duration. Generally, this "hydraulic jack-up framework system" is used for high-rise buildings in general contracting of china construction third engineer bureau or forth engineer bureau.

　　In order to cope with the pressure of the steel structure construction, a strategy of "tower crane liberation" is adopted currently. The main point is that the tower crane should do the lifting of some heavy component only, not to be occupied by some other small components or construction measures. Specifically, for the first a set of self-climbing operation platform, seen from Figure 9, should be designed to provide the installation operating platforms for the outer frame steel columns, then the tower crane will not be used to disassemble and turnover the operation platforms entirely. In addition, there is an assumption that several small tower crane, seen from Figure 10, should be designed attached to the pillars upon to solve those light elements like steel beams of floors and profiled steel sheet, and further liberate "the large tower crane".

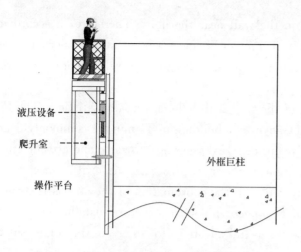
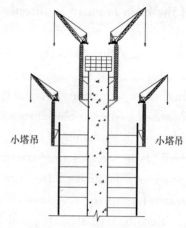

Fig 9 Self-climbing operation platform for columns installation

Fige 10 Tower crane attached to the columns

3.2 Issue of Security

"Tower crane attachment" should be talked about when it concerns the security issues. The task of tower crane attachment to the core-wall general belongs to general contractor originally. However, it is gradually assigned to the subcontractor of steel structure recently. Because large tower crane is mainly used in steel structure, and China Construction Steel Structure Company has purchased four Favco tower crane from M440D to M1280D, and has developed business in aspects of attachment, installment and demolition.

The form of the tower crane attachment hanging on the core-wall outside is considered to be an unreliable one. The one hanging on West Tower is M900D, seen from Figure 11. Now when the height increases, the amount of steel structure grows. M1280D have been used in Pingan Financial Center in Shenzhen and Gaoyin 117 building in Tianjin. In the past, when M900D was used, the bearing beam for the tower crane attachment in generally rooting directly on the core-wall and the rooting site is close to the hole in wall, sometimes only a few centimeters. Well, it is a big issue whether the wall has enough bearing capacity.

According to the previous analysis and practical, if tower crane M900D is used, the minimum thickness required of wall provided the capacity was 500mm, when the rooting site is near the hole. While the M1280D was employed, there was a big

Fig 11 Tower crane attached to the core-wall

risk if the main beam roots directly on the wall near the hole. Therefore, in accordance with the characteristics of the core-wall in Pingan Financial Center in Shenzhen, the position of rooting switched to the intersection of wing wall and web wall in core-wall structures.

In this way, the bearing capacity of the wall with hole is basically satisfied. The Pingan Financial Center in Shenzhen and Gaoyin 117 building in Tianjin are commonly conditional to use this scheme. Then the main idea of discussing "tower crane attachment of core-wall" is to stresses the "coordination of the steel structure and civil" concept what was previously mentioned. It not only refers to the coordination of the steel structure itself, also refers to the coordination of equipment for steel structure installation with the civil, so that the civil structure will be taken full advantage of to create the conditions for security protection of the equipment.

4. Conclusions

For the steel structure construction of the high-rise building, "structural deformation control + structure internal force control + welding quality control" are three keys of construction control. In addition, the coordination of civil engineering and steel structure is probably one of the most prominent characteristics in the high-rise building construction process. From the idea of coordination, the issue of "duration" and "security" should be focused to solve.

型钢高强钢筋混凝土结构柱在地下室逆作施工方案的应用与设计

刘志健，李志城

（新鸿基建筑设计有限公司，香港，中国）

摘 要：随着经济高速的发展与城市建设规模的不断扩大，城市地面空间拥挤、交通阻塞、土地资源匮乏等种种问题下，建筑工程日趋向地下空间纵深发展，已成为增强城市功能、改善城市环境的必要手段。通过规划建设各类型的地下空间，如建设地下停车场、商业街、休闲场、地下给（排）水设施和电力设施，有效地利用城市地下空间对发展城市经济、改善环境、方便人民生活已起到积极作用，也是城市走向可持续发展道路的重要途径。

高层建筑的地下室工程越来越多，且大量挖掘地下室工程集中在市区，工期紧张，施工场地狭小，施工条件复杂。为了减小地下室开挖对周围建筑物、道路和地下管线的影响、达到环境保护越来越高的要求以及要有效缩短工期等一系列施工难题时，地下室逆作法施工在建筑工程中得到了广泛应用。它特点在于可使建筑物上部结构的施工和地下室结构施工平行立体作业，地面上、下可同时进行施工，直至工程结束，可大大节省很多任务期，使在城市建设中具有良好的经济效应和实用价值。

为实现地下室逆作施工法，地下室结构设计常以型钢混凝土柱构件作为承受临时施工及永久的全部荷载。型钢混凝土结构是由型钢与钢筋混凝土结合起来，在互相发挥其优点且弥补各自的缺点，它主要是克服了钢材的稳定性缺陷和耐火的不足，较之钢结构可大大减少钢材用量，使它成为更安全与经济的结构体。

本文结合工程实例，主要目的在介绍型钢混凝土结构柱在地下室逆作施工方案的应用与设计方法。另外，对于型钢混凝土组合结构的发展情况，本文会介绍各国所制定的型钢混凝土柱的设计规范要求。经过探讨各国的设计规范和例题分析，说明采用适当设计的型钢混凝土构造更能够有效发挥其优点，从而达到安全与经济结合的目标。

关键词：型钢混凝土；地下室；逆作施工法

1. 引言

近年来，随着环境保护及可持续性发展的要求，高层建筑的地下室工程愈见普遍，地下室越来越深，当遇到周边施工场地狭窄、周围环境复杂、高水平的围护要求及紧缩工期等一系列施工难题时，采用地下室逆作法施工技术是首选。它的主要特点在于可使建筑物上部结构的施工和地下室结构施工平行立体作业，直至工程结束，使工期可大大节省，在城市建设中具有良好的经济效应和实用价值。

作为地下室逆作施工法的重点结构的型钢混凝土组合柱，在结构工程中得到了广泛的应用。由于它结合了型钢与钢筋混凝土两种构造，使得它兼具这两种构造的特色。因此，型钢混凝土组合结构都体现出了比钢结构或钢筋混凝土结构更加优越的特性。而型钢混凝土设计主要分为二类别：包覆型型钢混凝土结构和钢管混凝土结构。

由于包覆型型钢混凝土结构，特点在于混凝土之包覆有助于避免型钢断面发生局部屈曲之现象，它较之钢结构可大大减少钢材用量，从而大幅降低了造价与使用成本，使它成为更安全与经济的结构体。另外，混凝土之包覆并可作为型钢之防火被覆，节省了防火防锈等方面的费用，减少了结构维护保养费用。包覆型型钢混凝土结构相对于纯钢构造而言，由于它的断面刚度比较大，有助于减少建筑物之侧向位移，很适合用于地下室建筑中。而本文主要引述包覆型型钢混凝土结构柱在地下室的逆作施工法的应用与设计。

2. 地下室逆作施工工程实例

香港九龙牛池湾丰盛街住宅发展程项目——峻弦，建筑 5 座楼高 31 层住宅，大厦主楼裙房部分三层，高约 21m，塔楼部分 30 层，高度为 120m，地下三层，埋深约 14m。地盘面积约为 14,459 万 m²，地上部分总建筑面积 100,143m²，地下部分总建筑面积 11,900 m²。大楼地下部分采用全逆作法施工，柱为型钢高强钢筋混凝土结构柱。大厦地上部分采用钢筋混凝土框架剪力墙体系，柱和剪力墙为高强钢筋混凝土结构，楼板为钢筋混凝土结构。

图 1　峻弦　住宅发展程项目

本工程采用地下钻孔后注浆连续墙作为地下室外壁结构，起挡土、挡水、承重、围护等多功能作用，墙厚 1.2m、深约 30m，整个建筑由长 127m×36m 左右地下连续墙及在 29 根钻孔灌注桩基础上设置的型钢柱共同临时承重，随地下室 3 层结构完成时，将型钢柱灌注混凝土处理成型钢混凝土方柱。根据受力计算采用逆作法施工，该地下室 3 层结构完成时，上盖结构能同时施工至 16 层。该工程用 26 个月完成钢筋混凝土结构施工，工期缩短 4 个月。本工程的地下室逆作法主要施工工况介绍如下各图。

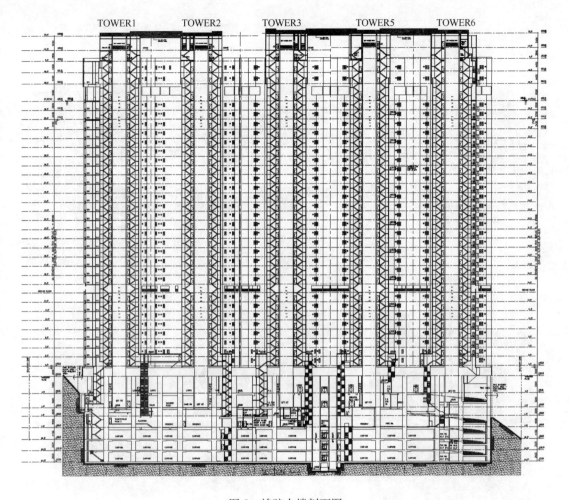

图 2　峻弦大楼剖面图

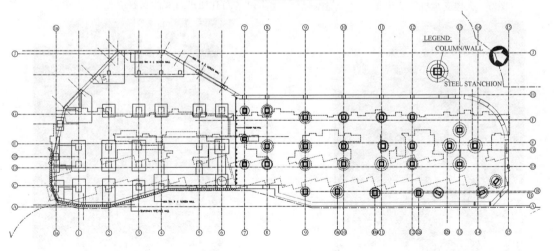

图 3　峻弦大楼地基平面图

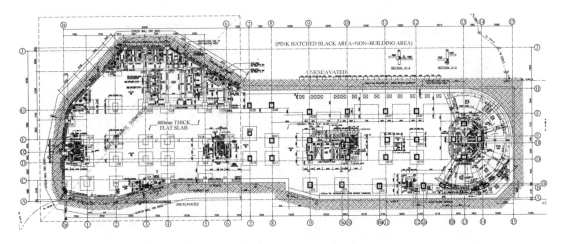

图 4 峻弦大楼负一层平面图

图 5 大楼支承柱安装施工

图 6 对孔底进行实测安放位置

图 7 格构柱垂直度实时纠偏调垂

图 8 钢柱垂直度监测

图 9 逆作施工的土方开挖工况

图 10 大楼支承柱的小量土方开挖工况

图 11 开挖后的大楼支承柱

图 12 大楼型钢钢筋混凝土结构支承柱施工

图 13 大楼楼面的钢筋构造施工

图 14 地上部分同时施工

3. 地下室逆作法施工特点

地下室建筑工程逆作法是预先以型钢柱构件作为中间逆作柱，利用地下连续墙和逆作柱作各层面的支撑承重，由首层或地下1层楼板作为水平分界层，同时向分界层上下各层组织施工，并把逆作柱逐层处理成型钢组合结构柱，桩及逆作柱承受临时施工的全部荷载，形成随地下室挖掘的同时，由分界层楼板向地上、地下同时组织施工的一种施工方法。

3.1 由于地下连续墙在施工中起挡土、挡水、承重和围护作用，地下室各层楼板起刚性水平支撑作用，逆作柱起中间承重作用，使整个地下室各层楼板、地下连续墙和逆作柱形成一个整体，在分层施工中起着围护和挡土、挡水作用。

3.2 地下室逆作法因不用进行深基坑支护与主体工程间施工空间的土方开挖，土体变形小，受力可靠，临时挡土结构与永久结构合二为一，并由于有层间楼面可作为工作平台，不必另外架设开挖工作平台与支撑支护结构，这样大幅度削减了支撑和工作平台等大型临时设施，减少了施工费用。

3.3 土方分层开挖深度可达到最小，从而节省了临时支护系统的费用。但如遇较大层高的地下室，有时需另设临时水平支撑或加大围护墙的断面及配筋。

3.4 浇筑完分界层楼板后，即可利用分界层楼面作施工场地分别向上向下同时施工，从而增大施工场地使用面积，缩短整体结构的施工工期。

3.5 地下室逆作支承柱在底板以下采用大直径钻孔灌注桩直接传力至地基，减少了大开挖时卸载对持力层的影响，降低了基坑内地基回弹量。

3.6 保证邻近建筑物及设施的安全，由于地下连续墙受地下室各楼层楼板连续支撑，土方分层开挖，故结构变形很小，避免了地下室施工中主体结构与支护结构间施工空间的抽排水，使连续墙外土体不受扰动，因而避免了地下室在土方开挖时对周围建筑结构物地基沉降的影响。

3.7 基础在封闭条件下施工，施工可少受风雨的影响，有利于工程围护、提升工程质量和减少工期。

4. 地下室逆作法施工工艺要点

地下室逆作法施工工艺中围护设计要点和逆作施工阶段的结构施工工艺要点和处理方法：

4.1 由于逆作法的施工工艺的特殊性，决定了地下室的竖向构件必须采用型钢混凝土柱，而型钢柱施工的垂直度是该工序的关键，钢柱垂直度监测与调垂系统利用先进的监测技术等对结构柱垂直度实时纠偏调垂。

4.2 挖土是逆作施工的重要环节，由于挖土是在顶部封闭状态下进行，基坑中还分布有一定数量的中间支承柱和降水用井点管，使挖土的难度增大。

4.3 挖土是产生变形的主要关键因素。挖土专项方案严格按照设计要求进行分区、分段、分层跳跃式开挖，严禁未按设计要求每次每层开挖厚度、分段长度过大。在基坑边缘地面严禁堆放超出设计允许荷载的材料、弃土和设备等。

4.4 另一关键工序是地下连续墙施工、支承柱施工、地下室梁板结构混凝土的浇筑及插墙体钢筋及柱插筋留设，需按设计及施工规范要求。由于逆施到正施完成时间较长，预留的墙、柱、梁、板钢筋外露时间很长，对其保护避免锈蚀很重要。此外，因为建筑方案修改使部分预埋钢筋位置偏差太大而失去作用，在实际施工中应用植筋的办法解决。

4.5 竖向构件墙柱混凝土后浇筑施工，考虑到顶部混凝土的密实性，采取梁板模板施工时，在平台楼板上预留洞作为下料口和振动棒的插入口。检查构件节点处的构造情况、梁、板的标高、轴线和起拱高度，标高、轴线的安装偏差需确保符合规定要求。

4.6 本工程监测重点为基坑侧向水平位移、沉降。要求在挖土施工期间每天监测二次。

如果基坑水平位移或者沉降出现异常或者变形增量超过设计确定的预警值，则要再增加观测次数。

为了保证工程工期、质量和安全均达到第一流水平，我们针对本工程地下室全逆作法的施工特点，在施工技术方案、质量措施、安全措施等各个方面进行了严密的管理组织，有完善的验收制度，有切实可行的建筑方案，真正形成完整的工程施工保证体系。对于地下室逆作法施工是由多种工艺组合而成的施工工法，由地下连续墙或带止水的排桩墙构成的围护结构是逆作法施工的前提条件，墙式支护结构的施工质量及止水效果是逆作法施工的第一关键。

5. 型钢混凝土柱的设计

在地下室逆作法施工中，大楼支承柱的作用是在地下室底板封底前，承受地上和地下各层结构的自身荷载和施工荷载。它承受的最大荷载是在地下室修到最下一层，而地面上已经修筑到相应的最高层数时的荷载。因此本工程的支承柱是采用包覆型型钢混凝土柱。型钢混凝土构件的外包混凝土可以防止钢构件的局部屈曲，并能提高钢构件的整体刚度，显著改善钢构件的平面扭转屈曲性能，使钢材的强度得以充分发挥，较之钢结构可大大减少钢材用量。此外，外包混凝土增加了结构的耐久性和耐火性，克服了钢材的稳定性缺陷和耐火的不足。与普通钢筋混凝土结构相比，由于配置了型钢，因此大大提高了构件的承载力，尤其是采用实腹型钢的型钢混凝土构件，其抗剪承载力有很大提高，并大大改善了受剪破坏时的脆性性质，这样做一方面可以减小截面尺寸，又能够满足承载力的要求，也增加建筑结构的使用面积和空间，节省混凝土用量、减轻地基荷载、使它成为更安全与经济的结构体。

另外，由于在逆作法施工中的开挖时期，型钢混凝土柱的钢骨构件未能灌注混凝土合成为型钢混凝土组合柱，钢骨在不同所逆作法施工程序下会承受不同的临时荷载，而这钢骨于挖掘地下室时所受的临时施工荷载是不可能被释放的内锁荷载。因此型钢混凝土柱设计需要考虑钢骨在各种施工模式下的内锁荷载组合对其作出承载力的影响、以满足大楼结构承载力的要求。

对型钢混凝土组合结构的发展情况，本文会下述介绍各国所制定的型钢混凝土柱的设计规范要求。经过探讨各国的设计规范，使型钢混凝土构造设计能够更有效地发挥型钢与钢筋混凝土的优点，达到安全与经济并存的目标。

5.1 各国设计规范回顾

本文介绍现行型钢混凝土结构设计规范主要有：(1)香港行业设计规范；(2)中国设计规范；(3)美国 AISC-LRFD 钢构造设计规范；(4)美国 ACI318 混凝土构造设计规范；(5)美国 NEHIRP 设计规范；(6)欧洲设计规范 Eurocode4；(7)日本建筑协会所发行之型钢混凝土结构规范设计和(8)中国台湾内政部营建署颁布之"型钢混凝土构造与解说"设计规范。

目前型钢混凝土构造相关设计规范中，其设计观念可以概分为以下三大类：

5.1.1 第一类为引用《钢结构构造的换算截面法》设计公式，将型钢混凝土构造中混凝土所提供的强度与刚度转换为相当的型钢来进行设计，例如美国 AISC-LRFD 钢结构设计规范，美国 NEHRP 合成构造设计规定，欧洲规范与香港规范。

5.1.2 第二类为使用《混凝土构造构造的换算截面法》的设计方法，将型钢混凝土构造中的型钢视为增加构材强度与刚度的普通钢筋来进行设计，如美国 ACI 结构混凝土设计规范。

5.1.3 第三类为采用"强度叠加法"，此法先将型钢及混凝土之强度个别分开计算，然后予以叠加以求得型钢混凝土构材之强度，如中国设计规范，日本建筑学会的型钢混凝土设计规范，台湾设计规范。

三种计算理论的主要差别之一是对型钢与混凝土之间的粘结滑移性能给予了不同的考虑。在上述的设计规范中，中国始于 20 世纪 80 年代中期展开了广泛的型钢混凝土结构的研究，在这一时期研究主要针对型钢混凝土构件的受力性能，进而对型钢混凝土构架之行为进行研究，经由试验研究与理论分析，初步形成了一套较完整的设计计算理论。目前中国设计规程基本上还是采用强度叠加法来计算受弯、受压和受剪承载力和构件刚度，但进一步提出了较为准确的轴力分配方法，称为改进简单的叠加方法。改进简单叠加方法与理论方法和一般叠加法基本吻合，即将型钢和钢筋混凝土所分别承担的弯矩叠加同时考虑构件轴向力的影响，且不考虑彼此间的变形的相互约束作用，因此得到的计算结果偏于安全。

美国 ACI 规范及 AISC-LRFD 规范中有关型钢混凝土构件设计规定均仅包含构材之设计，较不如日本的型钢混凝土及 NEHRP 之规定完整。目前美国规范的型钢混凝土构件设计规定则采用极限强度设计法。而日本的型钢混凝土规范之主要优点在于具有较多的经验及研究成果，该规范对型钢混凝土构造细则之规定较为明确。日本在型钢混凝土构造设计方法上采用工作应力设计法再辅以极限层剪力之检核。另外，台湾地区的型钢混凝土设计规范主要是根据美国和日本的设计方法，加以再作适当改良。

欧洲大部分国家很少使用型钢混凝土建筑，因此相关之研究甚少，尤其规范对梁柱接头区之剪力强度，并未有明确的规范，仅仅在混凝土包覆型钢柱断面之轴力强度与弯矩强度有比较详细规定。而香港的型钢混凝土设计规程，是根据欧洲设计规范方法，加以改良发展，而对梁柱接头区之剪力强度，根据欧洲和新西兰的设计方法有明确的规范。

针对中国香港、中国、欧洲、美国、日本等针对型钢混凝土结构设计条文之要求，各主要材料的最大容许设计计算应力如下列出：

表 1

设计规范	中国香港	中国	欧洲	美国	日本
混凝土最大抗压强度	60MPa	80MPa	50MPa	70MPa	60MPa
型钢最大屈曲强度	460MPa	390MPa	420MPa	500MPa	355MPa

5.2 计算例题分析

本文计算例题主要是考察在同一型钢混凝土柱截面和其他材料设计应力相同时，按各国对型钢混凝土柱的设计极限承载力作出比较，探讨各国的设计差异。

另外，因为高强混凝土是近年来混凝土材料发展的一个重要方向，以其抗压强度高、抗变形能力强，在高层建筑结构中得到广泛的应用。高强混凝土是抗压强度一般为普通混凝土的 2 倍，故可减小构件的截面，减轻自重，因而可获得较大的经济效益。因此世界各国对于采用钢筋高强混凝土的大楼设计越来越多应用于高层和超高层建筑工程中。针对此点，本例题会按照各国或地区对钢筋高强混凝土的设计参数，加以应用于相应的型钢混凝土设计规范中，说明高强混凝土对型钢混凝土设计的影响。

计算例题：构件截面尺寸取 2400mm × 2000mm，竖向受力型钢等级为 S355，对称布置（如下图），构件截面含钢率为 11.2%。材料特征

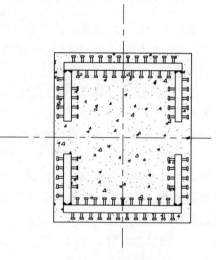

图 15 构件截面

参数均取自各国的设计规范，通过计算程序计算型钢混凝土柱的极限轴压承载力。

假定按香港地区规范计算分析中普遍采用 C45 混凝土抗压强度的型钢混凝土柱极限轴压承载力为 1.0 的情况下，比较按各国规范计算所得的型钢混凝土柱极限轴压承载力比对值，如下列出。

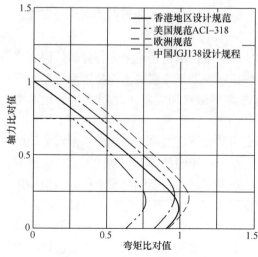

图 16 对比国设计规范构件断面 PM Curve 分析
（采用 C45 抗压强度混凝土的型钢混凝土柱）

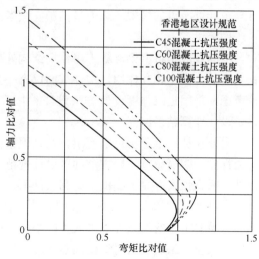

图 17 对采用不同抗压强度混凝土的构件断面 PM Curve 分析

表 2

混凝土抗压强度	极限轴压承载力比对值			
	中国香港	中国	美国	欧洲
45MPa	1.000	1.090	0.784	1.158
60MPa	1.112	1.196	0.912	1.292*
80MPa	1.262*	1.335	1.083*	1.501*
100MPa	1.412*	1.467*	1.255*	1.742*

* 按各设计规范对混凝土和钢的应力特性，和型钢混凝土设计原理加以计算。

  从以上分析可以看出，在其他截面构件条件不变的情况下，只考虑标准值的荷载，不考虑各国或地区不同的荷载组合值和其分项系数下，其对型钢混凝土柱的设计的极限轴压承载力的受力性能有明显影响。

  按香港地区设计规范，在忽略混凝土抗拉强度、遵从平截面假定及不考虑型钢与混凝土之间的粘结力等条件下，以强度叠加法作为理论基础。由于没有考虑型钢与混凝土之间的相互约束作用，设计计算结果偏于保守。

  目前中国设计规程 JGJ138 假定型钢与混凝土完全共同达到屈服状态，受压区钢骨与混凝土之间始终没有滑移，构件截面始终保持平面。若要满足平截面假定则，必须沿全构件焊接栓钉，并满足一定构造要求，在工程应用中则会增加造价与施工工序。因此设计计算结果对比香港地区的设计值最多高约 10%。另外，有试验结果表明，在达到极限承载力之前，钢骨和混凝土之间已经产生了相对滑移。并且，变形协调模型的公式比较复杂尚需要进一步的探讨。

  而美国规范 ACI-318 在考虑型钢和混凝土构件的变形协调和内力平衡下，再考虑型钢材料本身的残余应力和初始位移后。此法最突出的优点是容易计算得到构件的弯矩与构力，但由于它是以考虑初始位移和残余应力的纯钢结构为基础。型钢混凝土柱的截面轴压承载力最大比其他各国设计计算值为少，最大轴力的计算值比其他各国的设计计算值少约 20%，此时若不考虑各国对于型钢混凝土设计构件所采用的荷载组合分项系数下，美国设计规范计算所得的极限承载力是偏于安全考虑了，但断面之间破坏分界点的判定尚需要进一步的探讨。

  欧洲规范是将型钢混凝土当作钢结构处理，从而应用钢结构理论来进行分析，假定型钢与混凝土完全相互作用，构件截面仅有一个对称轴，将型钢与混凝土均按矩形应力块理论考虑，采用极限强度设计方法设计。由于没有考虑型钢骨和混凝土之间产生的相对滑移，设计计算结果偏高。

  另外，从以上分析可以看出，在其他条件不变的情况下，混凝土的抗压强度对截面的受力性能有明显影响。以香港地区设计规范为例，对于在相同条件下配制的型钢高强混凝土柱，因高强度混凝土提供了有利条件下，采用高强混凝土的型钢混凝土柱的截面轴压承载力最大可增加为 41.2%。表明它在截面面积不变的条件下可大大节约钢材，或在轴压承载力要求不变的条件下可大大减小柱截面，也增加建筑结构的使用面积和空间，促进了经济效益。虽然国内对型钢高强混凝土柱有了一定的试验研究，但在计算机数值模拟方面研究相对较少，相关知识结构比较松散。变形协调模型的公式比较复杂尚需要进一步的探讨。因此使用高强度混凝土作为型钢混凝土柱的材料需要进一步的研究。

6. 结束语

6.1 本文结合工程实例，主要目的在介绍型钢混凝土结构柱在地下室逆作施工方案的应用与设计方法。在施工过程中，对支撑体系、土方、模板、钢筋、混凝土、地下室变形位移监测等各分项采取了有效控制措施，避免了地下室变形对周边建筑物、道路和地下管线的安全影响。

6.2 地下室逆作施工方案与传统的深基坑施工方法相比，逆作法具有保护环境、缩短建设周期等诸多优点。因为围护墙的支撑体系由地下室楼盖结构代替，它克服了常规临时支护存在的诸多不足之处，还可省了内支撑体系造价。而且楼盖结构即支撑体系，还可以解决特殊平面形状建筑或局部楼盖缺失所带来的布置支撑的困难，并使受力更加合理。

6.3 随着高层建筑越来越多，基坑深度越来越大，对于具有多层地下室的高层建筑，采用逆作法施工具有明显的经济效益，相信逆作法施工技术必将得到越来越广泛的应用。

6.4 对型钢混凝土组合结构的发展情况，本文比较各国所制定的型钢混凝土柱的设计规范。目的在使我国的型钢混凝土柱设计规范能够互相配合，并具有一贯性。一个经过适当设计的型钢混凝土构造，可以有效发挥型钢与钢筋混凝土的优点，达到安全与经济的目标。

6.5 由于型钢混凝土构材力学行为相当复杂，因此假设型钢与混凝土不具有合成作用，因此个别计算构材之强度后，再予以叠加视为型钢混凝土构材之强度，设计偏向保守。

6.6 从本文工程例题表明在结构的受压杆件中采用型钢高强混凝土结构，可大大节约钢材，或可大大减小柱截面，促进了经济效益。但在试验研究和计算机数值模拟方面研究相对较少，因此使用高强度混凝土作为型钢混凝土柱的材料需要进一步的研究探讨。

6.7 型钢混凝土的出现为建筑行业带来了巨大的经济效益。如何加强混凝土在型钢混凝土中的应用是一个综合性问题，从设计、施工到监理各方都应密切配合，认真研究分析，才能保证建筑质量的要求。

参考文献

[1] Hong Kong Code of Patrice for the Structural Use of Steel 2011，Hong Kong，2011
[2] Hong Kong Code of Patrice for the Structural Use of Concrete 2004，Hong Kong，2004
[3] 中华人民共和国国家标准 GB 50010—2010，混凝土结构设计规范，中国，2010
[4] 中华人民共和国行业规程 JGJ 138—2001，型钢混凝土组合结构技术规程，中国，2001
[5] 中华人民共和国设计规程 YB 9082—2006，型钢混凝土结构设计规程，中国，2006
[6] 中华人民共和国行业标准 CECS 230：2008，高层建筑钢－混凝土混合结构设计规程，中国，2008
[7] 中华人民共和国行业标准 CECS 104：99，混凝土结构技术规程，中国，1999
[8] NEHRP(2012)，"Recommended Provisions for the Development of Seismic Regulation for NewBuilding,"National Earthquake Hazards Reduction Program，Building Seismic Safety Council，Washington，D. C.，2012
[9] IBC(2012)，"International Building Code"International Conference of Building Officials，Whittier，California，2012

[10] ACI(2011), Building Code Requirements for Structural Concrete, ACI 318-11, American Concrete Institute, Farmington Hills, MI

[11] AISC(2010), Specification for Structural Steel Buildings, ANSI/AISC 360-10, American Institute of Steel Construction, Inc., Chicago, IL. August3rd, 2009

[12] Eurocode 2: Design of Concrete Structures Part1-1: General Rules and Rules for Buildings, EN1992-1-1: 2004, CEN, Brussels, Belgium

[13] Eurocode 4: Design of Composite Steel and Concrete Structures Part1-1: General Rules and Rules for Buildings, EN1994-1-1: 2004, CEN, Brussels, Belgium

[14] 日本建筑学会,「铁骨铁筋コンクリート造配筋指针・同解说（2010）」,东京,2010

[15] 日本建筑学会,「铁骨铁筋コンクリート构造计算规准・同解说：一许容应力度设计と保有水平耐力一」,东京,2001

[16] Johnson, R. P., and Anderson, D. (2004), Designers' Guide to EN1994-1-1: Eurocode 4, Thomas Telford Publishing, London, England

[17] Leon, R. T., Kim, D. K., and Hajjar, J. F. (2007), "Limit State Response of Composite Columns and Beam Columns PartI: Formulation of Design Provisions for the 2005 AISC Specification,"Engineering Journal, AISC, Vol. 44, No. 4, 4th Quarter, pp. 341-358

[18] Leon, R. T., and Hajjar, J. F. (2008), "Limit State Response of Composite Columns and Beam Columns PartII: Application of Design Provisions for the 2005 AISC Specification,"Engineering Journal, AISC, Vol. 45, No. 1, 1st Quarter, pp. 21-46

[19] Bao-Jun Sun and Zhi-Tao Lu, DESIGN AIDES FOR REINFORCED CONCRETE COLUMNS, Journal of Structural Engineering, Vol. 118, No. 11, November, 1992

[20] 施素萍；浅谈逆作法施工工艺在建筑工程的应用,山西建筑,2010,36(32).

[21] 邓远征；型钢一钢管混凝土轴压短柱承载力和组合刚度研究．大连：大连理工大学土木水利学院,2005

[22] 吕帝雄；高层建筑"逆作法"施工的探讨；国外建材科技,2005,26(5),pp30-31

[23] 李惠；混凝土及其组合结构；北京：科学出版社,2004,pp163-164

[24] 李少泉,沙镇平；钢骨混凝土柱正截面承载力计算的叠加方法；建筑结构学报,2002,23(3),pp27-31

[25] 阿里浦江；钢骨混凝土柱承载能力的研究,南京：河海大学,2002

[26] 陈洪涛．各种截面钢管混凝土轴压短柱基本性能连续性的理论研究．哈尔滨：哈尔滨工业大学土木工程学院,2001

[27] 翁正强；型钢混凝土结构(SRC)设计规范在台湾的发展,钢结构会刊,第十一期,2000年,台北

[28] 翁正强、颜圣益、林俊昌；包覆型型钢混凝土柱型钢对混凝土围束箍筋量之影响,中国土木水利工程学刊,第十卷,第二期,第193-204页

日本 E-DEFENSE 五层楼实尺寸含制震斜撑钢构架振动台试验反应预测

蔡克铨[1,2]，游宜哲[2]，李昭贤[2]，翁元滔[2]，蔡青宜[1]

(1. 台湾大学 土木工程学系，台湾；
2. 国家地震工程研究中心，台湾)

摘 要：日本 E-Defense 振动台于 2009 年进行五层楼实尺寸含制震斜撑钢构架受震试验，并举办实验反应预测竞赛。本文介绍参赛所采之分析模型及仿真方法，包含了试验前以 OpenSees 对加装金属阻尼器构架，及以 PISA3D 对加装黏性阻尼器构架的预测分析。分析模型中对于结构含混凝土楼板之复合钢梁构件行为、箱型钢柱及梁柱交会区、各式阻尼器之劲度、强度、阻尼力等，提出模拟方法并与试验结果比较，讨论模型分析结果及对结构反应预测的影响。预测模型中考虑了混凝土楼板对梁构件反应的影响、以动力试验结果校正金属阻尼器模型、以 Maxwell model 仿真黏性阻尼器，分析结果对加装金属及黏性阻尼器钢构架之楼层绝对最大反应及阻尼器最大反应皆能准确的预测。

关键词：挫屈束制支撑；黏性阻尼器；结构非线性分析

EARTHQUAKE RESPONSE PREDICTIONS OF A FULL-SCALE 5-STORY STEEL FRAME EQUIPPED WITH DAMPERS

K. C. Tsai[1,2], Y. J. Yu[2], C. H. Li[2], Y. T. Weng[2], C. Y. Tsai[1]

(1. Department of Civil Engineering, National Taiwan University, Taiwan;
2. National Center for Research on Earthquake Engineering, Taiwan)

Abstract: The seismic performance tests of a full-scale five-story passively-controlled steel building were conducted on the E-Defense shaking table in Japan in March 2009. Before the tests, a blind prediction contest was held to allow researchers and practitioners from all over the world to construct analytical models and predict the dynamic responses of the steel frame specimen equipped with buckling-restrained braces (BRBs) or viscous dampers (VDs). This paper presents the details of two refined prediction models made before the tests. A three-dimensional shell finite element subassembly model was constructed to gain insight into the BRB end joint stiffness. The dynamic effects on the BRBs' strength were incorporated into the BRB frame prediction model. For the viscous damper frame model, the viscous dampers were represented using the Maxwell model, which connects the damping, the stiffness provided by both the viscous damper and the steel brace in series. The fibered beam elements for both two frame response prediction models were considered to account for the steel and concrete full-composite actions. The frame response prediction results show

---

通讯作者：蔡克铨（1955—），男，博士，台湾大学土木工程学系教授，国家地震工程研究中心顾问，E-mail：kctsai@ntu.edu.tw

good agreement with the test results. This suggests that the proposed modeling techniques may be considered for the structural analysis of similar types of value-aided frame.

Keywords：nonlinear analysis；buckling-restrained brace；viscous damper

1. 介绍

2009年3月，日本E-Defense研究中心利用全世界最大振动台，进行了实尺寸五层楼含制震斜撑钢构架受震试验。此试验以相同的抗弯梁柱构架，分别加装置换四种不同的阻尼器进行测试，包含了金属阻尼器（steel damper）、油压阻尼器（oil damper）、黏性阻尼器（viscous damper）及黏弹性阻尼器（viscoelastic damper），最后再测试无加装阻尼器的抗弯构架。此试验测试研究了构架加装四种不同阻尼器之受震性能，并将其与抗弯构架反应比较[1]。在试验进行前，主办单位举办了一个实验反应预测的国际竞赛[2]，参与此竞赛的学者专家针对此五层楼钢构架分别加装金属阻尼器及黏性阻尼器之受震反应进行预测。本文介绍国家地震中心组队以OpenSees[3]及PISA3D[4]分析软件于参赛中进行的预测分析、对未融入比赛项目的含油压阻尼器构架之模拟结果，并在试验后另建置简化模型，以便探讨数值模拟方法。本文提出数种分析模型及仿真技巧，希望透过本文的介绍及模拟分析经验的分享，提供工程师及结构分析者有效的结构仿真方法。

2. 日本E-DEFENSE五层楼实尺寸含制震斜撑钢构架振动台受震试验

2.1 试体介绍

图1显示该五层楼钢构架加装金属阻尼器试体之情形。图2、图3为构架的平面图及

图1 含金属阻尼器五层楼钢构架

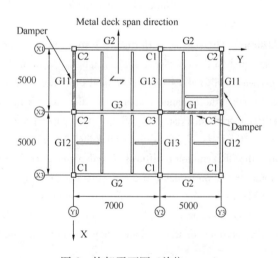

图2 构架平面图（单位：mm）

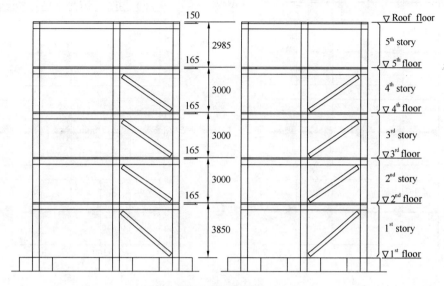

图 3　构架立面图（单位：mm）

立面图。此试体构架之长向定义为 Y 向，结构短向为 X 向，长向为双跨共 12m 短向为双跨共 10m。此五层楼钢结构一楼楼高为 3.85m，二至四楼楼高为 3m，五楼楼高为 2.985m。混凝土楼板厚度二至五楼皆为 165mm，顶楼则为 150mm。试体的重量组成包括钢梁、钢柱、钢承板、混凝土楼板、内外墙、楼梯、量测仪器等。分配到每层楼的载重，从二楼到顶楼分别为 867、842、835、790、1451 kN。钢材部分，梁材质为 SN490B，冷弯箱型柱材质为 BCR295。构架的梁柱尺寸详见表 1 至表 3，梁断面为宽翼断面，柱断面为中空方管断面。如图 2 及图 3 所示，斜撑装置在一楼到四楼的 X2、Y1 及 Y3 构架在线，共 12 支斜撑。

2.2　试验反应预测竞赛

此五层楼试验构架于振动台上以 1995 年 Kobe 地震，取 Takatori 测站数据，进行三向地震振动台试验，以研究此结构受震下之行为[5]。主办单位并在构架试验前进行阻尼器构件试验，以试验结果供参赛者建置构架模型时有较精确的参考依据。此试验反应预测比赛依分析的形式分为两个类型，三维分析组（3D-analysis）及二维分析组（2D-analysis）。而在这两组中，又依阻尼器的性质分为金属阻尼器组（Steel damper）及黏性阻尼器组（Viscous damper）。主办单位将 Takatori 测站加速度历时，以逐渐增大的方式重复施加于试体上。参考图 4（a），含金属阻尼器构架受力之放大系数为 0.15、0.4、0.5、0.7 及 1，参赛者需预测在 0.15 倍及 1 倍 Takatori 三向地震历时下之试体反应；参考图 4（b），含黏性阻尼器构架为 0.4、0.5、0.6、0.7、0.83 及 1，参赛者需预测在 0.4 倍及 1 倍 Takatori 三向地震历时下之试体反应。需预测的反应包括[2]：每层楼的绝对最大位移反应、绝对最大加速度反应、绝对最大层间剪力、绝对最大层间位移角，位于 X2 及 Y1 构架线之一楼及四楼斜撑最大受拉与受压轴力及轴向变形，以及位于梁及柱上各一位置之绝对最大应变。

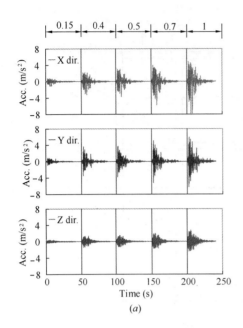
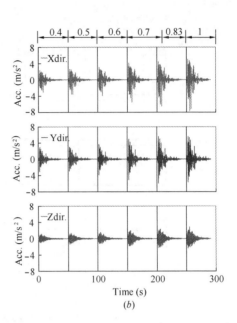

图 4 地震加速度历时
(a) 含金属阻尼器构架；(b) 含黏性阻尼器构架

梁尺寸之一 表1

Floor	G1 (full length)	G2 (end portion)	G2 (center portion)	G3 (end portion)	G3 (center portion)
RF	H 400×200×9×12	BH 400×200×9×12	H 400×200×9×12	BH 400×200×9×16	H 400×200×9×12
5F	BH 400×200×12×16	BH 400×200×9×16	H 400×200×9×12	BH 400×200×9×16	H 400×200×9×12
4F	BH 400×200×12×19	BH 400×200×9×19	H 400×200×9×16	BH 400×200×9×19	H 400×200×9×16
3F	H 400×200×12×22	BH 400×200×9×19	H 400×200×9×16	BH 400×200×9×19	H 400×200×9×16
2F	H 400×200×12×22	BH 400×200×9×19	H 400×200×9×16	BH 400×200×9×19	H 400×200×9×16
1F	BH 900×500×16×28	BH 900×500×16×28		BH 900×500×16×28	

梁尺寸之二 表2

Floor	G11 (full length)	G12 (end portion)	G12 (center portion)	G13 (end portion)	G13 (center portion)
RF	H 400×200×9×12	BH 400×200×9×12	H 400×200×9×12	BH 400×200×9×12	H 400×200×9×12
5F	BH 400×200×12×16	BH 400×200×9×16	H 400×200×9×12	BH 400×200×9×16	H 400×200×9×12
4F	BH 400×200×12×16	BH 400×200×9×19	H 400×200×9×16	BH 400×200×9×19	H 400×200×9×16
3F	H 400×200×12×19	BH 400×200×9×19	H 400×200×9×16	BH 400×200×9×19	H 400×200×9×16
2F	H 400×200×12×22	BH 400×200×9×19	H 400×200×9×16	BH 400×200×9×19	H 400×200×9×16
1F	BH 900×500×16×28	BH 900×500×16×28		BH 900×500×16×28	

柱尺寸   表3

Story	C1	C2	C3
5	□350×350×12×12	□350×350×12×12	□350×350×12×12
4	□350×350×12×12	□350×350×12×12	□350×350×12×12
3	□350×350×16×16	□350×350×16×16	□350×350×19×19
2	□350×350×16×16	□350×350×19×19	□350×350×19×19
1	□350×350×19×19	□350×350×22×22	□350×350×22×22

3. 含金属阻尼器构架反应预测

在含金属阻尼器构架反应预测比赛中，国家地震中心以 OpenSees 为分析软件建置数值分析模型，参赛获得三维分析组第二名及二维分析组第三名。此分析模型以刚性楼板仿真楼板效应，质量及梁线高程设定均在每层楼的混凝土楼板表面（top of concrete），并于楼板质心处设置节点，每层楼的水平向平移质量 Mx 与 My 与绕 z 轴旋转的质量惯性矩（Iz）集中施加于刚性楼版质心，每层楼的垂直向质量 Mz 则将各柱构件承载面积范围之质量集中施加至九个柱节点上，使垂直地震力对每个柱造成个别的影响。结构阻尼比视为2%。此模型在进行非线性动力分析时，是以逐步积分法（step-by-step integration method）来求取结构历时反应，选用的数值积分法为 Newmark Method 中的等平均加速度法（constant average acceleration method）。

3.1 金属阻尼器分析模型

主办单位所用之金属阻尼器其实就是挫屈束制支撑，在受拉及受压时均能发展出降伏强度的轴力构件[6]，具有高劲度及高韧性，挫屈束制下的特性可完全展现钢材良好的迟滞消能能力。装置于五层楼钢构架的金属阻尼器尺寸如图5及表4所示。在 OpenSees 分析模型中，以 Truss 元素搭配 Giuffre-Menegotto-Pinto steel 材料来仿真金属阻尼器。此材料进入非线性后，其应力应变关系呈现曲线软化，并可定义由线性段进入非线性段的转换曲线，对受拉及受压之等向硬化（isotropic strain hardening）参数亦可各自定义。与简化的双线性材料模型相比，此模型可更精确地仿真钢材降伏后变形与等向硬化之应力应变关系。

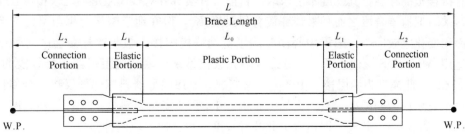

图 5 金属阻尼器示意图

金属阻尼器尺寸　　　　　　　　　　表4

	Story	Total Length L (mm)	Plastic Portion		Elastic Portion		Connection Portion	
			Length L_0 (mm)	Section Area A_0 (mm²)	Length L_1 (mm)	Section Area A_1 (mm²)	Length L_2 (mm)	Section Area A_2 (mm²)
X2 Frame Y2-Y3	4	5728.4	2600	2794	393.3	7568	1170.6	9856
	3	5728.4	2600	2794	393.3	7568	1170.6	9856
	2	5728.4	2450	4400	469.5	12276	1169.7	12276
	1	6121.8	3300	4400	434.4	12276	976.5	12276
Y1, Y3 Frame X1-X2	4	5728.4	2900	1408	277.3	3840	1136.9	7264
	3	5728.4	2900	1408	277.3	3840	1136.9	7264
	2	5728.4	2750	2204	303.3	5909	1185.9	8569
	1	6121.8	3600	2204	258.1	5909	1002.8	8569

3.1.1　金属阻尼器劲度

挫屈束制支撑是由三个部分所组成，即核心单元、围束单元及滑动机制[7]。核心单元由钢板组成，为主受力单元，围束单元负责提供束制机制防止核心单元受压局部挫屈，滑动机制则须在核心及围束单元间提供脱层单元避免核心受压后断面增大与围束单元间产生过大的摩擦力而使轴压力大量增加。在实际应用中，此种金属阻尼器可视为一单纯轴力构件，仅核心部分提供构件劲度。如图5所示，金属阻尼器构件整体的断面积并非完全一致，而是包含核心消能段（Plastic Portion）、转换段（Elastic Portion）及接合段（Connection Portion），因此金属阻尼器构件轴向弹性劲度须以此三段串联后的等效劲度来决定，如下式[8]：

$$K_{\text{eff}} = \frac{1}{\frac{1}{K_p} + \frac{1}{K_e} + \frac{1}{K_c}} \quad (1)$$

其中 K_{eff} 为金属阻尼器工作点至工作点（work point-to-work point）之等效劲度，K_p、K_e 及 K_c 分别为核心塑性消能段、弹性转换段及接合段之劲度。然而，由于接合段部分有接合板的存在，其劲度不易求取。目前工程实务界对金属阻尼器之模拟，在接合段部分，采阻尼器接合段之面积来计算接合段劲度 K_c，保守地忽略接合板与梁柱交会区附近断面积变化之效应[9]。国震中心团队为了更精确模拟接合板与梁柱交会区对劲度之效应，运用 ABAQUS[10] 建立了一个三维有限元素模型来求取金属阻尼器于接合段的劲度。如图6（a），此模型为两层楼，由薄壳元素组成，金属阻尼器斜撑部分仅建立金属核心单元。模型之所以为两层楼，是因为加装于构架上的12支制震斜撑，仅有两种长度（参考表4、图2及图3），此模型即包含了此两种长度，可一次求取两种斜撑之实际劲度。此模型考虑走动硬化双线性钢材，共37601个薄壳元素组成，如图6（a）所示，构架柱底为固接，并在梁柱接头部分施与面外束制。对此模型三楼梁形心进行位移控制侧推静力分析，侧推方向为使斜撑受拉的方向。参考图 6 (b)，分析完成后，可由接合段的实际轴向变形及通过斜撑的轴力来求得 K_c，再利用（1）式求得 K_{eff}，应用至 OpenSees 构架模型代表金属阻尼器的桁架元素定义中。

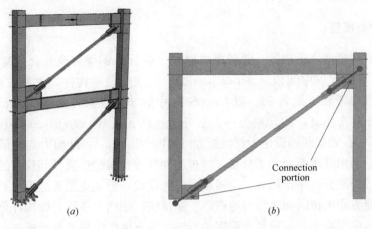

图 6　薄壳元素模型

3.1.2　金属阻尼器强度

于构架试验前，主办单位进行了实尺寸金属阻尼器构件试验，并将试验结果提供给参赛者，使其建置构架模型时有较精确的参考依据[11]。受测的金属阻尼器轴向反复受载，施力方式包含了静力受载及四种不同施力频率的动力受载（0.2、0.5、1 及 2 Hz），施加了三种不同的金属阻尼器冲程（12mm、24mm 及 36 mm），组合起来共有 15 组试验反应。如图 7 所示，金属阻尼器在动力受载下的强度反应比静力受载明显为高，因此为了正确预测金属阻尼器在动力历时作用下的反应，金属阻尼器的速度相依特性也应被考量。为了模拟此一特性，国震参赛团队根据此构件试验结果，调高数值模型中金属阻尼器之材料降伏强度，使其计算结果接近动力受载下之反应。另于整体构架的模态分析中，得到五层楼含金属阻尼器构架之第一模态周期为 0.5s，因此在含金属阻尼器构件的构架分析模型中，选定加载频率为 2Hz 的构件试验结果作为建模的依据。图 7 为

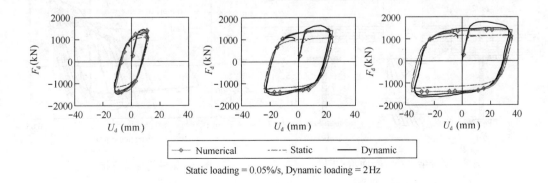

图 7　金属阻尼器构件试验与模拟反应比较

构件试验受静力加载、2Hz 频率动力加载，及国震参赛团队考虑金属阻尼器速度相依特性的构件模拟结果，由图中可看出以此方法调整之单一构件分析模型可相当准确地仿真金属阻尼器受动力加载之下的反应，并能以同一组材料模型参数，有效地模拟受不同冲程下金属阻尼器之反应。由此构件试验模拟所采之参数，即应用于构架反应预测模型中之每个金属阻尼器构件。

3.2 梁构件分析模型

此五层楼构架模型之梁构件，皆考虑成混凝土楼板与钢梁完全复合之复合梁（composite beam），并且仅以中间段（center portion）代表整段梁构件，未考虑端部段（end portion）影响（参考表1及表2）。以 OpenSees 中的非线性梁柱元素（Nonlinear Beam-column Element）来仿真，此元素为一力法分布塑性元素（force-based distributed plasticity element）[12]。元素断面设定为纤维断面（Fiber Section），断面力与变形关系由此纤维断面来求得。利用纤维断面元素分析，透过不同纤维元素的性质及其位置的安排，虽比一般常用之塑铰模型分析来得耗时，也需较多时间建模，但能更精确地模拟复合构件的劲度及强度，尤能方便模拟断面之轴力与弯矩互制反应。如图8所示，梁构件分析模型考虑了钢承板走向造成混凝土与钢梁复合程度的影响，此五层楼钢构架钢承板走向是沿着结构长向，因此考虑结构长向的大梁在钢承板下缘以上之所有混凝土（图8a）均对复合梁构件有贡献效果；而在结构短向的大梁，则只考虑钢承板上缘部分之混凝土才对复合梁有贡献（图8b）。钢梁部分由40个走动硬化双线性钢材（OpenSees 中之 Steel01 Mateial）纤维所组成，此 Steel01 材料于线性及非线性阶段，其应力应变关系皆为直线段，使用者仅需定义 E、F_y 及降伏后劲度比。此处对钢梁的模拟，未考虑钢材的等向硬化效应，仅定义降伏后劲度比 0.01 来模拟钢材走动硬化特性。混凝土楼板由仅受压不受拉之混凝土材料（Open Sees 中之 Concrete01 Material）纤维组成，结构长向梁之混凝土楼板断面为50个混凝土材料纤维，结构短向则是25个。参考图8，混凝土等效板宽 b_E 之指定根据先前有关 2007 年抗弯构架受震试验反应预测竞赛的研究[13]，模拟结果如图9所示，此时 b_E 为 30cm。依此模拟所得出之最佳混凝土楼板等效板宽，便应用于五层构架模型中的每根梁构件。

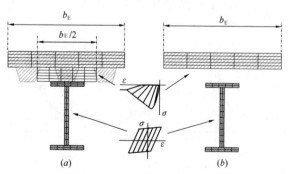

图8 复合梁之纤维断面示意图
(a) y 向；(b) x 向

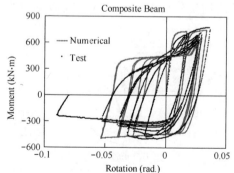

图9 梁柱构件抗弯接合试验与模拟反应比较

3.3 柱构件及梁柱交会区分析模型

此五层楼构架模型之柱构件亦以 OpenSees 中的非线性梁柱元素搭配纤维断面来模拟，钢柱纤维断面由 40 个走动硬化双线性钢材纤维所组成。此次参赛预测模型不考虑任何梁柱交会区之剪力变形，梁端及柱端亦未指定任何刚域。

3.4 反应预测结果

模态分析所得参赛构架模型的第一模态方向为结构长向，自然周期为 0.5s；第二模态方向为结构短向，周期为 0.481s；第三模态为扭转模态，周期为 0.343s。以图 4（a）所示之加速度历时进行三向动力分析，每层楼的绝对最大反应见图 10（a）至 10（d）。由图中可看出不论在 15% 或 100% Takatori 地震历时下，预测的结构楼层绝对最大反应（位移、加速度、层间剪力、层间位移角）皆与实验值[1]很接近，且楼层绝对最大位移角小于 1%（图 10d）。图 11 为 100% Takatori 地震历时下，预测模型 X2 及 Y1 构架线之一楼及四楼斜撑轴力对轴向变形关系图，由图中可看出斜撑均已进入大量非线性反应，且其迟滞循环饱满，大量能量由斜撑构件所消散。图 12 显示在 15% 及 100% Takatori 地震历时下，试体一楼及四楼斜撑最大受拉与受压轴力及轴向变形皆被准确预测。分析结果显示，于 15% 地震历时下，梁、柱构件及金属阻尼器，皆仍处于弹性范围；而于 100% 地震历时下，梁及柱构件仍维持弹性，但所有金属阻尼器皆进入塑性范围。此分析结果与试体设计要求及试验结果一致，地震能量皆由阻尼器消散。

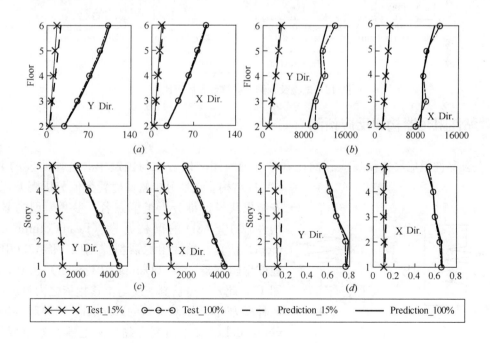

图 10 试验及预测之绝对最大反应比较

(a) Ma. Reletive Disp（mm）；(b) Max. Abs. Acc.（mm/sec^2）；
(c) Max. Story Shear（kN）；(d) Max. Story Drift（%rad.）

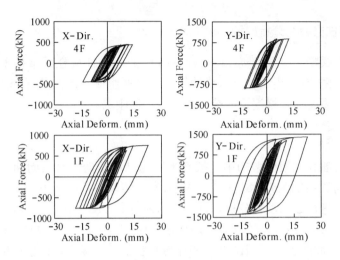

图 11 预测模型斜撑反应

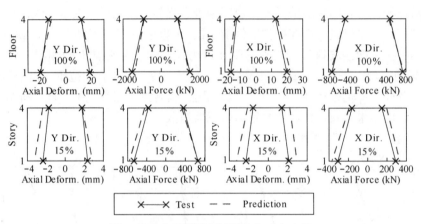

图 12 试验及预测之斜撑最大反应比较

4. 含黏性阻尼器构架反应预测

在含黏性阻尼器构架反应预测比赛中，国震团队以台大与国震中心合作研发的 PISA3D 结构非线性分析软件建置数值分析模型，参赛结果获得三维分析组第三名。参赛模型的梁柱构件以 PISA3D 的梁柱元素（BeamColumn element）仿真、黏性阻尼器斜撑则以 PISA3D 中的速度型阻尼器元素（Damper Element）仿真，见图 13，此元素可仿真速度与位移相依之力量反应，两种反力可串联（Maxwell model）或并联（Kelvin model）。在本研究含黏性阻尼器构架模型中，是以 Maxwell model 来模拟，PISA3D 中 Maxwell model 之详细计算处理方式详见参考文献[14]。如图 14（a）所示，梁线的高程设定在钢梁中心。每

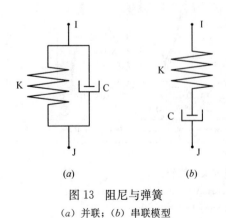

图 13 阻尼与弹簧
(a) 并联；(b) 串联模型

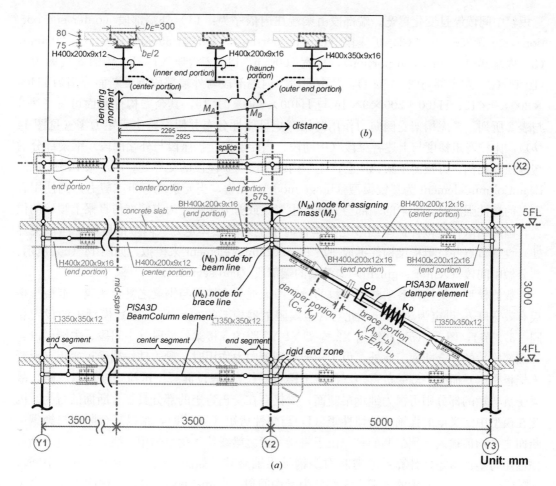

图 14 （a）含黏性阻尼器构架之模拟细节（以 X2 Frame 四楼为例）；
（b）采扩座细节之构件梁弯矩分布图及各部分断面图

层楼的质心位置均设置节点，质心点的高程设定在混凝土楼板表面，此外，柱线与混凝土楼板表面相交处皆设置节点（图 14（a）中 NM），在每层楼质心节点与同一水平面上九个柱上节点 NM 之间设定有刚性楼板约束条件（rigid diaphragm constraint），每层楼的水平向平移质量 M_x 与 M_y 与绕 z 轴旋转的质量惯性矩（I_z）集中施加于刚性楼板，每层楼的垂直向质量 M_z 则将各柱构件承载面积范围之质量集中施加至九个柱节点 NM 上。为了精确仿真阻尼器的位置，参赛模型在柱线与阻尼器中心线交会处设置节点 N_b（见图 14（a））做为阻尼器元素的端点，可注意到阻尼器元素端点 N_b 与柱线、梁线交点 N_B（见图 14（a））并非同一点。

4.1 梁构件之模拟

如图 14（a）所示，试验构架在未装置斜撑的跨度中皆以梁端扩座（Haunch）的方式来确保梁柱接合的耐震性能，因此梁跨可分作中间段（center portion）与端部段（end portion），端部段以全断面焊接的方式与柱构件相连；中间段选用 H400×200 的热轧型钢、以栓接方式与端部段相连。端部段为变断面构件，其梁深和腹板厚度与中间段相同但翼板较中间段厚且变化翼宽。端部段由梁端往内依序为：(1) 外端部段（outer end portion），翼宽 350mm；(2) 扩座段（haunch portion），翼宽由 350mm 渐变小至 200mm；(3) 内端部段（inner end portion），翼宽 200mm。以 X2 构架 Y1-Y2 跨的 5 楼梁为例（图 14 (a) 左半部与图 14 (b)），其中间段、内端部段与外端部段钢梁断面分别为 H400×200×9×12、H400×200×9×16 与 H400×350×9×16，其余梁构件断面可参考表 1 与表 2 所列。考虑构架受侧向力作用时梁弯矩呈线性分布并假设跨中弯矩为零（见图 14 (b)），比较弯矩梯度与上述各梁段（中间段、内端部段、扩座段与外端部段）钢梁部分的塑性强度比例，可预测梁塑铰应会产生于内端部段与扩座段的交界处。由于采 PISA3D 的 BeamColumn element 为塑铰模型（hinge model）元素，当元素之弯矩或剪力达到使用者定义的塑性强度时，元素的端部会分别产生弯矩或剪力塑铰。为求精准仿真梁上塑铰的位置，国震团队在模拟含扩座细节之梁构件时于上述塑铰预定处（内端部段与扩座段交界处）设置节点并将单根梁构件拆成三段 BeamColumn element 来模拟，如图 14 (a) 所示，共分做中间段（center segment）与两端的端部段（end segment）。

数值模型中梁构件各项断面参数的设定考虑混凝土楼板与钢梁之复合效应，混凝土楼板有效宽度 b_E 设定为 300mm，此与含金属阻尼器构架的 OpenSees 参赛模型中的设定相同（3.2 节），图 14 (b) 亦绘出 X2 构架 Y1-Y2 跨的 5 楼梁各梁段（中间段、内端部段与外端部段）的复合断面。此外，由于构架受到侧向力作用时，梁构件一端受正弯矩（混凝土楼板受压）而另一端受负弯矩（混凝土楼板受拉），假设混凝土抗拉强度可以忽略，故梁受负弯矩的部分则可视为纯钢梁断面；梁仅有在受正弯矩的部分具复合断面的行为。因此在模型中定义梁的强轴旋转惯性矩（I）与塑性弯矩（M_p）时，系以复合断面与纯钢梁断面之平均值输入。须强调的是，由于将梁元素之梁线位于钢梁的中心线，前述复合断面惯性矩与塑性弯矩之计算系考虑断面绕钢梁强轴旋转。如图 14 (a) 所示，由于中间段（center segment）元素的范围内大部分为梁中间段（center portion），故中间段元素之强轴旋转惯性矩系由梁中间段决定，以求模拟到梁中间段的弹性劲度；而中间段元素之强轴塑性弯矩则由梁内端部段的断面决定（塑铰预定处之断面），以求能模拟到梁的塑铰强度。另一方面，端部段元素的强轴旋转惯性矩（I）与塑性弯矩（M_p）则由梁外端部段的断面决定。梁柱交会区之行为以刚域（Rigid end zone）来模拟，如图 14 (a) 所示，梁端刚域之大小为柱深之一半，柱上刚域则由钢梁底部延伸至混凝土楼版表面之半高处，此番设定系因考虑到柱两侧混凝土楼版在构架侧位移时一侧受拉一侧受压之故。

阻尼器斜撑规格与阻尼器元素之参数　　表 5

构架	楼层	阻尼器部分 (Damper Portion)		斜撑部分 (Brace Portion)			PISA3D Maxwell 阻尼器元素 (PISA3D Damper Element)	
		C_d (kN/(mm/s)$^{0.38}$)	K_d (kN/mm)	L_b (mm)	A_b (mm^2)	K_b (kN/mm)	C_D (kN/(mm/s)$^{0.38}$)	K_D*** (kN/mm)
X2 Frame	4F	98	193*	2104	8380	797	98	155
	3F	98	193*	2104	8380	797	98	155
	2F	196	438*	1542	15323	1987	196	359
	1F	196	438*	2322	15323	1320	196	329

续表

构架	楼层	阻尼器部分 (Damper Portion)		斜撑部分 (Brace Portion)			PISA3D Maxwell 阻尼器元素 (PISA3D Damper Element)	
		C_d (kN/(mm/s)$^{0.38}$)	K_d (kN/mm)	L_b (mm)	A_b (mm^2)	K_b (kN/mm)	C_D (kN/(mm/s)$^{0.38}$)	K_D*** (kN/mm)
Y1、Y3 Frame	4F	49	119**	2429	9121	751	49	103
	3F	49	119**	2429	9121	751	49	103
	2F	98	193*	2104	8380	797	98	155
	1F	98	193*	2864	8380	585	98	145

* 由构件试验所得

** 由外插法所得

*** $K_D = 1/(1/K_d + 1/K_b)$

4.2 黏性阻尼器

黏性阻尼器属于速度型之被动消能装置，一般做成圆管活塞型，管内填充高分子黏滞性流体（silicon oil），当活塞于圆管内前后移动时，即会挤压流体通过节流孔（orifices）而产生黏滞阻尼力，地震能量经由阻尼器之运动过程中转换成热能消散出去。阻尼力 F_D 与速度 V 关系如式（2），因高分子液体材料不同而分成线性（$\eta=1$）与非线性（$\eta \neq 1$），线性黏性阻尼器之阻尼力与位移迟滞循环为椭圆，非线性型则为介于椭圆与长方形之间。当结构作简谐运动时，结构体在最大位移，即速度为零时，黏性阻尼器受力最小，此与结构杆件受力最大时恰呈 90°之相位差。此阻尼器在频率小于 2Hz 之情形下几乎不具劲度，且在消能的过程中虽会产生热能而使温度上升，但除非是在高速之作用下，黏性阻尼器之效能并不会有太大之影响，温度之因素几乎可以忽略。

$$if(V \geqslant 0) \quad \text{sign} = 1$$
$$if(V < 0) \quad \text{sign} = -1$$
$$F_D = C \times |V|^\eta \times \text{sign} \tag{2}$$

如图 14（a）所示，黏性阻尼器斜撑系由阻尼器部分（damper portion）以及钢斜撑部分（brace portion）所组成。主办单位在振动台试验之前进行了阻尼器部分（damper portion）的构件试验，以提供参赛者运用 Maxwell model 数值模型来仿真阻尼器受力变形行为时所需要的各项参数。Maxwell model 数值模型为阻尼器元素（阻尼系数 C_d）与轴向弹簧（弹性劲度 K_d）的串联。主办单位对于三种阻尼器（厂商提供之规格分别为：阻尼系数 $C_d=98$、131 与 196kN/（mm）0.38；速度指数皆为 0.38）进行了多种频率与多种振幅的正弦波型反复荷载试验，由试验结果求算出此三种阻尼器最适当的 K_d 值，依序（阻尼系数由小至大）分别为 193、222 与 438kN/mm。然而，$C_d=131$kN/（mm）0.38 的阻尼器并未安装于五层楼试验构架中，实际安装于试验构架的阻尼器除了 $C_d=98$ 与 196kN/（mm）0.38 两种阻尼器之外，尚有阻尼系数 $C_d=49$kN/（mm）0.38 的一种。由于并无 $C_d=49$kN/（mm）0.38 的阻尼器构件试验数据，国震团队参考前述构件试验所得的三种阻尼器 C_d-K_d 关系，求算出 C_d-K_d 关系的指数函数回归曲线（如图 15 所示），再以外插法求算出 $C_d=49$kN/（mm）0.38 阻尼器的 $K_d=119$kN/mm。黏性阻尼器出现

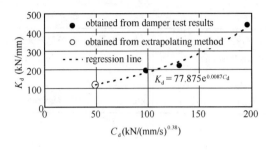

图 15 运用外插法求算 $C_d=49$kN/(mm)$^{0.38}$ 阻尼器之 K_d

弹性储存劲度 K_d 的原因可能是在高频反复载重下校正室调整阀反应不及所致[15]，不同规格之阻尼器的 K_d 值与 C_d 值之间理应不存在特定关系，上述国震团队利用主办单位给知三支阻尼器 C_d-K_d 关系，采指数函数回归曲线进行外插来求算未知规格阻尼器 K_d 值的方法并无任何理论根据。所有安装于试验构架中的阻尼器规格与所搭配之斜撑的长度（L_b）与断面积（A_b）皆列于表 5 中。每组阻尼器斜撑（含阻尼器部分与钢斜撑部分）皆以单根 PISA3D Maxwell model Damper element 模拟，考虑到斜撑部分的柔度对整体阻尼器斜撑构件的影响，阻尼器元素所定义之弹性劲度 K_D 为阻尼器劲度 K_d 与斜撑劲度 $K_b=EA_b/L_b$ 的串联；阻尼器元素之阻尼系数 C_D 则为阻尼器之 C_d。表 5 亦详列阻尼器元素所设定之参数。

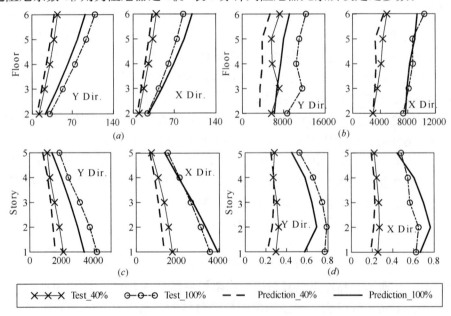

图 16 试验及预测之绝对最大反应
(a) Max. Reletive Disp (mm); (b) Max. Abs. Acc. (mm/sec^2);
(c) Max. Story Shear (kN); (d) Max. Story Drift (%rad.)

4.3 反应预测结果

此加装黏性阻尼器构架参赛模型经模态分析，第一模态周期为 0.681s，方向为结构长向；第二模态周期为 0.662s，方向为结构短向；第三模态为扭转模态，周期为 0.459s。图 16a 至 16d 为此模型受图 4b 所示之三向加速度历时动力分析所得每层楼绝对最大反应与实验值[1]之比较。由图 16a 及图 16d 可看出在 40% Takatori 地震下，参赛模型对结构绝对最大位移及层间位移角之预测很接近试验结果；于 100% Takatori 下，结构 Y 向之位移反应略小于试验结果，于结构 X 向则是略大于试验结果，而楼层绝对最大位移角皆小于1%。而由图16b及图16c，预测的结构X向绝对最大加速度及层间剪力皆与实验值很接近；但结构Y向预测之绝对最大加速度不论于40%或100% Takatori地震下，反应皆小于实验结果。图17显示PISA3D参赛模型在40%及100% Takatori地震历时下，对试体一楼及四楼斜撑最大受拉与受压轴力及轴向变形之预测很准确。图18为100% Takatori地震历时下，预测模型X2及Y1构架线之一楼及四楼斜撑轴力对轴向变形关系图，由图中看出阻尼器迟滞循环饱满，达到消散能量的效果，并且显示出阻尼反力与劲度反力串联的迟滞循环反应。

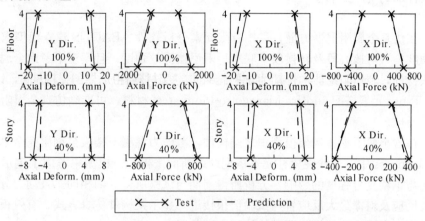

图17 试验及预测之斜撑最大反应

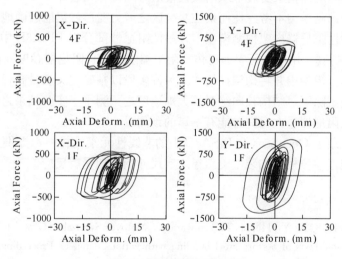

图18 预测模型斜撑反应

5. 实尺寸五层楼固定基础之抗弯构架及含隔震器之隔震构架反应预测

于2011年8月，NEES/E-Defense再次利用此五层楼实尺寸钢构架，进行受震反应试验。此次试验分别对固定基础之抗弯构架及含隔震器之隔震构架进行测试，研究构架加装三层摩擦单摆隔震器（Triple Friction Pendulum）之受震性能，并将其抗震反应与固定基础之抗弯构架反应比较。在试验进行前，主办单位亦举办了试验反应预测国际竞赛。国震中心利用 OpenSees 程序建置数值模型，运用本文第三章介绍之模拟方法，参赛获得抗弯构架组第一名及含隔震器之隔震构架组第二名[16]。此次参赛再次验证利用本文提出之模拟方法，不仅能准确模拟含制震斜撑之钢构架，亦可准确预测钢抗弯构架及隔震构架之受震反应。

6. 分析结果讨论与结论

（1）在仿真金属阻尼器方面，运用有限元素分析软件 ABAQUS 建立模型求得阻尼器接合段劲度，再与核心段及转换段串联求得阻尼器等效劲度；且模型中考虑金属阻尼器的速度相依特性，利用阻尼器构件动力试验结果来定义材料性质。由以上方法，以 OpenSees 中之 Truss 元素搭配 Giuffre-Menegotto-Pinto steel 材料，可准确仿真金属阻尼器动态受力变形反应。

（2）在含金属阻尼器构架模型中之梁构件，考虑为混凝土楼板与钢梁完全复合之复合梁，以 OpenSees 中的非线性梁柱元素搭配纤维断面来模拟。模型中考虑了钢承板走向造成混凝土与钢梁复合程度的影响，分析所得周期与试验所得周期相同。且动力分析后楼层绝对最大反应及斜撑最大反应皆与试验值很接近，显示梁构件仿真方式、有效板宽定义及阻尼器劲度定义皆准确地模拟了试体构架实际状况。

（3）含金属阻尼器构架受震下双向最大楼层层间位移角皆小于 0.8%，因构架加装金属阻尼器，参赛预测模型不考虑任何梁柱交会区之剪力变形，仍能准确地预测试验反应。

（4）加装黏性阻尼器构架预测模型中，黏性阻尼器部分以 PISA3D 中 Damper 元素之 Maxwell model 仿真，用单一元素即可仿真阻尼器阻尼力、阻尼器劲度及阻尼器斜撑串联下阻尼器的行为。模型中并将构架柱构件分为许多段以精确定义质量高程位置、梁高程及斜撑角度。此模型除了对结构长向绝对最大加速度之预测小于实验结果之外，对其余结构绝对最大反应及阻尼器最大反应皆能有不错的预测。

（5）本文介绍了 OpenSees 及 PISA3D 中的非线性元素及材料，提供了数种分析模型及方法，能准确地预测模拟加装制震斜撑之实尺寸钢构架受震反应。

参考文献

[1] Kasai K, Ito H, Ooki Y, Hikino T, Kajiwara K, Motoyui S, Ozaki H, and Ishii M (2010). Full-scale shake table tests of 5-story steel building with various dampers. Proceedings of the 7th CUEE and 5th ICEE Joint Conference, Tokyo, Japan, March 3-5

[2] Hikino T, Ohsaki M, Kasai K, and Nakashima M (2010). Overview of 2009 E-Defense blind analysis contest results. Proceedings of the 7th CUEE and 5th ICEE Joint Conference, Tokyo, Japan, March 3-5

[3] McKenna F (1997). Object oriented finite element programming frameworks for analysis, algorithms and parallel computing. Ph. D. thesis, Univ. of California, Berkeley. Calif

[4] Lin BZ, Chuang MC, Tsai KC. (2009). Object-oriented development and application of a nonlinear structural analysis framework. Advances in Engineering Software. 40：66-82

[5] Nakashima M, Inoue K, Tada M. (1998). Classification of damage to steel buildings observed in the

1995 Hyogoken-Nanbu earthquake. Engineering Structures. 20 (4-6): 271-281

［6］ Lin PC, Tsai KC, Wang KJ, Yu YJ, Wei CY, Wu AC, Tsai CY, Lin CH, Chen JC, Schellenberg AH, Mahin SA, Charles CW. (2012). Seismic Design and Hybrid Tests of A Full-scale Three-story Buckling-Restrained Braced Frame Using Welded End Connections and Thin Profile. Earthquake Engineering and Structural Dynamics, Vol. 41, No. 5, pp 1001-1020

［7］ 蔡克铨、吴安杰、魏志毓、庄明介（2012），"槽接式挫屈束制支撑与脱层材料性能研究"结构工程，in review

［8］ Tsai KC and Hsiao PC (2008). Pseudo-dynamic test of a full-scale CFT/BRB frame - Part 2: Seismic performance of buckling-restrained braces and connections. Earthquake Engineering and Structural Dynamics. 37: 1099-1115

［9］ 蔡克铨、黄彦智、翁崇兴（2004），"含挫屈束制消能支撑构架耐震性能试验与分析研究"结构工程，第十九卷，第一期，第3-40页

［10］ ABAQUS (2006). ABAQUS analysis user's manual. Version 6.6. ABAQUS Inc

［11］ Ooki Y, Kasai K, Motoyui S, Kaneko K, Kajiwara K, and Hikino T. (2009). Full-scale tests of passively-controlled 5-story steel building usinf E-Defense shake table, Part 3: Full-scale tests for dampers and beam-column subassemblies. Proceedings of 6th International Conference on Behavior of Steel Structures in Seismic Areas, Philadelphia, Pennsylvania, USA

［12］ Neuenhofer A and Filippou FC (1997). Evaluation of nonlinear frame finite-element models. Journal of Structural Engineering

［13］ Yu YJ, Tsai KC, Weng YT, Lin BZ, and Lin, JL (2010). Analytical Studies of a Full-Scale Steel Building Shaken to Collapse. Engineering Structures. 32(10), 3418-3430

［14］ Hatada T, Kobori T, Ishida M, and Niwa N. (2000). Dynamic analysis of structures with Maxwell model. Earthquake Engineering and Structural Dynamics. 29: 159-176

［15］ Constantinou MC, Tsopelas P, and Hammel W (1997). Testing and Modeling of an Improved Damper Configuration for Stiff Structural Systems. Center for Industrial Effectiveness, State University of New York, Buffalo, NY

［16］ Yu YJ, Tsai KC, Chen CH, Li CH, and Tsai CY (2012). Blind Analyses of a Full-Scale 5-Story Steel Building with the Fixed-Base Configuration. Proceedings of the 9th International Conference on Urban Earthquake Engineering and 4th Asia Conference on Earthquake Engineering, Tokyo, Japan, March 6-8

# 脚手架和看台的二阶直接分析*

刘耀鹏[1]，陈绍礼[1]，Stephen McCrory[2]
（1. 香港理工大学 土木及结构工程系，香港九龙红磡；
2. McCrory Scaffolding Limited，U. K.）

摘 要：本文提出一种简单高效的弹簧单元，可以按规范输入弹簧的刚度模型、精确地模拟出扣件式钢管脚手架的扣件的结构行为。与传统的设计脚手架考虑扣件半刚性的方法不同，本文的弹簧单元不仅可以精确地反映扣件的行为，还可以反映杆件的连续性行为。该单元在二阶分析软件 NIDA 中进行了实现，进而提出了一种完善的用于脚手架二阶直接设计的方法。该方法同时考虑了 P-Δ 和 P-δ 效应，考虑了结构整体和构件局部的初始缺陷，符合了直接分析的要求，因而不再需要假设计算长度，其计算结果可以直接作为设计的依据，可达到更安全、更经济的设计目的。

关键词：二阶分析；直接分析；脚手架；非线性弹簧单元
中图分类号：TP391

SECOND-ORDER DIRECT ANALYSIS OF STEEL SCAFFOLD AND STAND STRUCTURES

Y. P. LIU[1], S. L. CHAN[1], S. McCrory[2]
(1. Department of Civil and Structural Engineering, The Hong Kong Polytechnic University;
2. McCrory Scaffolding Limited, U. K.)

Abstract: In this paper a nonlinear spring element for modeling of the coupler of tube-and-fitting scaffolds is proposed. By using the moment-rotation curve from specified design code or from experimental test, the spring element can precisely reflect the behavior of the coupler. Unlike the traditional modeling method which sets the coupler as a semi-rigid connection at the two ends of the beam-column element and breaks the continuous member into two parts, the spring element maintains the continuous nature of the member as the fact as it is. Incorporating the spring element to the second-order analysis software NIDA, a comprehensive design method for design of scaffold structures is proposed. This method considers P-Δ and P-δ effects, frame and member imperfection, and therefore the effective length assumption for stability design can be skipped and a safer and more economical design can be achieved.

Keywords: second-order analysis; direct analysis; scaffold; nonlinear spring element

---

* 基金项目：香港特别行政区 RGC 项目
第一作者：刘耀鹏（1978—），男，博士，主要从事钢结构二阶直接分析的研究，E-mail：ceypliu@polyu.edu.hk。
通讯作者：陈绍礼（1957—），男，博士，教授，主要从事钢结构二阶直接分析的研究，E-mail：ceslchan@polyu.edu.hk。

1. 引言

脚手架是为安装施工或建筑施工而搭设的临时结构。钢管脚手架由于易安装、易拆卸、承载能力高、可重复利用等诸多优点而得到广泛应用。同时，相对于传统支撑系统（如竹结构、木结构）标准化程度高，在选材、模块化方面，容易把各种不确定因素控制下来，从而降低使用风险。

根据脚手架的用途，可以分为两大类。第一类是安装脚手架，有侧向连墙件，承担较小的荷载，如工人、装修材料、装修设备、拆卸物等，给工人提供一个安全施工的地方。第二类是施工脚手架，承担较大的荷载，如建筑结构材料、模板、施工设备等，起到临时支撑作用等结构完成形成一个有效的承载体系后再拆卸走。

引起两类脚手架的破坏因素也不尽相同。对于安装脚手架，连墙件破坏导致立杆系统长度增大、临时拆卸物堆积过多过于集中、斜撑布置不够是常见的因素。对于施工脚手架，浇筑混凝土产生的动力放大效应、顶部承托构件破坏增大了 P-Δ 效应[1]、支座不均匀沉降等是常见因素。

从设计的角度而言，由于脚手架属于临时结构，缺乏像永久结构那样的严格设计流程，重视程度不够。更应该指出的是，脚手架的构件长细比通常都比较大，极易发生弹性失稳（见图1），而目前许多国家通行的基于构件的线性设计方法（包括 GB 50017—2003[2]，JGJ 130—2011[3]）在预测结构稳定性方面表现不佳，不能为这类结构的设计提供很好的依据。

近十多年来，欧美规范（如 Eurocode3—2005[4]，AISC360—2010[5]）得到了很大的发展，尤其是二阶直接分析在规范的详细引入，表明了规范发展的潮流与方向。

现行的用于稳定性设计的计算长度法是基于线性分析的，在系数取值上倍受争议，常常只有在简单的情况下才可以取得较为合理的值，更多的情况下只能凭经验估计，一方面给结构安全带来很大的隐患，另一方面也给结构设计带来很多浪费。计算长度法是基

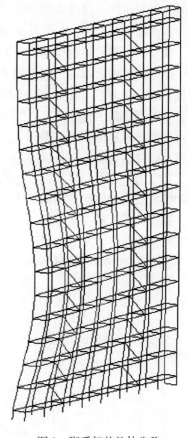

图1 脚手架的整体失稳

于单根构件的失稳模态，并假设所有构件同时失稳，在理论上就存在不合理之处。因而，欧美规范对这种方法的应用给出了严格的条件。总的而言，计算长度法是历史产物，已经逐渐无法去满足现代化设计的要求了。

二阶直接分析法是基于体系的，可充分考虑结构变形和材料屈服问题，理论合理，是结构设计的必然趋势。这种方法可保证系统的整体和局部稳定性，可适用于任意结构形式。一个显著的特点是，二阶直接分析不再需要假设计算长度了，所谓"设计"也仅限于

进行截面强度设计而已，从而避免了因计算长度而引入的不确定因素，给工程师们带来省时、经济且安全的设计体验。二阶直接分析也可视为各种非线性分析（如火灾、地震、施工分析、连续性倒塌等）的基础，具有很高的可扩展性，为工程师们分析、研究结构物的各种真实行为提供了很好的手段。

脚手架结构是细长结构体系的典型代表，二阶直接分析法可为这类结构的设计提供可靠的依据，因而引起了众多研究人员和工程师们的重视。然而，人们往往只关注了脚手架结构整体的二阶 $P\text{-}\Delta$ 效应，忽略了构件自身的二阶 $P\text{-}\delta$ 效应，其结果是仍旧不能反映结构的真实受力和变形，不能预测出结构真实的失稳模态，不能达到"直接分析"的目的。

脚手架的可重用性非常高，造成构件的初始缺陷尤其是几何缺陷就要比一般钢构件的高。在二阶分析中正确地考虑构件的初始缺陷，就可以全过程追踪到结构的受力变形情况，从而真正实现了直接分析，不再需要传统冗长、繁琐的构件稳定性设计。本文采用的二阶直接分析法，同时考虑了 $P\text{-}\Delta$ 和 $P\text{-}\delta$ 两种效应，并用特征值屈曲模态来模拟结构整体和构件局部的初始缺陷，从而可以精确分析、解决脚手架这类细长结构的稳定性问题。

常见的钢管材料制作的脚手架有扣件式钢管脚手架、碗扣式钢管脚手架、承插式钢管脚手架、门式脚手架等。钢管脚手架由于要易安装、易拆卸，造成节点的半刚性行为非常明显，以至于节点刚受力就呈非线性，这跟普通的钢结构非常不同。扣件式钢管脚手架除具有一般脚手架的优点外，还可以保持构件的连续性，承受一定的弯矩，节省斜撑的应用，因而在全球范围内都得到了广泛的应用。但这类脚手架的节点也比较复杂，因而本文将重点研究这类脚手架，提出了一个非线性弹簧单元，不仅可以考虑扣件的拉、压、弯、扭、剪行为，还可以显式地考虑偏心情况，配合二阶直接分析，可以非常精准地分析、研究扣件式钢管脚手架的结构行为。

扣件式钢管脚手架不仅可用于一般的安装、建筑施工中，还广泛地应用到临时演出台、观众看台等。本文将以一个安装施工钢管脚手架和伦敦奥运的一个体育看台为例，证明本文方法的精确性和可用性。

2. 二阶直接分析法

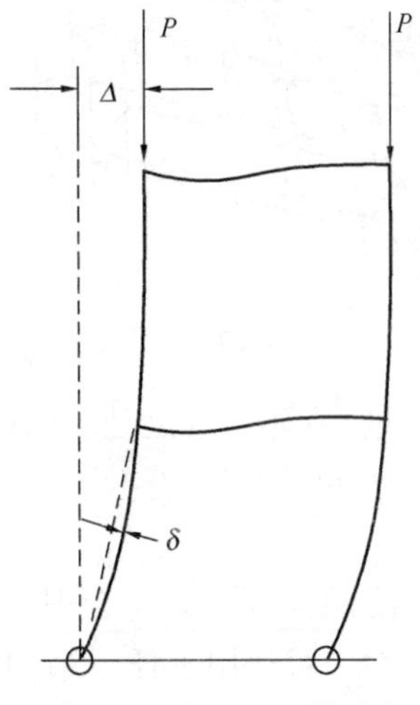

二阶直接分析必须将各种影响结构变形的因素都考虑进去，欧盟规范 Eurocode3[4]、美国 AISC360[5] 和香港规范 CoPHK[6] 都显式地要求在二阶分析中考虑 $P\text{-}\Delta$ 效应、$P\text{-}\delta$ 效应（见图2）以及初始缺陷。扣件式钢管脚手架还必须考虑节点半刚性，弯矩-转角曲线可以参考 EN12811-1[7]。

本节将简述二阶分析法的主要观念，更详细的说明可以参考欧盟规范 Eurocode3[4]、美国 AISC360[5] 和香港规范 CoPHK[6]。

2.1 初始缺陷

工程中使用的构件都不可能是完美的，初始弯曲以及残余应力等不可避免，而这些因素会影响构件的真实响应，进

图2 $P\text{-}\delta$ 和 $P\text{-}\Delta$ 效应

而影响了结构系统的响应。基于一阶线性分析的设计方法虽然也认识到了这一点，比如对不同截面进行柱子失稳曲线分类，但仅在线性分析后在构件设计中考虑，即这种影响是局部的，而不是像二阶分析一样，在分析中引入这个因素，不仅构件自身得已考虑，同时也考虑了构件对系统整体的影响。

欧盟规范 Eurocode3[4]、美国 AISC360[5] 和香港规范 CoPHK[6] 都明确规定了框架以及构件的初始缺陷。初始缺陷效应从两方面予以考虑：（1）框架整体分析中，P-Δ 效应；（2）构件单独分析中，P-δ 效应。

（1）框架初始缺陷

框架初始缺陷为安装过程中不可避免的误差所引起的结构整体或框架的不垂直度，在竖向荷载下，结构的 P-Δ 效应将放大，是结构整体失稳的一大来源。模拟该类缺陷的方法常用有两种，一种是特征值屈曲模态法，另外一种假想水平力法。

①特征值屈曲模态法

在二阶直接分析前，先采用特征值屈曲分析，获得结构整体失稳的第一阶模态，然后依此模态作为结构的初始缺陷分布形式，最大的缺陷值可取为：

$$\Delta = \frac{h}{200} \tag{1}$$

式中，h 代表建筑高度或者结构的最大尺寸，Δ 表示初始变形或不垂直缺陷值。

在确定了结构的初始缺陷分布形式后，再开始二阶非线性分析，获得结构的全过程的荷载变形曲线。

香港规范 CoPHK[6] 对临时性及需拆除结构的初始缺陷值进行适当放大，如对于临时结构，初始缺陷值取为 $h/100$；对于需要拆除的结构，初始缺陷值取为 $3h/100$。规则的多层框架结构其初始缺陷形状可简单近似为一根倾斜的直线。本文将采用 NIDA[8] 软件进行二阶直接分析。

②假想水平力法

对于规则框架结构，如果屈曲模态是明显的侧移模态，则可以在水平方向施加如图 3 所示的 0.5% 的竖向荷载值来模拟结构的整体初始缺陷。采用该方法时，这种附加的假想水平力应加在引起结构最不利反应的方向上，不可起到抵消水平外载的作用。因而，假想水平力通常仅应用在规则结构中，在复杂的三维结构应用中会陷入无法确定结构最不利方向的困境。此外，

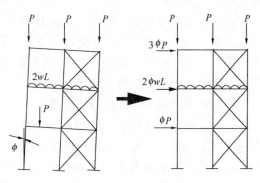

图 3 假想水平力法

由于这种假想水平力是附加的，并不是真实的设计荷载，这就增大了支座反力，使得基础和底部连接的设计偏于保守。

（2）构件的初始缺陷

初始弯曲和残余应力作为两种主要影响构件性能的因素，在制作、运送和安装过程中都不可避免地引入，因此计算构件的屈曲荷载时应该对它们充分加以考虑。传统方法把这

两种因素统一放在几条柱子曲线（如图 4）来考虑，对于单根构件而言可以达到设计的目的，但没有考虑到构件对系统的影响。

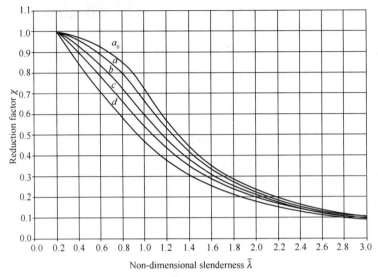

图 4　柱子失稳曲线（Eurocode3）

当采用二阶直接分析时，以欧盟 Eurocode3[4] 和美国 AISC360[5] 为代表的两大规范系统在处理初始弯曲和残余应力问题上有所不同。

欧盟规范将残余应力等效成几何缺陷，再叠加到初始弯曲上，即可认为是对原有几何缺陷的放大。实际操作上，规范所给出的杆件初始缺陷值是通过拟合构件的柱子曲线（屈曲强度-长细比曲线）得到的，这些值使构件在二阶分析中得到比试验拟合的柱子曲线低 5％的下界曲线。因为柱子曲线跟截面形状有关，杆件的初始缺陷值也就随着跟截面形状相关，如欧盟规范 Eurocode3 的表 5.1 和香港规范 CoPHK[6] 的表 6.1。

美国规范则对初始弯曲和残余应力作了分别处理，几何缺陷一律取为 1/1000 的杆件长度（这跟美国只采用一条柱子曲线有关）；而残余应力则采取折减弹性模量 E 的方式，不论受压还是受拉构件，均取 $0.8E$，从而起到降低结构刚度的效果，客观上是杆件的 $P\text{-}\Delta$ 效应和系统的 $P\text{-}\delta$ 效应将会增大。

欧美两大规范都在结构分析中考虑了杆件初始弯曲和残余应力对结构系统的影响，方式不一样，但预期效果是一致的。总的而言，欧盟规范从柱子曲线出发，采用等效几何缺陷的方式，可操作性强，某种程度上讲实现了从线性设计到直接分析较为平滑的过渡。美国规范将两种因素分开考虑，概念清晰，但弹性模量的折减系数一律取为 0.8 缺乏说服力，还需按极限状态和正常使用状态来考虑是否折减刚度，使得直接分析法在实际使用过程中变得繁琐。

本文所采用的软件 NIDA 支持上述两大规范的分析方法，因所设计项目要满足欧盟规范，故杆件的初始缺陷统一为几何缺陷处理。NIDA 所采用的二阶分析单元是 Chan 和 Gu[9] 提出的可考虑初始弯曲的稳定函数，Liu, Chan 和 Lam[10] 将这种稳定函数运用在半刚性框架的二阶分析设计中。值得指出的是，一般软件所采用的梁柱单元没有在单元层次上考虑杆件的初始缺陷，因而是不能满足规范直接分析的要求的，事实上也就无法精确地预测出结构的失稳过程。

2.2 截面承载力

本文所采用的二阶直接分析法，充分考虑了 P-Δ 和 P-δ 效应以及初始缺陷，因此传统的单独构件的稳定性检验就不再需要了，否则是对这些因素作了双重考虑，使得设计过于保守。根据欧盟规范 Eurocode3 中 5.2.2（7）节中所述，公式（2）的截面承载力检验能够充分保证结构的安全性，结构以及构件的强度和稳定性可以通过进行逐根构件的截面强度检验得到保证：

$$\frac{N_{Ed}}{Af_y} + \frac{(M_{y,Ed} + P\Delta_y + P\delta_y)}{M_{y,Rd}} + \frac{(M_{z,Ed} + P\Delta_z + P\delta_z)}{M_{z,Rd}} = \phi \leqslant 1 \quad (2)$$

式中，A 为横截面面积；f_y 为设计强度；N_{Ed} 为轴力；$M_{y,Ed}$ 和 $M_{z,Ed}$ 分别为不考虑二阶效应的关于 y 轴和 z 轴的弯矩；$M_{y,Rd}$ 和 $M_{z,Rd}$ 分别为关于 y 轴和 z 轴的弯矩承载力（等于 $f_y \cdot S$ 或者 $f_y \cdot Z$，S 和 Z 分别表示塑性截面模量和弹性截面模量），如果截面还受到侧向扭转屈曲的影响，对 Y 轴（强轴）的弯矩承载力 $M_{y,Rd}$ 就应该用 $M_b = f_b \cdot S_y$ 或者 $f_b \cdot Z_y$ 代替，f_b 为考虑侧向扭转屈曲的设计强度；Δ 为节点位移，由框架不垂直缺陷加上荷载引起的框架侧移组成；δ 为由初始缺陷以及荷载引起的构件弯曲；ϕ 为截面承载力系数，如果 $\phi > 1$，则意味着构件会发生破坏。

在二阶直接弹性分析法中，任一根构件的截面承载力系数 $\phi > 1$ 都表明整体结构达到了设计极限荷载。在二阶直接塑性分析法中，若截面承载力系数大于 1 则构件将产生一个塑性铰，直到整个结构因塑性铰过多变成机构为止。

3. 非线性弹簧单元

扣件式钢管脚手架所采用的扣件主要有直角扣件、旋转扣件和对接扣件。直角扣件用于连接两根互相垂直交叉的钢管；旋转扣件用于连接两根任意角度交叉的钢管；对接扣件则用于连接两根对接的钢管。

直角扣件除了能抵抗滑移、保证构件之间的结合点不会偏离原来的位置外，还可以承受扭矩，这种抗扭作用就为垂直交叉处节点提供了半刚性，从而杆件之间可以像一般梁柱连接一样传递弯矩。由于直角扣件的这种能力，扣件式钢管脚手架就不用像桁架结构那样把所有节点都作为铰接点来设计了，这也就节省了斜撑的应用。旋转扣件不提供抗扭能力，相交杆件可以绕结合点自由转动，但要保证相交杆件不脱离。对接扣件主要为两对接杆件提供轴向抗拉和抗压能力。

直角扣件的受力行为最为复杂，如图 5a 所示。欧洲脚手架规范 EN12811-1 对直角扣件给出了双线性关系的扭矩-转角曲线（见图 5b），用于节点的半刚性分析。EN12811-1 还对直角扣件给出了抗滑移、抗破坏的承载力。

本文根据直角扣件的受力特征（如图 5a 所示），提出了一种非线性弹簧单元，用于精确地模拟直角扣件的结构行为。常见的直角扣件实物图可见图 6a，本文提出的弹簧单元用于连接两相交钢管的形心，计算模型见图 6b。

每根弹簧单元由两个节点定义，每个节点拥有 6 个自由度，方便与考虑初始弯曲的稳

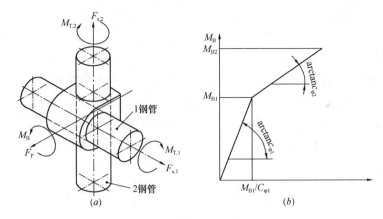

图 5　直角扣件受力行为（EN12811-1）
(a) 直角扣件受力图；(b) 直角扣件扭矩

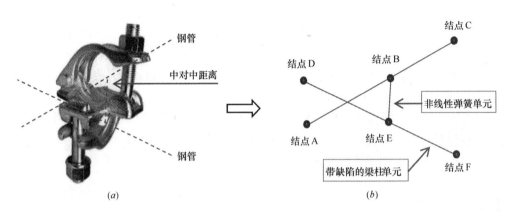

图 6　直角扣件及计算模型
(a) 直角扣件实物图；(b) 直角扣件计算模型

定函数连接。对于直角扣件，弹簧的单元刚度矩阵由 6 根相互独立的子弹簧组成：一根扭转弹簧，满足图 5b 所示的半刚行为，用于模拟直角扣件的受扭行为，对应图 5a 所示的 M_B；一根轴向弹簧，用于模拟直角扣件的受拉破坏，对应图 5a 所示的 F_P；两根剪切弹簧，用于模拟直角扣件的滑移，对应图 5a 所示的 $F_{s,1}$ 和 $F_{s,2}$；两根受弯弹簧，用于模拟直角扣件对钢管的自旋约束，对应图 5a 所示的 $M_{T,1}$ 和 $M_{T,2}$。因此，直角扣件的主要行为可以精确地反映，并满足规范的要求。本文的分析主要采用了 EN12811-1 所给出的直角扣件的属性，对于其他规范或相应产品，只要替换掉弹簧属性即可。对于旋转扣件和对接扣件，可根据其相应的受力特征，忽略相关的不起作用的弹簧即可。换而言之，本文提出的弹簧单元模型，可以适用于任意产品。

本文提出的用于考虑扣件式钢管脚手架的扣件的半刚行为的非线性弹簧单元，模型简单、清晰、直接，不仅可以精确地模拟扣件的真实行为，还保证了相连钢管的连贯性，使得钢管行为与真实结构相符，与传统半刚性计算把半刚节点放在杆件两头导致连续杆件打断不同，如 Prabhakaran 等[11]的研究工作。本文的工作很好地解决了长期困扰工程师的问题，即连续杆件的行为得不到真实反映。此外，采用本文的计算模型时，钢管之间的偏心也得到了显式的考虑，使得计算结果更接近结构真实响应，符合直接分析法的精神。

4. 例子

本文所提出的非线性弹簧单元已经整合到 NIDA 软件，并用于以下的验证算例和工程实例中。

4.1 两层扣件式钢管脚手架

本例题对一个两层扣件式钢管脚手架进行了研究。该脚手架为一常见脚手架的一部分，结构尺寸如图 7a 所示，除侧向横杆采用 CHS48.3×3.2 外，其余截面尺寸包括斜撑皆为 CHS48.3×4，钢的材料规格均为 S275，即设计强度为 275 MPa。

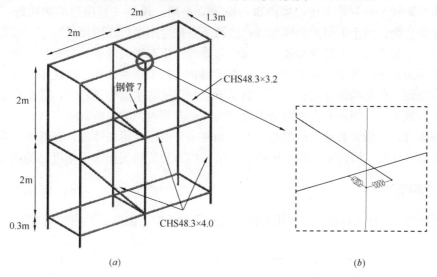

图 7 例 1——两层扣件式钢管脚手架
(a) 结构尺寸和构件截面；(b) 本文弹簧模型

本例旨在测试本文所提弹簧单元对结构反应的影响，故为简单起见，此脚手架所受荷载仅为竖向均布荷载和侧向节点力，如图 8 所示，柱底设为刚接。本文对该扣件式脚手架采取了三种模拟方式进行分析：

模型一：刚接模型，即忽略脚手架的扣件，认为杆件之间是刚接的且没有偏心；

模型二：弹簧模型，即采用本文所提弹簧模型对扣件进行模拟，如图 7b 所示，杆件直接存在偏心；

模型三：杆端半刚模型，即把扣件的行为放在了杆件的两端进行考虑，认为杆件之间是半刚连接的且没有偏心。

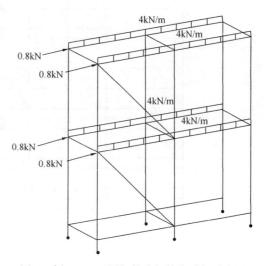

图 8 例 1——两层扣件式钢管脚手架受力图

对以上三种计算模型分别采用了二阶直接分析，取钢管 7（见图 7a）作为研究对象，其弯矩图如图 9 所示。

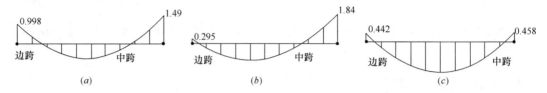

图 9 例 1——不同模型的弯矩图
(a) 刚接模型；(b) 本文弹簧模型；(c) 杆端半刚模型

图 9 所示弯矩图表明，模型一和模型二的弯矩图趋势跟两跨连续梁类似，中间弯矩较大，两边弯矩较小，与简支连续梁两边弯矩为 0 不同，模型一和模型二两边跨有抗弯刚度，故存在弯矩。由于采用弹簧模型时，边跨节点抗弯刚度远比刚接模型小，故弯矩也较小。同时，模型二由于边跨变形较大，对中跨处的弯矩有增大效果，故模型二的中跨弯矩大过刚接模型的中跨弯矩。对于模型三，由于将扣件的半刚性放在了梁柱单元两端，其两端弯矩就呈现出跟两端固支梁相似，而由于单元两端抗弯刚度毕竟较小，弯矩也就较小且两端弯矩差别不大（弯矩差别主要是因为结构一侧有节点力的影响）。

总的而言，采用弹簧单元模型时，不仅扣件本身的行为得以体现，钢管的结构响应也得到了准确地反映，因而可以得到更精确的结构内力和位移用作脚手架设计的依据。

4.2 装修脚手架

本例题为一英国实际装修工程用的扣件式钢管脚手架。该脚手架的结构布置如图 10 所

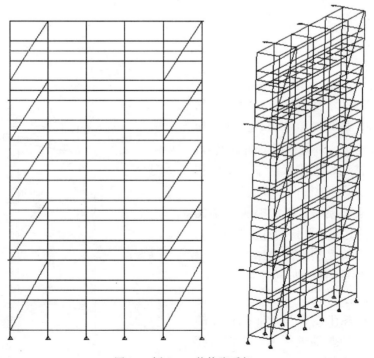

图 10 例 2——装修脚手架

示。所采用的钢管截面主要有两种，CHS48.3×3.2 和 CHS48.3×4，钢的材料规格均为 S275，即设计强度为 275MPa。同时采用了直角扣件和旋转扣件，均用本文所提弹簧单元进行模拟，对于旋转扣件，只需要把扭转弹簧忽略掉即可。

该脚手架的柱底设计为铰接，连墙件只约束轴向行为。荷载组合主要为两种情况，组合一为施工活荷载控制，风荷载较小；组合二为风荷载控制，施工活荷载较小。

对该结构进行特征值类型的屈曲分析表明，第一阶失稳模态对应的弹性临界荷载因子系数为 2.76，根据欧洲规范 Eurocode3，必须采用二阶分析因该系数小于 5，此结构为侧向失稳敏感结构。

对该脚手架进行二阶分析，以第一阶整体失稳模态（见图 11）作为缺陷分布，受压构件的初始缺陷取为 $L/500$，L 为杆件长度，此缺陷包含了残余应力的影响。

由于采用了直接分析，故不用再采用计算长度法进行杆件稳定性设计，只需进行截面承载力效应。因所采用的直接分析为二阶全过程非线性分析，不允许进行荷载效应组合，要对所有组合工况进行分析，所有杆件的所有截面在任一组合工况下，其按公式（2）所得的截面承载因子都不能大过 1，满足此条件时，则表明了该结构是满足设计要求的。

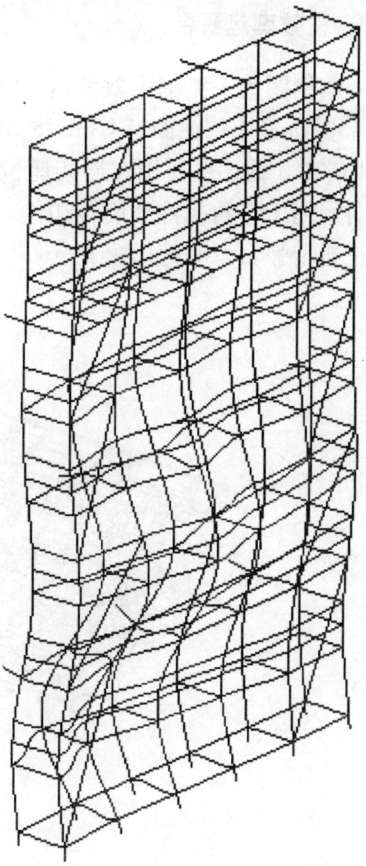

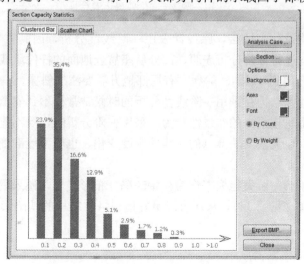

图 11 例 2——第一阶失稳模态（2.76）

本结构的截面承载因子的统计图见图 12，该图表明了所有构件的截面承载因子都小于 1，约 0.3% 的构件处于 0.8~0.9 水平，大部分构件的承载因子都较低。

图 12 例 2——截面利用率

4.3 伦敦奥运看台

本例题为一英国伦敦奥运的临时看台,采用了扣件式钢管系统,方便安装与拆卸,见图 13。所采用的截面有 CHS48.3×3.2,CHS48.3×4 和 RHS60×40×2.5,其中 RHS60×40×2.5 截面主要用在连接座位处,钢的材料规格均为 S275,即设计强度为 275MPa。同时采用了直角扣件和旋转扣件,均用本文所提弹簧单元进行模拟,对于旋转扣件,只需要把扭转弹簧忽略掉即可。

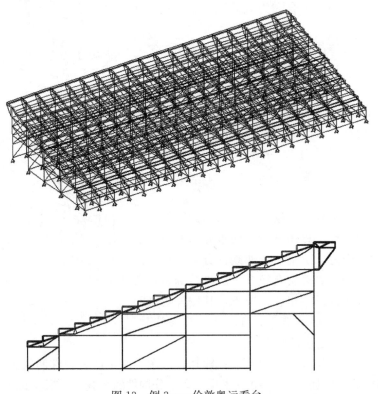

图 13 例 3——伦敦奥运看台

该看台的柱底均设计为铰接,原设计采用一阶线性分析,因其重要性要求了二阶分析。对原设计的结构进行二阶分析表明,部分后座靠近地面的杆件承载力不够,需要把截面从 CHS48.3×3.2 调到 CHS48.3×4,有局部地方需要添加斜撑。

该案例表明,此类结构在采用一阶线性分析的时候常常得不到安全的设计,这类结构无论杆件还是节点都呈现很高的非线性行为。采用一阶分析时,既不能获得合理的计算长度进行稳定性设计,也无法反映节点的非线性刚度变化,也就不能准确预测结构的受力和变形了。

本工程除采用了第一阶失稳模态作为初始缺陷分布形式外,还采用了荷载作用下的变形作为初始缺陷分布(见图 14),从而可以更好地包络住结构的缺陷形式,达到安全设计的目的。

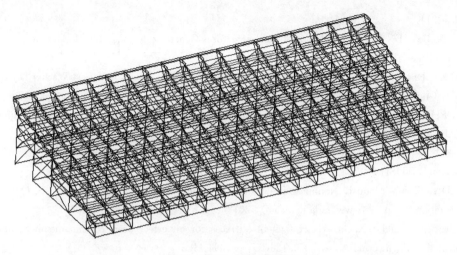

图 14 例 3—结构变形图

5. 结论

本文提出一种简单高效的弹簧单元，可以按规范输入弹簧的刚度模型、精确地模拟出扣件式钢管脚手架的扣件的结构行为。本文进而提出一种新的设计脚手架的方法，不仅可以精确地反映扣件的行为，还可以反映杆件的连续性行为，以及脚手架结构的钢管之间常见的偏心行为。此外，本文采用二阶分析软件 NIDA，同时考虑了结构整体和构件局部的初始缺陷，考虑了 $P\text{-}\Delta$ 和 $P\text{-}\delta$ 效应，符合了直接分析的要求，获得了精确的结构内力和位移，因而不再需要假设计算长度，其计算结果可以直接作为设计的依据。

本文认为传统的将节点半刚性放在梁柱单元两端的做法不能很好地反映扣件式钢管脚手架的结构行为。本文提出的单元和模拟方法很好地解决了此类结构的设计问题，有很好的应用前景。

ACKNOWLEDGEMENT

The authors acknowledge the financial support by the Research Grant Council of the Hong Kong SAR Government on the projects "Stability and second-order analysis and design of re-used and new scaffolding systems (PolyU 5116/11E)" and "Second-order and Advanced Analysis and Design of Steel Towers Made of Members with Angle Cross-section (PolyU 5115/08E)".

参考文献

[1] Peng, J. L., et al., Structural modeling and analysis of modular falsework systems. Journal of Structural Engineering-Asce, 1997. 123(9): p. 1245-1251

[2] GB 50017, Code for design of steel structures. Zhong Hua Ren Min Gong He Guo Jian She Bu, 2003

[3] 130-2011, J., 建筑施工扣件式钢管脚手架安全技术规范. 中华人民共和国住房和城乡建设

部,2011
[4] Eurocode3, EN 1993-1-1: Design of steel structures-General rules and rules for buildings. European Committee for Standardization, 2005
[5] AISC-LRFD, Specification for Structural Steel Buildings. AISC, Inc., One East Wacker Driver, Suite 700, Chicago, Illinois 60601-1802, 2010
[6] COPHK, Code of Practice for the Structural Use of Steel 2005. Buildings Department, Hong Kong SAR Government, 2005
[7] 12811-1, E., Temporary Works Equipment, Part 1: Scaffolds-Performance requirements and general design. BRITISH STANDARDS INSTITUTION, London, 2003
[8] NIDA, User's Manual, Nonlinear Integrated Design and Analysis. NIDA 9.0 HTML Online Documentation. (http://www.nida-naf.com), 2012
[9] Chan, S. L. and J. X. Gu, Exact tangent stiffness for imperfect beam-column members. Journal of Structural Engineering-Asce, 2000. 126(9): p. 1094-1102
[10] Liu, Y. P., S. L. Chan, and D. Lam, Case Study for semi-rigid design, Section V, in Semi-rigid Connections Handbook, edited by Wai-Fah Chen, Norimitsu Kishi, and Masato Komuro. 2011
[11] Prabhakaran, U., R. G. Beale, and M. H. R. Godley, Analysis of scaffolds with connections containing looseness. Computers & Structures, 2011. 89(21-22): p. 1944-1955

带竖向加劲肋钢板剪力墙设计研究

范 重，黄彦军，刘学林，肖 坚，王义华

(中国建筑设计研究院，北京 100044)

摘 要：钢板剪力墙是一种新型结构抗侧力构件，设置竖向加劲肋可以有效抑制墙板的面外变形，延缓内嵌钢板的屈曲，但目前在我国现行的结构设计规范中尚缺乏可操作的具体规定。本文主要针对带竖向加劲肋的钢板剪力墙，对竖向加劲肋的作用机理进行了较为深入的探讨，并全面分析了墙板厚度、加劲肋刚度、间距等参数对钢板墙受力形态的影响。本文对初始缺陷、边界条件等对钢板墙屈曲承载力的影响进行了详细比较，对是否设置竖向加劲肋对钢板墙应力与变形的分布情况、轴压比对骨架曲线与滞回性能的影响进行了对比分析。最后，提出了钢板墙竖向加劲肋设置的原则、简化模型与相应的设计建议。

关键词：钢板剪力墙；竖向加劲肋；屈曲；承载力；参数分析；滞回性能

RESEARCH ON THE DESIGN OF STEEL PLATE SHEAR WALL WITH VERTICAL STIFFENERS

Fan Zhong, Liu Xuelin, Huang Yanjun, Li Li, Cao He, Xiao Jian, Wang Yihua

(China Architecture Design & Research Group, Beijing 100044, China)

Abstract: Steel plate shear wall (SPSW) is an innovative lateral loading-resisting system. Vertical stiffeners are adopted to restrain the out-plane deformation and buckling of infill plate of SPSW effectively. However, concrete provisions are not seen in current structural design specification. This paper mainly focuses on SPSW with vertical stiffeners, the action mechanism of vertical stiffeners is deeply studied; the influence of some parameters on SPSW force pattern is overall analyzed including plate thickness and stiffener rigidity and distance between stiffeners. The effect of initial imperfection and boundary conditions on the buckling capacity of SPSW is compared in detail. The distribution of stresses and deformation of SPSW and SPSW with stiffeners is studied. The influence of axial compression ratio on SPSW seismic behavior is also analyzed. Finally, principle of vertical stiffeners and simplified model and is proposed. Corresponding design suggestion is presented.

Keywords: steel plate shear wall; vertical stiffener; buckling; bearing capacity; parameter analysis; hysteretic behavior

1. 引言

1.1 概述

钢板剪力墙是 20 世纪 70 年代发展起来的一种新型抗侧力结构体系，其主要作用是提供结构的侧向刚度、抗剪强度和抗震延性。钢板剪力墙由周边框架和内嵌钢板组成，具有自重轻、安装方便等特点。研究表明，钢板剪力墙可以充分发挥钢材延展性好、耗能能力强的特点，结构侧向刚度大，构件延性性能好，具有出色的抗震性能，是一种具有广阔发展前景的超高层建筑抗侧力构件。

钢板剪力墙可以按照是否设置加劲肋分为加劲钢板墙和非加劲钢板墙。非加劲钢板墙中包括墙板两侧与柱脱开形式、带竖缝钢板墙、低屈服点钢板墙、开洞钢板墙等形式。带加劲肋钢板剪力墙有竖向均匀布置、十字或井字形布置、对角交叉布置等形式。设置加劲肋的主要作用是提高钢板剪力墙的屈曲临界荷载，使其在弹塑性范围内具有稳定饱满的滞回曲线，克服无加劲肋时钢板滞回曲线的"捏拢"现象。防屈曲钢板混凝土剪力墙是在内嵌在型钢边框钢板的单侧或两侧放置钢筋混凝土预制板，通过抗剪连接件将钢板与钢筋混凝土板结合在一起。预制混凝土墙板不但可以起到阻止钢板弹性屈曲的作用，使钢板到达剪切屈服强度，还可以起到防火、保温、隔声等效果，可有效降低工程造价。

鉴于钢板剪力墙的优越性能，加拿大规范（Limit States Design of Steel Structures，CAN/CSA S16-01）[1]与美国规范（Seismic Provisions for Structural Steel Buildings，AISC-2005）[2]均增加了钢板剪力墙的相关内容。

1970年建于东京的日本钢铁大厦（NIPPON STEEL BUILDING）是世界第一栋采用钢板剪力墙的建筑，其后钢板剪力墙在日本、北美等高烈度地震区得到许多应用。国内对于钢板剪力墙的相关研究虽然起步较晚，但是也取得了一些进展，特别是在工程应用方面尤为突出，2010年竣工的天津津门津塔是目前世界上应用钢板剪力墙抗侧力体系最高的建筑，目前正处于施工阶段的天津国际金融会议酒店工程则是钢板剪力墙在大型复杂建筑中的最新应用[3]。

1.2 钢板剪力墙研究现状

近几十年来，各国特别是北美和日本地区的许多学者对于钢板剪力墙这一新型抗侧力体系进行了系统的试验和理论研究。由于钢板剪力墙板件的宽厚比很大，在受力较小时就会发生局部屈曲，并且随着侧向力的逐渐增大在钢板墙对角线方向形成拉力场，沿对角线方向的拉力分布将对柱产生很大的作用。因此在设计钢板墙时，对其周边构件应相应加强，以保证钢板墙能够充分发挥拉力场的作用。钢板剪力墙的滞回曲线有不同程度的捏拢现象，捏缩的程度主要取决于墙板的宽厚比及周边框架的抗侧刚度[4]。Thorburn L J 等[5]利用钢板墙屈曲后强度的概念，提出了钢板墙的铰接斜拉杆模型，Driver R G 等[6,7]根据最小能量原理给出的拉杆倾角计算公式，为钢板墙的分析与设计提供了理论依据。Alinia M M 等[8]研究了薄钢板剪力墙屈曲后的非线性受力性能。

Rezai M[9]对钢板墙模型分别进行了拟动力试验和振动台试验，结果表明钢板墙具有良好的承载能力和延性，柱子破坏是影响钢板墙极限承载力的主要因素。Astaneh-Asl A 等[10]对钢管混凝土柱框架-钢板剪力墙进行的理论与试验研究表明，钢管混凝土柱抗侧刚度大，能够使内嵌钢板的拉力场效应得到充分发挥。

郭彦林、郝际平等人[11-15]分别对无加劲与有加劲板钢板剪力墙以及带竖缝钢板剪力墙做了大量试验和理论研究工作。Alinia M M、Hughes O F 等[16-18]在钢板剪力墙的加劲肋设计方面进行了大量试验研究与有限元分析，得到了一些在加劲肋设计方面有益的结论。蔡克铨等[19]对低屈服强度钢板剪力墙结构进行了系统试验和理论研究，发现采用低屈服钢的钢板剪力墙结构抗震性能更加优越。张素梅等人[20]对钢板剪力墙滞回性能进行了分析，提出了混合杆系模型。郑悦等[21]研究了四边简支、两侧边固支、上下边简支和周边固支条件下矩形钢板剪力墙在剪压作用下的弹性屈曲性能。

钢板墙自身延性非常好，延性系数在 8～13 之间[22]，不会发生钢板剪力墙承载力明显下降的情况，外框架分担的水平力不会相应发生很大变化，有利于实现结构多道抗震设防的设计理念。由于钢板剪力板的厚度比钢筋混凝土墙体小得多，可有效降低结构自重，减小地震响应与基础工程费用，增加建筑有效使用面积。由于钢板剪力墙具有很强的变形能力，与钢结构、钢管混凝土结构的变形能力相匹配，可以共同构成以钢结构为主要抗侧力构件的结构体系，具有很大的应用前景。

1.3 本文的研究内容

虽然钢板剪力墙具有初始弹性刚度大、抗震延性好的优点，但也存在不少问题。国外在工程中主要采用无加劲肋的钢板剪力墙，墙板面外刚度很小，容易发生屈曲变形，在使用过程中可能会发出响声。由于钢板剪力墙主要承受水平剪力，不承担竖向重力荷载，在施工过程中需要滞后安装。但由于经常会与施工进度要求以及室内装修发生矛盾，墙板难以避免承担部分重力荷载，此时在钢板墙设置竖向加劲肋可以避免钢板过早地出现面外屈曲变形。目前国内外对带有竖向加劲肋的钢板墙的受力机理与设计方法还缺乏系统的理论分析与试验研究。在我国现行结构设计规范中尚无对钢板剪力墙设计的具体规定，对钢板剪力墙设置竖向加劲肋的形式与间距等目前无具体规定。本文的主要内容如下：

（1）对钢板剪力墙竖向加劲肋的作用机理进行较为深入的探讨；
（2）对加劲肋刚度、间距、钢板厚度等多种参数的影响进行全面分析；
（3）对墙板边界条件、初始缺陷等对钢板墙屈曲承载力与屈服承载力的影响进行研究；
（4）对带竖向加劲肋钢板墙应力与变形的分布情况、轴压比对骨架曲线与滞回性能的影响等进行了分析与对比；
（5）提出了钢板墙竖向加劲肋设置的原则、简化模型与相应的设计建议。

2. 钢板剪力墙与加劲肋的作用

2.1 钢板剪力墙的一般规定

根据 AISC 341—05 和 FEMA 450 的规定，边框梁、边框柱之间的内嵌钢板墙的宽高比一般应在 0.8～2.5，且不应大于 3:1，即

$$0.8 \leqslant \frac{L}{h} \leqslant 2.5 \tag{1}$$

式中 L——钢板剪力墙竖向边缘构件中心线之间的距离；
h——钢板剪力墙水平边缘构件中心线之间的距离。

国外规范对钢板剪力墙内嵌钢板的厚度较低，高宽厚比满足下式要求即可：

$$200 \leqslant \frac{\min(L_{cf}, h_{cf})}{t_w} \leqslant 25\sqrt{\frac{E}{f_{wy}}} \tag{2}$$

式中　L_{cf}——钢板剪力墙竖向边缘构件之间的净距离；

　　　h_{cf}——钢板剪力墙水平边缘构件之间的净距离；

　　　t_w——内嵌钢板墙厚度。

　　　E、f_{wy}——分别为剪力墙钢板的弹性模量与屈服强度。

对于 Q235 钢材，$25\sqrt{\dfrac{E}{f_y}} = 25 \times \sqrt{\dfrac{206000}{235}} = 740$。

依据通用的划分标准，将屈曲发生在弹性阶段的墙板定义为薄板[4]。因此钢板的材料强度也将会成为影响钢板墙划分的影响因素。根据目前的研究现状，以高厚比 λ 限值作为标准，同时考虑钢材强度的影响，其分类如表 1 所示[20]：

钢板剪力墙高厚比分类　　　　　　　　　　表 1

宽厚比	λ＞300	100≤λ≤300	λ＜100
板件分类	薄板	中厚板	厚板

由此可见，对于一般工程来说，钢板剪力墙的宽厚比绝大部分处于中厚板的范畴。为了避免钢板剪力墙过早屈曲，通常可以考虑采取设置竖向加劲肋的方式。

2.2　加劲肋的作用

当板件高厚比 λ 在 300～100 的范围内，由于钢板厚度很小，在水平荷载作用下墙体很早便会出现面外"鼓曲"，即进入整体屈曲失稳状态，出现墙板屈曲先于屈服的情况。Thorburn L J 等[23]提出利用钢板墙屈曲后强度的概念，现已被各国钢板剪力墙设计规范普遍采用。但这并不意味着提高钢板剪力墙的屈曲荷载变得没有意义。屈曲荷载越低，钢板剪力墙的面外变形越大，且薄钢板剪力墙在侧向荷载作用下过早进入屈曲状态，造成了其在往复荷载作用下的滞回曲线出现明显的"捏拢"，滞回环不够饱满，因此其耗能能力不强。Alinia M M[16]的研究表明，提高钢板剪力墙的屈曲荷载，可以有效控制钢板的面外变形，克服薄钢板滞回曲线的"捏拢"效应，能够有效地改善薄钢板剪力墙的滞回耗能能力。

增加钢板厚度可以有效提高钢板剪力墙的屈曲荷载，改善其滞回性能，但是钢板厚度的增加会造成成本的显著增加。为在控制建造成本的前提下改善钢板墙的受力性能，在钢板上布置加劲肋是一种非常实用的方式。加劲肋的作用主要体现在以下几个方面：

（1）提高钢板剪力墙的水平屈曲临界荷载，延缓剪切屈曲的发生，减小屈曲后钢板面外的变形量，避免伴随钢板发生弹性屈曲变形时发出的响声；

（2）提高钢板剪力墙的竖向屈曲临界荷载，可以提供部分竖向承载能力，避免钢板在安装时出现屈曲，满足施工进度方面的实际需求；

（3）改善钢板剪力墙的滞回性能，克服无加劲肋时钢板滞回曲线的"捏拢"现象，提高构件的延性与耗能能力；

（4）提高钢板剪力墙的抗侧刚度，控制高层建筑结构的侧向变形。

2.3　加劲肋的形式

加劲肋有 I 形截面、T 形截面、[型截面和□形截面等多种形式，不同截面形式的加劲肋其效能不尽相同，Alinia MM[17]、Hughes O F[18]分别对"I"形截面和"T"形截面加劲肋的设计进行了研究。合理的加劲肋截面形式，可以使钢板剪力墙受力更加合理，施工方便，可以达到节约钢材的经济性要求。在实际工程中，双槽形加劲肋具有实用方便，稳定性能好等诸多优点，如图1所示。

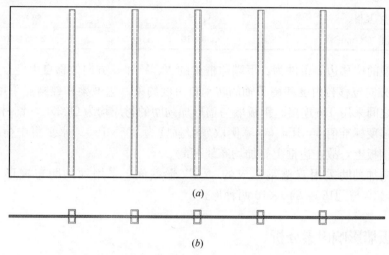

图1 带竖向加劲肋的钢板剪力墙
(a) 钢墙板立面；(b) 钢墙板剖面

为了有效发挥加劲肋的作用，需要控制加劲肋的抗弯刚度与钢板厚度的关系。在进行钢板剪力墙设计时，定义竖向加劲肋的肋板刚度比如下：

$$\gamma = \frac{EI_s}{Db} \tag{3}$$

式中 E——钢板的弹性模量；

D——墙板的刚度，$D = \dfrac{Et^3}{12(1-\nu^2)}$；

I_s——单个加劲肋的截面惯性矩；

b——竖向加劲肋的间距。

2.4 算例模型

算例模型采用单层单跨框架内嵌钢板剪力墙结构，主要研究中间楼层钢板墙的受力性能。对于中间层钢板墙，边框梁上、下所受到拉力近似平衡，故可以近似将边框梁近似视为刚性杆，即边框梁的抗压刚度 EA 和抗弯刚度 EI 均为无穷大。对于边框柱，当边框柱截面尺寸远大于钢板厚度时，其抗弯刚度可视为无穷大。故此，假定钢板墙的边界条件为周边固接。

在 Abaqus 软件中建立有限元模型，边框梁、边框柱均采用 B31 单元，边框梁刚度无穷大，钢板采用 S4R 单元。本文所取钢板的长度、高度和厚度，加劲肋的截面形式、尺寸等参数如表2所示。

算例模型的几何与物理参数 表 2

构 件	钢 板 墙	加 劲 肋
截面尺寸（长×宽×厚度）(mm)	7200×3200×12, 20	□120×120×10，□120×240×10，
单元名称	S4R	S4R
材料规格	Q235	Q235
弹性模量（N/mm^2）	2.06×10^5	2.06×10^5

算例模型的位移边界条件为：下端边框梁的 X、Y 和 Z 方向位移自由度全部约束，上端边框梁 Z 方向位移自由度和绕 Z 轴的转动自由度约束。边框梁和柱结点采用铰接连接，钢板和边框之间采用 Tie 连接。钢板墙与槽形加劲肋的材质均为 Q235C，钢材最大屈服强度不得高于强度标准值 50MPa（本算例取屈服强度为 205MPa）。水平集中荷载由边框梁均匀传递到钢板上，故在边框上施加均布线荷载。

本文竖向加劲肋采用双槽钢形成的"□"形截面，采用加劲肋布置形式，尺寸为 □120×120×10 与 □120×240×10 两种形式。

3. 加劲钢板墙影响因素分析

3.1 钢板墙屈曲模态分析

本文算例钢板剪力墙的肋板刚度比如表 3 所示。

钢板墙的肋板刚度比 $\gamma=\dfrac{EI_s}{Db}$ 表 3

板厚（mm）	$t_w=12$		$t_w=20$	
加劲肋间距 b	□120×120×10	□120×240×10	□120×120×10	□120×240×10
7200	0	0	0	0
3600	15.97	88.38	3.45	19.09
2400	24.80	137.22	5.36	29.64
1800	33.65	186.22	7.27	40.22
1440	42.83	237.01	9.25	51.19
1200	52.35	289.68	11.31	62.57

20mm 厚的钢板剪力墙，当分别采用 □120×120×10 与 □120×240×10 时，其在受剪、受压和受弯状态时的一阶屈曲模态如图 2~图 4 所示。

从图 2 中可以看出，对于厚度为 20mm 承受剪切作用的钢板墙，当设置 1~2 道 □120×240×10 的加劲肋时，屈曲模态主要以板件的屈曲变形为主，加劲肋未发生变形或变形量很小；当设置 3~4 道 □120×240×10 的加劲肋时，屈曲模态以板件与加劲肋均发生屈曲变形为特征，但板件的屈曲变形量明显大于加劲肋的变形量；当设置 5 道 □120×240×10 的加劲肋时，屈曲模态为板件与加劲肋均发生屈曲变形，但此时加劲肋的屈曲变形量与板件基本相同。钢板墙受剪时屈曲波形的数量与倾斜角度与加劲肋数量密切相关。由上述现象可知，肋板刚度比 λ 为 30 左右时，竖向加劲肋对钢板剪切屈曲变形的约束作用较

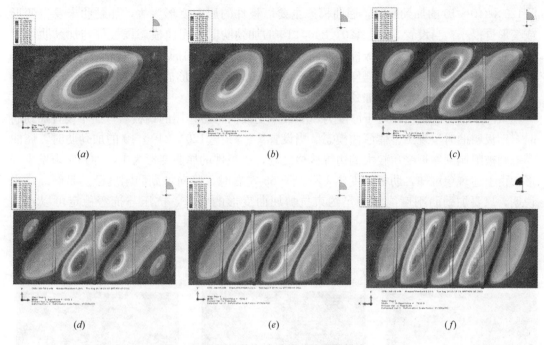

图 2 钢板剪力墙受剪时的屈曲模态
(a) 无加劲肋；(b) 加劲肋间距 3600mm；(c) 加劲肋间距 2400mm；
(d) 加劲肋间距 1800mm；(e) 加劲肋间距 1440mm；(f) 加劲肋间距 1200mm

为适中。

从图 3 中可以看出，对于厚度为 20mm 承受竖向压力作用的钢板墙，当设置 1~4 道

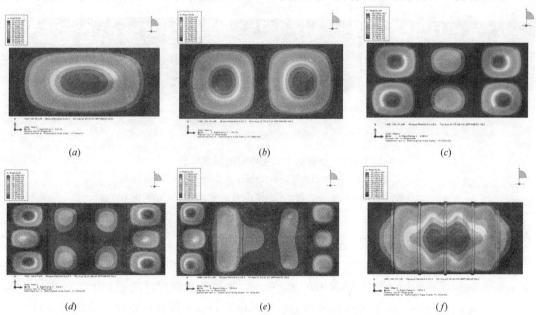

图 3 钢板剪力墙受压时的屈曲模态
(a) 无加劲肋；(b) 加劲肋间距 3600mm；(c) 加劲肋间距 2400mm；
(d) 加劲肋间距 1800mm；(e) 加劲肋间距 1440mm；(f) 加劲肋间距 1200mm

□120×240×10 的加劲肋时，屈曲模态主要以板件的屈曲变形为主，加劲肋未发生变形或变形量很小；当设置 5 道□120×240×10 的加劲肋时，屈曲模态以板件与加劲肋均发生屈曲变形为主要特征，但板件的屈曲变形量略大于加劲肋的变形量。由上述现象可知，肋板刚度比 λ 为 50 左右时，竖向加劲肋对钢板受压屈曲变形的约束作用较为适中，钢板墙受压时屈曲波形的数量与加劲肋数量密切相关。

从图 4 中可以看出，对于厚度为 20mm 承受面内弯矩作用的钢板墙，当不设置加劲肋时，钢板屈曲时呈现出整体屈曲模态。当设置 1～5 道□120×240×10 的加劲肋时，屈曲模态则被限制到靠近受压较大的边缘区格之内，以板件的屈曲变形为主，加劲肋未发生变形。由上述现象可知，肋板刚度比 λ 不大于 55 左右时，竖向加劲肋刚度已经足够。这也说明，只要加劲肋刚度能够满足约束抗剪屈曲变形的要求，抗压与抗弯稳定可以自然满足。

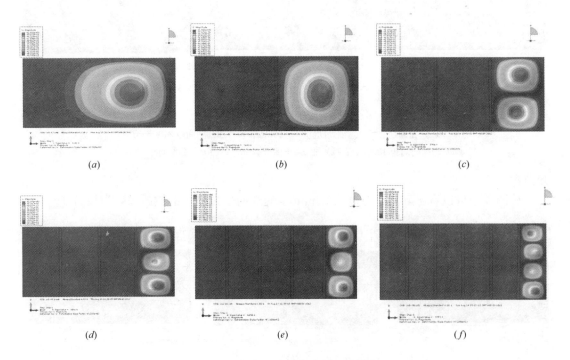

图 4　钢板剪力墙受弯时的屈曲模态
(a) 无加劲肋；(b) 加劲肋间距 3600mm；(c) 加劲肋间距 2400mm；
(d) 加劲肋间距 1800mm；(e) 加劲肋间距 1440mm；(f) 加劲肋间距 1200mm

3.2　屈曲承载力分析

本文对带加劲肋钢板剪力墙在受剪、受压及受弯时的弹性屈曲应力进行了详细的有限元分析，钢板墙平面尺寸为 7200×3200×20mm，加劲肋截面尺寸为□120×240×10，采用周边固定边界条件。由于带肋钢板剪力墙的弹性屈曲应力没有解析解，故此与按照文献 [24] 中相关公式的计算结果进行了对比。钢板墙在受压状态下的屈曲应力分别如表 4 与图 5 所示。

钢板墙在受压状态下的屈曲应力（MPa）　　　　　表 4

肋间距（mm）	$t_w=12$ □120×120×10		$t_w=12$ □120×240×10		$t_w=20$ □120×120×10		$t_w=20$ □120×240×10	
	有限元	文献[24]	有限元	文献[24]	有限元	文献[24]	有限元	文献[24]
1200	124.84	119	126.05	113	134.75	81	368.90	314
1440	95.62	76	90.20	76	127.43	67	259.96	210
1800	64.08	47	61.86	47	97.79	52	174.41	130
2400	39.61	25	38.41	25	75.75	39	108.89	70
3600	20.66	11	20.46	11	52.00	26	56.08	29
∞	12.15	4.62	12.14	4.62	33.70	12.83	33.70	12.83

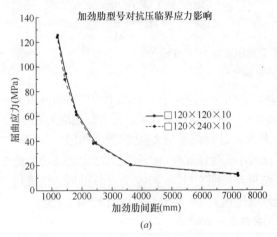
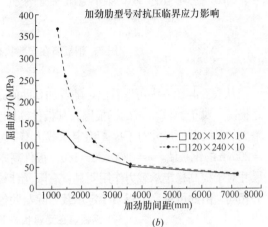

图 5　钢板墙在受压状态下的屈曲应力（MPa）
(a) $t_w=12$；(b) $t_w=20$

从表 4 与图 5 中可以看出，随着加劲肋间距逐渐减小，钢板在受压状态下的屈曲承载力随之提高。对于 $t=12$mm 的钢板墙，其板件高厚比较大，$\lambda=3200/12=267$，受压屈曲应力较小，计算结果与文献［24］比较接近，增大加劲肋截面尺寸对于提高稳定承载力作用不大。由此可见，钢板墙稳定承载力处于肋板刚度比有关外，主要与板件的宽厚比有关。对于 $t=20$mm 的钢板墙，$\lambda=3200/20=160$，板件高厚比较小，受压屈曲应力较大，增大加劲肋截面尺寸对于提高稳定承载力作用显著。由此可见，钢板墙稳定承载力除与肋板刚度比相关外，主要与板件的宽厚比有关。有限元计算结果与按照文献［24］中公式计算得到的结果具有较好的一致性。

钢板墙在受剪状态下的屈曲应力（MPa）　　　　　表 5

肋间距（mm）	$t_w=12$ □120×120×10		$t_w=12$ □120×240×10		$t_w=20$ □120×120×10		$t_w=20$ □120×240×10	
	有限元	文献[24]	有限元	文献[24]	有限元	文献[24]	有限元	文献[24]
1200	116.12	57	148.98	36	209.44	229	323.67	205
1440	92.73	51	106.36	26	181.47	222	247.77	195
1800	65.69	41	75.07	18	136.42	212	187.74	179
2400	49.91	23	49.13	22	115.93	192	136.53	149
3600	33.22	32	32.35	28	90.97	90	91.37	78
∞	26.37	28.16	26.37	28.16	73.19	54.46	73.19	54.46

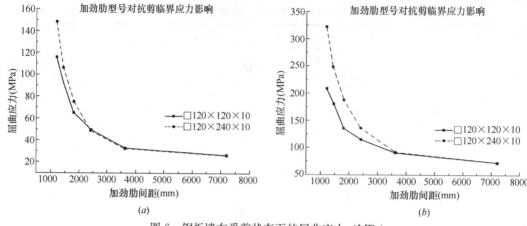

图 6 钢板墙在受剪状态下的屈曲应力（MPa）
(a) $t_w=12$；(b) $t_w=20$

从表 5 与图 6 中可以看出，随着加劲肋间距逐渐减小，钢板在受压状态下的屈曲承载力随之提高。对于 $t=12$mm 的钢板墙，其板件高厚比较大，$\lambda=3200/12=267$，受剪屈曲应力较小，增大加劲肋截面尺寸对于提高稳定承载力有一定作用，计算结果与文献[24]差别很大。对于 $t=20$mm 的钢板墙，$\lambda=3200/20=160$，板件高厚比较小，受压屈曲应力较大，增大加劲肋截面尺寸对于提高稳定承载力作用明显，有限元计算结果与按照文献[24]中公式计算得到的结果趋势相同。由此可见，钢板墙稳定承载力除与肋板刚度比相关外，主要与板件的宽厚比有关。

钢板墙在受弯状态下的屈曲应力（MPa）   表 6

肋间距 (mm)	$t_w=12$ □120×120×10		$t_w=12$ □120×240×10		$t_w=20$ □120×120×10		$t_w=20$ □120×240×10	
	有限元	文献[24]	有限元	文献[24]	有限元	文献[24]	有限元	文献[24]
1200	169.27	97	162.01	109	315.55	89	463.31	378
1440	124.84	76	120.55	76	294.86	77	340.10	241
1800	88.83	50	89.12	50	223.31	65	247.04	139
2400	61.00	31	60.51	31	161.99	55	167.90	85
3600	41.44	18	41.86	18	118.81	45	113.67	49
∞	37.47	14.1	37.47	14.1	104.03	39.2	104.03	39.2

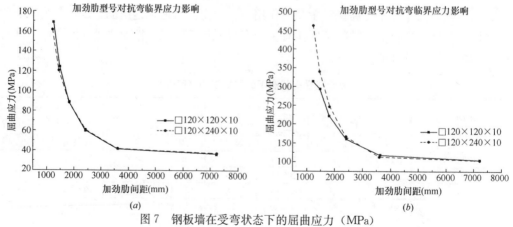

图 7 钢板墙在受弯状态下的屈曲应力（MPa）
(a) $t_w=12$；(b) $t_w=20$

从表6与图7中可以看出，随着加劲肋间距逐渐减小，钢板在受弯状态下的屈曲承载力随之提高。对于 $t=12mm$ 的钢板墙，其受弯屈曲应力与受压屈曲应力相比有较大幅度提高，增大加劲肋截面尺寸对于提高稳定承载力作用明显。对于 $t=20mm$ 的钢板墙，板件高厚比较小，受弯屈曲应力较大，增大加劲肋截面尺寸对于提高稳定承载力作用非常显著。由此可见，钢板墙稳定承载力除与肋板刚度比相关外，主要与板件的宽厚比有关。

4. 边界条件影响分析

4.1 钢板墙边界条件的模拟

钢板墙的边界条件与其边框梁、边框柱的刚度有关。对于中间层的钢板墙，边框梁上、下层钢板所受到的拉力近似平衡，一般来说，边框梁自身的刚度不需要很大。对于边框柱，为了满足抗弯刚度 $I_c \geqslant \dfrac{0.00307 th^4}{L}$ 的要求，其截面尺寸一般较大。

为了考察边界条件对钢板墙屈曲稳定的影响，分别研究了墙板周边固接、周边铰接以及边框柱固接、边框梁铰接三种情况。

4.2 钢板墙受剪状态

各种边界条件时钢板墙的受剪屈曲应力分别如表7与图8所示。从表与图中可以看出，加劲肋间距是影响钢板墙在受剪状态时稳定性的主要因素，随着加劲肋间距减小，三种边界条件下的稳定性均可逐渐迅速提高。其中周边固接稳定承载力最高，周边铰接最低，上下铰接-左右固接介于两者之间。当加劲肋间距较大（即肋板刚度比较小）时，边界条件影响较大。

钢板墙在各种边界条件时受剪状态下的屈曲应力（MPa） 表7

肋间距 (mm)	$t_w=12$ □120×120×10			$t_w=12$ □120×240×10			$t_w=20$ □120×120×10			$t_w=20$ □120×240×10		
	周边 固支	周边 简支	上下 简支	周边 固支	周边 简支	上下 简支	周边 固支	周边 简支	上下 简支	周边 固支	周边 简支	上下 简支
1200	116.12	96.10	102.31	148.98	114.71	147.06	209.44	162.44	171.46	323.67	273.05	287.87
1440	92.73	76.65	82.61	106.36	82.73	104.43	181.47	141.62	149.11	247.77	207.48	220.45
1800	65.69	53.43	56.98	75.07	57.47	71.12	136.42	101.15	105.33	187.74	153.89	164.72
2400	49.91	38.72	41.82	49.13	37.73	45.74	115.93	81.29	84.03	136.53	106.21	118.27
3600	33.22	23.33	26.49	32.35	22.87	26.05	90.97	63.30	70.69	91.37	63.79	72.11
∞	26.37	16.40	17.13	26.37	16.40	17.13	73.19	45.52	47.55	73.19	45.52	47.55

4.3 钢板墙受压状态

各种边界条件时钢板墙的受压屈曲应力分别如表8与图9所示，从中可以看出，加劲肋间距是影响钢板墙在受压状态时稳定性的主要因素，随着加劲肋间距减小，三种边界条

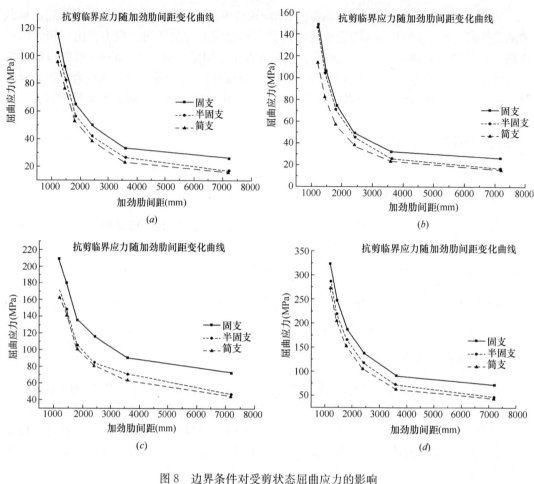

图 8 边界条件对受剪状态屈曲应力的影响
(a) $t_w=12$, □120×120×10; (b) $t_w=12$, □120×240×10;
(c) $t_w=20$, □120×120×10; (d) $t_w=20$, □120×240×10

件下的稳定性均可逐渐迅速提高。其中周边固接稳定承载力最高，周边铰接最低，上下铰接-左右固接介于两者之间。当肋板刚度比较小时，边界条件影响较大，减小加劲肋间距的作用也逐渐减弱。

钢板墙在各种边界条件时受压状态下的屈曲应力（MPa） 表 8

肋间距 (mm)	$t_w=12$ □120×120×10			$t_w=12$ □120×240×10			$t_w=20$ □120×120×10			$t_w=20$ □120×240×10		
	周边固支	周边简支	上下简支	周边固支	周边简支	上下简支	周边固支	周边简支	上下简支	周边固支	周边简支	上下简支
1200	124.84	93.03	101.40	126.05	86.60	113.98	134.75	76.36	79.98	368.90	252.25	309.83
1440	95.62	64.65	85.74	90.20	60.30	80.46	127.43	73.36	77.94	259.96	176.14	232.44
1800	64.08	41.54	54.45	61.86	40.19	51.57	97.79	53.93	57.65	174.41	112.55	151.56
2400	39.61	24.62	30.70	38.41	23.74	30.00	75.75	40.36	45.01	108.89	67.12	83.77
3600	20.66	10.07	14.13	20.46	10.00	14.10	52.00	23.19	29.71	56.08	26.84	37.38
∞	12.15	3.76	4.38	12.14	3.76	4.38	33.70	10.43	12.17	33.70	10.43	12.17

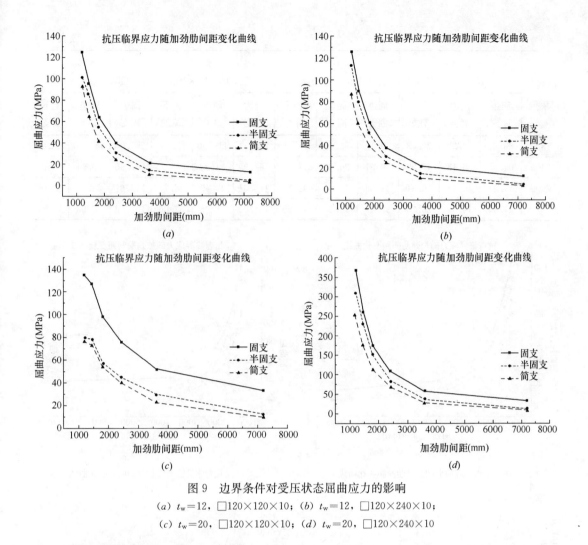

图 9　边界条件对受压状态屈曲应力的影响

(a) $t_w=12$，□120×120×10；(b) $t_w=12$，□120×240×10；
(c) $t_w=20$，□120×120×10；(d) $t_w=20$，□120×240×10

4.4 钢板墙受弯状态

各种边界条件时钢板墙的受弯屈曲应力分别如表9与图10所示。从表与图中可以看出，边界调节的影响与钢板墙在受压状态时非常接近，随着加劲肋间距减小，三种边界条件下的稳定性均可逐渐迅速提高。其中周边固接稳定承载力最高，周边铰接最低，上下铰接-左右固接介于两者之间。当肋板刚度比较小时，边界条件影响较大，减小加劲肋间距的作用也逐渐减弱。

钢板墙在各种边界条件时受弯状态下的屈曲应力（MPa）　　表9

肋间距 (mm)	$t_w=12$ □120×120×10			$t_w=12$ □120×240×10			$t_w=20$ □120×120×10			$t_w=20$ □120×240×10		
	周边固支	周边简支	上下简支	周边固支	周边简支	上下简支	周边固支	周边简支	上下简支	周边固支	周边简支	上下简支
1200	169.27	115.84	154.93	162.01	110.59	147.77	315.55	165.42	218.16	463.31	313.71	421.12
1440	124.84	83.18	113.75	120.55	79.71	109.34	294.86	159.13	214.27	340.10	226.42	307.69

续表

肋间距 (mm)	$t_w=12$ □120×120×10			$t_w=12$ □120×240×10			$t_w=20$ □120×120×10			$t_w=20$ □120×240×10		
	周边固支	周边简支	上下简支	周边固支	周边简支	上下简支	周边固支	周边简支	上下简支	周边固支	周边简支	上下简支
1800	88.83	56.32	76.43	89.12	56.65	75.35	223.31	119.90	171.60	247.04	155.53	212.01
2400	61.00	36.43	47.99	60.51	36.20	47.91	161.99	84.38	126.90	167.90	99.91	131.21
3600	41.44	19.38	30.31	41.86	19.81	31.19	118.81	49.49	75.77	113.67	52.65	82.16
∞	37.47	14.05	20.55	37.47	14.05	20.55	104.03	38.98	57.06	104.03	38.98	57.06

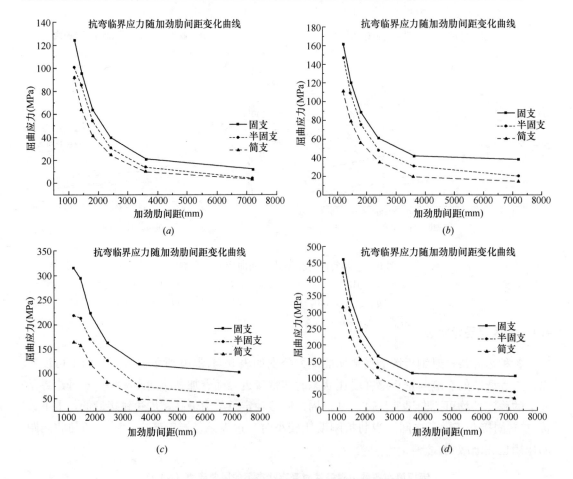

图 10 边界条件对受弯状态屈曲应力的影响
(a) $t_w=12$,□120×120×10；(b) $t_w=12$,□120×240×10；
(c) $t_w=20$,□120×120×10；(d) $t_w=20$,□120×240×10

考虑到工程中边框柱的截面尺寸相对较大，对钢板墙面外变形约束效果显著，而边框梁相对的侧向刚度较小，故此，钢板墙上、下边采用铰接、左右采用固接的边界条件，与实际情况比较接近。

5. 初始缺陷对钢板墙的屈曲影响分析

在实际工程中，钢板剪力墙在加工、运输和安装过程中出现墙板面外变形难以避免，对于正常使用阶段而言，相当于结构的初始缺陷[25]。由于钢板墙板件的宽厚比很大，有必要对初始缺陷的形态以及大小对于钢板剪力墙抗剪承载力和弹塑性变形性能的影响进行全面深入的研究。

钢板墙的平面尺寸为 3200mm×7200mm（高×宽），钢板厚度为 20mm，材质为 Q235，加劲肋尺寸为 □120×240×10。在计算分析时，采用在钢板墙周边设置刚性边缘构件的方式，分别考虑无加劲肋与加劲肋间距为 1440mm 时的情况。

在墙板中引入初始缺陷时，首先进行钢板墙的屈曲特征值分析，得到钢板墙的低阶屈曲模态，将该模态或几个模态的组合作为构件的初始变形，最大变形值则根据施工控制条件或制作安装容许偏差确定。

在本文研究中，初始缺陷分别选取了受剪、受压及受弯状态下的屈曲模态，计算模型编号和面外变形幅值规定如表 10 所示。

钢板剪力墙的初始缺陷类型与幅值 表 10

缺陷类型	无	a	a	a	b	c	d	a+b
屈曲模态描述	—	受剪	受剪	受剪	受剪	受压	受弯	组合
缺陷幅值（mm）	0	16	64	128	64	64	64	64

各类初始缺陷对钢板墙在水平剪力作用下屈服荷载与抗剪承载力的影响如表 11 所示，钢板墙在水平剪力作用下的荷载位移曲线如图 11 所示。从表 11 与图 11 中可以看出，初始缺陷对钢板剪力墙的屈服荷载和抗剪承载力均有一定影响。初始缺陷的影响与其面幅值有关，随着初始缺陷幅值的增大，钢板墙的剪切屈服荷载与抗剪承载力均有所降低。对于初始缺陷模态 a，当初始缺陷幅值为 64mm、即钢板墙高度的 1/50 时，带肋钢板墙屈曲荷载降低为 4.81%，当初始缺陷幅值达到 128mm、即钢板墙高度的 1/25 时，带肋钢板墙屈曲荷载降低为 10.1%，说明受剪屈服荷载对于初始缺陷幅值的敏感程度不高。钢板墙对缺陷类型不是非常敏感，当初始缺陷幅值相同时，各类缺陷相应的钢板墙剪切屈服荷载与抗剪承载力接近。此外，初始缺陷对钢板墙初始刚度以及发生屈服后性能的影响均不显著。从分析结果可以看出，对于带有加劲肋的钢板墙，初始缺陷对其力学性能的影响较小。

初始缺陷对钢板墙在水平剪力作用下屈服荷载与抗剪承载力的影响 表 11

屈曲模态	缺陷幅值（mm）	屈服荷载（kN）		抗剪承载力（kN）	
		无加劲肋	带加劲肋	无加劲肋	带加劲肋
无	0	17091	17091	18301	18301
a	16	15845	16874	16279	17859
a	64	14736	16293	16238	17535
a	128	13623	15387	15919	17261

续表

屈曲模态	缺陷幅值（mm）	屈服荷载（kN）		抗剪承载力（kN）	
		无加劲肋	带加劲肋	无加劲肋	带加劲肋
b	64	—	16337	—	17375
c	64	14609	16684	16261	17936
d	64	14626	16610	16294	18088
a+b	64	—	16026	—	17390

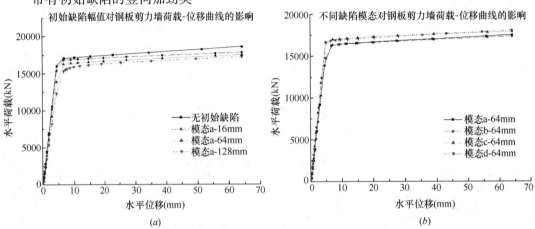

图 11　有初始缺陷加劲钢板墙的水平剪力-顶点位移曲线
(a) 缺陷幅值的影响；(b) 初始缺陷类型的影响

6. 钢板剪力墙抗震性能分析

6.1 钢板墙的拉力场与变形

与钢板墙受剪承载力分析相同，钢板墙平面尺寸为 7200×3200，板厚 20mm，当引入初始缺陷后，在钢板剪力墙上端刚性横梁施加往复强制位移，考察其受力特点与变形性能。

无加劲肋与加劲肋间距为 1800mm 的钢板墙在水平剪力作用下的面外变形与应力 S12 分别如图 12 与图 13 所示。从图 12 可以看出，设置加劲肋之后，钢板剪力墙的面外变形变形态由整体波形转换为局部波形，加劲肋可以有效地约束钢板的面外变形。从图 13 中可以看出，设置加劲肋之后，钢板墙在水平剪力的作用下也可以形成明显的拉力场，但拉力场与边框梁的角度明显增大。

不同加劲肋间距时钢板剪力墙的面外变形量（mm）如表 12 所示。当无加劲肋时，钢板剪力墙在发生 1/50 层间位移时，其"外凸"和"内凹"变形最大值的平均值达 76mm，当设置一个加劲肋后，其值迅速下降到 58mm，随着加劲肋个数的增加，由于受到初始缺陷的影响，其面外变形值具有一定的波动性，大致为 50mm 左右。当加劲肋个

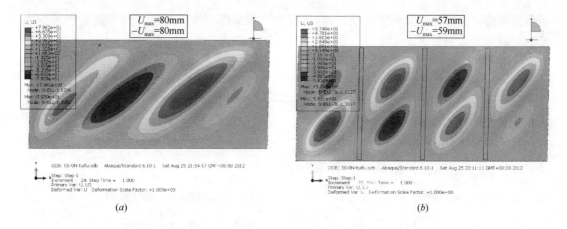

图 12 带竖向加劲肋钢板墙的变形分布
(a) 无加劲肋；(b) 加劲肋间距 1800mm

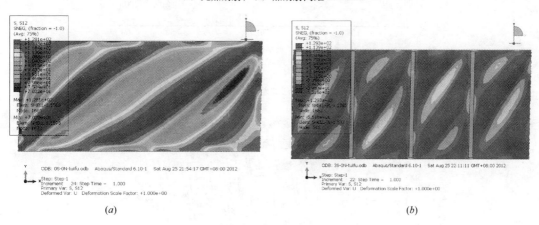

图 13 带竖向加劲肋钢板墙的剪应力 S12 分布
(a) 无加劲肋；(b) 加劲肋间距 1800mm

数为 1200mm 时，其值又迅速下降为 42mm。总体来看，加劲肋对于约束钢板剪力墙的面外变形的作用非常明显。

不同加劲肋间距时钢板剪力墙的面外变形量（mm）　　　　表 12

加劲肋间距（mm）	—	3600	2400	1800	1440	1200
最大面外变形 U_{max}	72.27	69.33	51.70	54.96	57.35	45.16
最大面外变形 U_{min}	−79.55	−46.55	−49.86	−50.02	−49.81	−39.41
$\frac{\|U_{max}\|+\|U_{max}\|}{2}$	76	58	51	53	54	42

6.2 钢板墙滞回性能

加劲肋间距对钢板墙抗侧力性能的影响如图 14 所示。从图 14 (a) 中可以看出，荷载-位移滞回曲线，设置竖向加劲肋对于钢板墙的抗侧力刚度与抗剪承载力有利，但随着加劲肋间距减小的改善作用幅度较小。从图 14 (b) 中可以看出，设置竖向加劲肋后，钢板剪力墙滞回曲线的饱满程度明显提高，"捏拢"现象得到大大缓解，对于提升钢板剪力墙的耗能能力效果显著。

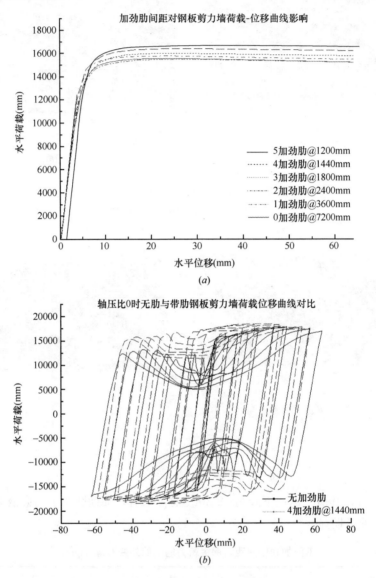

图14 加劲肋间距对钢板墙抗侧力性能的影响
（a）荷载-位移骨架曲线；（b）荷载-位移滞回曲线

6.3 轴压比影响分析

钢板剪力墙作为结构的抗侧力构件，主要承受水平荷载作用。在钢板剪力墙工程应用中，边框柱承受大部分的竖向荷载。虽然通过在施工阶段钢板墙滞后安装可以有效减小墙板承担的结构自重，但完全避免钢板墙承担室内装修等附加竖向荷载难度很大，施工进度计划往往需要钢板墙提前安装完毕。此外，由于钢板剪力墙镶嵌与边框柱与边框梁之间，在使用阶段变形协调，钢板墙实际上处于受压与受剪复合受力状态。因此，需要考察竖向压力对其力学性能的影响。

钢板剪力墙的轴压比可按下式计算：

$$\mu = \frac{N}{2N_c + N_w} \tag{4}$$

式中　N——各边框柱上的竖向荷载设计值之和；

　　　N_c——单个框架柱的竖向承载力，根据材料的强度设计值与截面面积计算得到；

　　　N_w——墙板的竖向承载力，$N_w = L f_y$。

轴压比对钢板剪力墙水平剪力作用下的屈服荷载与极限荷载的影响如表 13 所示，对钢板墙滞回性能的影响如图 15 所示。对于无加劲肋钢板墙，不设在轴压比为 0.4 的竖向压力作用下，钢板墙的屈曲荷载略有降低，板件较早进入屈曲状态。设置加劲肋后，可以有效抑制钢板的面外变形，防止过早发生屈曲，使滞回曲线更加饱满。但总起来说，对于钢板剪力

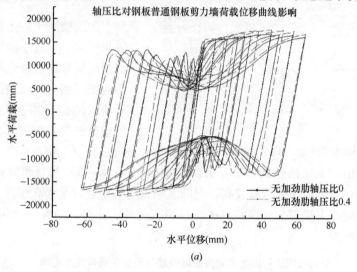

(a)

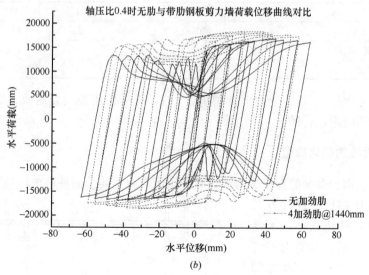

(b)

图 15　轴压比对钢板剪力墙抗震性能的影响
(a) 无加劲肋；(b) 带竖向加劲肋

墙结构，边框柱作为主要的承重构件，竖向力对钢板墙受力性能的影响是有限的。

轴压比对钢板剪力墙屈服荷载与极限荷载的影响　　　　表 13

轴压比	无加劲肋		4 加劲肋		备　　注
	0	0.4	0	0.4	
屈服荷载（kN）	14800	13800	16900	15300	相差 6.76%～9.47%
极限荷载（kN）	16200	15600	17400	16400	相差 3.70%～5.75%

7. 加劲肋作用分析

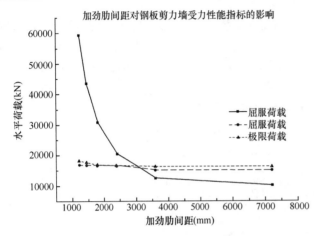

图 16　受剪屈曲荷载与屈服荷载与加劲肋间距的关系

7.1　受剪屈曲荷载与受剪屈服荷载

加劲肋对于防止面外屈曲变形效果明显，但过小的加劲肋间距将引起用钢量的显著增加。加劲肋间距对钢板墙受剪屈曲荷载与受剪屈服荷载的影响如表 14 所示，受剪屈曲荷载与受剪屈服荷载随加劲肋间距的变化情况如图 16 所示。从表与图中可以看出，无加劲肋钢板墙受剪屈曲承载力很小，随着加劲肋间距减小迅速增大；无加劲肋钢板墙受剪屈服荷载较高，但随着加劲肋间距的减小提高缓慢。当加劲肋间距小于 3000mm 时，其受剪屈曲荷载高于受剪屈服荷载。此外，随着加劲肋间距的减小，结构用钢量随之增大。

加劲肋间距对钢板墙受剪屈曲荷载与受剪屈服荷载的影响　　　　表 14

加劲肋间距（mm）	—	3600	2400	1800	1440	1200
屈曲荷载（kN）	6847	10383	17030	23719	31744	41453
屈服荷载（kN）	14117	14637	15198	15227	15418	15874
用钢量增加（%）	0	8.33	12.5	16.7	20.8	25.0

因此，在考虑抗震设防的竖向加劲肋设计时，应考虑遵照"先屈服、后屈曲"的原则，对加劲肋的间距进行最优。

7.2　带肋钢板墙的简化模型

对于设置竖向加劲肋的钢板剪力墙，其受压承载力主要受面外稳定性的控制，可以参照竖向轴心受压构件的设计方法进行简化计算，确定其可以承受的竖向荷载。在进行计算时，将加劲肋及在其两侧宽度 $20\varepsilon t$ 范围内的钢板视为"中"字形截面组合受压构件，如图 17 所示。可根

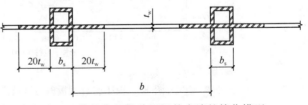

图 17　带竖向加劲肋钢板剪力墙的简化模型

据其有效面积为 A_e 和相应的面外方向截面惯性矩 I_{ey} 计算其面外方向的回转半径 $i_y = \sqrt{I_{ey}/A_e}$，组合截面受压构件面外方向的长细比 λ_y 按下式计算，

$$\lambda_y = \frac{h}{i_y} \tag{5}$$

根据《钢结构设计规范》(GB 50017—2003) 中的 B 类截面，可以得到组合受压构件面外方向的稳定系数 φ。带竖向加劲肋钢板墙的竖向压应力应满足下式要求：

$$\sigma < \varphi \frac{A_e}{t_w(b+b_s)} f_w \tag{6}$$

钢板墙简化模型受压时的竖向稳定应力与应力比如表 15 所示，此时钢板强度按照 Q235 考虑，有限元计算结果如表 16 所示。简化模型与有限元分析在钢板墙受压状态时的应力比较如图 18 所示。

钢板墙简化模型受压时的竖向稳定应力（MPa）与应力比　　　表 15

板厚 (mm)	加劲肋间距（mm）	@3600		@2400		@1800		@1440		@1200	
	加劲肋截面	应力	应力比	应力	应力比	应力	应力比	应力	应力比	应力	应力比
$t_w=12$	□120×120×10	28.08	0.119	42.11	0.179	56.15	0.239	70.19	0.299	84.24	0.358
$t_w=12$	□120×240×10	34.71	0.148	52.06	0.222	69.42	0.295	86.77	0.369	104.13	0.443
$t_w=20$	□120×120×10	18.41	0.078	27.61	0.117	36.82	0.157	46.03	0.196	55.23	0.235
$t_w=20$	□120×240×10	53.55	0.228	80.33	0.342	107.1	0.456	133.9	0.570	160.7	0.684

7.3 竖向加劲肋的设计原则

竖向加劲肋对于改善钢板墙受力性能的作用非常显著，对初始缺陷产生面外变形的抑制作用明显，对于改善钢板剪力墙的耗能性能效果显著。但由于其对用钢量存在一定影响，对加劲肋的设置进行优化非常重要。故此，应考虑加劲肋与墙板的相对刚度，通过调整加劲肋的截面尺寸与间距，确定适当的肋板刚度比，使加劲肋的屈曲应力接近于各区格钢板的屈曲应力，满足钢板剪力墙"先屈服、后屈曲"的性能指标。

为了保证板件在受剪时的稳定性，现行国家标准《钢结构设计规范》(GB 50017—2003) 与文献 [25] 中，对梁腹板宽厚比限值做出明确规定，对于不设中间加劲肋或仅按构造配置少量横向加劲肋的梁，$h_0/t \leqslant 80\sqrt{235/f_y}$。在钢板剪力墙设计时，参照我国现行《钢结构设计规范》(GB 50017—2003) 对直接承受动力荷载吊车梁腹板设置构造加劲肋的规定，钢板剪力墙设置竖向加劲肋的最大间距按下式确定：

$$\frac{b}{t_w} = 80\sqrt{\frac{235}{f_y}} \tag{7}$$

式中　b——钢板剪力墙加劲肋之间的净距。

对于本文研究的钢板剪力墙，加劲肋间距应满足 $b \leqslant 80\sqrt{235/f_y} \cdot t_w = 80 \times 20 = 1600\text{mm}$。前述多种分析结果表明，在此加劲肋间距条件下，能够充分满足钢板墙的屈曲承载力、屈服承载力以及抗震性能等方面的要求，并且具有一定的储备量。

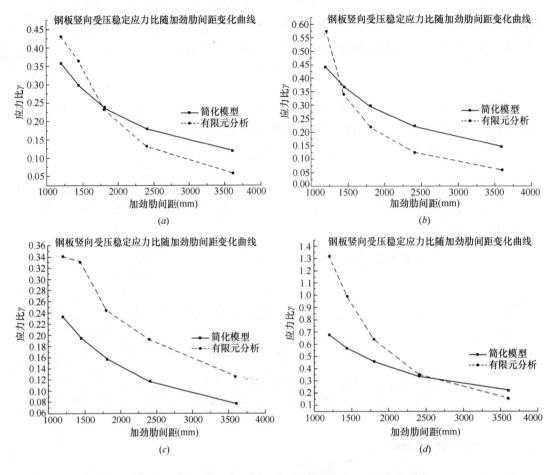

图 18 在钢板墙受压时简化模型与有限元计算结果的比较
(a) $t_w=12$，□120×120×10；(b) $t_w=12$，□120×240×10；
(c) $t_w=20$，□120×120×10；(d) $t_w=20$，□120×240×10

8. 结论

（1）竖向加劲肋可以显著改变钢板墙的屈曲变形模态，模态波形的数量与倾斜角度与加劲肋间距密切相关；

（2）带有加劲肋的钢板墙在竖向荷载作用下的稳定性显著提高，加劲肋刚度、间距、钢板厚度等参数对钢板墙受剪、受压与受弯状态下的屈曲应力有很大影响；

（3）墙板边界条件对屈曲承载力影响明显，说明边框梁、边框柱的刚度与钢板墙的受力形态关系较大；

（4）通过对不同类型、不同幅值的初始缺陷计算分析表明，钢板墙抗剪屈曲承载力与屈服承载力对初始缺陷不敏感；

（5）设置带竖向加劲肋对于钢板墙的面外变形有显著的抑制作用，使应力分布趋于均匀，但对拉力场角度有一定影响；

（6）设置竖向加劲肋对钢板墙滞回性能有一定的改善，可以有效缓解其"捏拢"现象，耗能能力得到提高；

（7）提出在竖向压力下钢板剪力墙的简化模型，可以在估算竖向加劲肋间距时作为参考；

（8）应通过调整加劲肋的截面尺寸与间距，确定适当的肋板刚度比，使加劲肋的屈曲应力接近于各区格钢板的屈曲应力，满足钢板剪力墙"先屈屈服、后屈曲"的性能指标；

（9）钢板墙焊接残余应力的影响不可忽视，还需要进行深入研究。钢板剪力墙合理的安装与焊接顺序对于保证期良好的工作性能至关重要。

致谢

在本文完成过程中，得到了浙江大学童根树教授的热情指导与帮助，在此谨表示衷心的感谢。

参考文献

[1] Canadian Standards Association (CSA). CAN/CSA S16-2001, Limit States Design of Steel Structures[S]. 2001

[2] ANSI/AISC 341-2005 American Institute of Steel Construction, Seismic Provisions for Structural Steel Buildings[S]. 2005

[3] 范重，刘学林，李丽，等. 天津金融会议酒店结构抗震超限审查报告[R]，北京：中国建筑设计研究院，2010

[4] 郭彦林，周明. 钢板剪力墙的分类及性能[J]. 建筑科学与工程学报，2009，26(3)：1-13

[5] Thorburn L J, Kulak G L, Montgomery C J. Analysis of steel plate shear wall: Structural Engineering Report No 107[R]. Edmonton, AB: Department of Civil Engineering, University of Alberta, 1983

[6] Driver R G, Kulak G L, Laurie K D J, et al. Seismic behavior of steel plate shear wall: Structural Engineering Report No 215[R]. Edmonton, AB: Department of Civil Engineering, University of Alberta, 1997

[7] Shishkin J J, Driver R G, Grondin G Y. Analysis of steel plate shear walls using the modified strip model: Structural Engineering Report No 261[R]. Edmonton, AB: Department of Civil Engineering, University of Alberta, 2005

[8] Alinia MM, Habashi H. R., Khorram A. Nonlinearity in the post buckling behavior of thin steel shear panels. Thin-Walled Structures 2009; 47: 412-420. doi: 10.1016/j.tws.2008.09.004

[9] Rezai M. Seismic behavior of steel plate shear walls by shake table testing [D]. Vancouver: Univ. of British Columbia, 1999

[10] Astaneh-Asl A, Zhao Q. Cyclic tests of steel shear walls[R]. Volume I-Final Report. Final Report to the Sponsors, Report Number UCB/CEE-STEEL-01/01, Dept. Of Civil and Env. Engineering, Univ. of California, Berkeley, 2002

[11] 陈国栋，郭彦林，范珍等. 钢板剪力墙低周期反复荷载试验研究[J]. 建筑结构学报，2004，25(2)：19-26

[12] 郭彦林，陈国栋，缪友武. 加劲钢板剪力墙弹性抗剪屈曲性能研究[J]. 工程力学，2006，23(2)：

84-91

[13] 侯蕾，郝际平，董子建等．十字加劲肋钢板剪力墙低周反复荷载的试验[J]．钢结构，2006，21(2)：12-16

[14] 王迎春，郝际平，李峰等．钢板剪力墙力学性能研究[J]．西安建筑科技大学学报，2007，39(2)：181-186

[15] 李峰，郝继平．钢板剪力墙抗震性能的试验与理论研究[D]．西安：西安建筑科技大学，2011

[16] Alinia MM, Dastfan M. Cyclic behavior, deformability and rigidity of stiffened steel shear panels. Int J Const Steel Research 2007；63：554-563. doi：10.1016/j.jcsr.2006.06.005

[17] Alinia MM, Dastfan M. On the design of stiffeners in steel plate shear walls. Int J Const Steel Research 2009；65：2069-2077. doi：10.1016/j.jcsr.2009.06.009

[18] Hughes OF, Ghosh B, Chen Y. Improved prediction of simultaneous local and overall buckling of stiffened panels. Thin-Walled Structures 2004；42：827-856. doi：10.1016/j.tws.2004.01.003

[19] 蔡克铨，林盈成，林志翰．钢板剪力墙抗震行为与设计[C]// 第四届海峡两岸及香港钢结构技术交流会议文集，2006

[20] 李然，郭兰慧，张素梅．钢板剪力墙滞回性能分析与简化模型[J]．天津大学学报，2010，43(10)：919-927

[21] 郑悦，赵伟．剪压荷载作用下钢板剪力墙弹性屈曲性能研究[J]．工业建筑，2011，41(8)：105-109

[22] SABOURI-GHOMI S, GHOLHAKI M. Ductility of thin steel plate shear walls[J]. Asian Journal of Civil Engineer：Building and Hoursing，2008，9(2)：153-166

[23] Thorburn L J, Kulak G L, Montgomery C J. Analysis of steel plate shear wall：Structural Engineering Report No 107[R]. Edmonton, AB：Department of Civil Engineering, University of Alberta，1983

[24] 高层民用建筑钢结构技术规程[S](JGJ99，报批稿)，2012年6月

[25] 陈绍蕃．钢结构稳定设计指南(第二版)，中国建筑工业出版社，2004年

面内挫屈斜撑之耐震行为与设计

陈诚直,汤伟干

(国立交通大学 土木工程学系,台湾新竹 30010)

摘　要:为使斜撑于受压时挫屈为面内挫屈,将斜撑端部翼板切削,使其产生塑铰,达成斜撑面内挫屈,并探讨其迟滞循环及轴向强度。研究方法采有限元素分析建立数值模型,研究不同参数对于斜撑行为之影响;进而以有限元素分析之参数研究结果规划试验,设计四组试体,进行往覆载重实验。试验结果显示四组试体皆具有稳定非线性行为与消散能量之能力,试体迟滞循环达 4%~5%弧度之层间位移角,斜撑产生面内挫屈并最终于斜撑中点发生断裂。切削量小时,斜撑端部劲度大,集中于斜撑之变形量就会越大,以致斜撑中点产生之裂缝较早,导致较早的断裂破坏。从面内变形量也发现,双接合板也比单接合板有着较大的斜撑面内变形量。整体而言,斜撑采端部切削形式可达成斜撑面内挫屈,具有典型的斜撑构材之迟滞循环行为。

关键词:特殊同心斜撑构架;面内挫屈;接合板;切削
中图分类号:TU391

SEISMIC BEHAVIOR AND DESIGN OF BRACES BUCKLED IN-PLANE

C. C. Chen, W. C. Tang

(Department of Civil Engineering, Chiao Tung University, Hsinchu, 30010, Taiwan)

Abstract: In order to achieve in-plane buckling of the braces while subjected to compression, and to study the hysteretic behavior and axial strength of the braces, the brace sections at both ends are reduced by cutting the flanges that results in forming plastic hinges at both ends. Finite element analysis was conducted to establish the numerical model for parametric study. On the basis of the analytical results, four specimens were designed and tested to validate the hysteretic behavior of the specimens. The test results demonstrated that all specimens attained stable nonlinear behavior, achieving 4 to 5% rad of the interstory drift angle, buckling in-plane, and fracturing at the middle of the brace. While the depth of the cut was small, specimen behaved premature fracturing failure due to the crack occurred at the middle of the brace because of the less reducing section resulting in higher rotational stiffness. Moreover, the in-plane deformation of the braces with double gusset plates is larger than those with single gusset plate. In summary, the braces designed by reducing section at both ends can achieve in-plane buckling and possess typical inelastic behavior.

Keywords: special concentrically brace frame; buckle in-plane; gusset plate; reduce section

---

作者:陈诚直(1956—),男,博士,教授,钢结构与型钢混凝土结构,E-mail:chrischen@mail.nctu.edu.tw.
　　　汤伟干(1987—),男,硕士生,钢结构研究,E-mail:ahei40832@gmail.com.

1. 引言

同心斜撑构架主要以斜撑构材抵御地震力，其非弹性侧向反应仰赖着斜撑构材、接合、与梁与柱构材，而构架侧位移可有效被控制，此为其主要优点；并且梁与柱构件的尺寸将可减小，亦有经济上的优点。AISC[1]对同心斜撑构架的分类区分为二种，为特殊同心斜撑构架（Special Concentrically Braced Frame，SCBF）与普通同心斜撑构架（Ordinary Concentrically Braced Frame，OCBF）。两者的区别在于特殊同心斜撑构架于抵御地震力时，期望能有可观的非弹性变形能力。因而 SCBF 之斜撑构材受压挫屈后则有比 OCBF 较少的强度衰减与较好的韧性行为。

斜撑构架于地震力作用下，斜撑构材将承受拉、压反复轴力，受压时斜撑会产生挫屈。因斜撑构架之斜撑与梁、柱构材的接合一般采用接合板（Gusset plate）连接，此连接方式使得斜撑构材于受压挫屈时将往面外变形，将可能造成非结构构材的破坏。因此，使斜撑构材产生面内挫屈为研究之课题。

同心斜撑构架的研究始于上世纪六十至七十年代，主要研究者有 Popov 等人[2]、Ghanaat[3]、Black 等人[4]、Thornton[5]、Astaneh-Asl and Goel[6]、Astaneh-Asl 等人[7]。北岭地震后已有学者重视同心斜撑构架之耐震行为，研究有 Tremblay 等人[8,9]、Richards and Uang[10]、Tremblay[11]。接合板作为传递斜撑构材与梁、柱构件之力量传递媒介，其重要性可想而知，较为重要的为 Whitmore[12]与 Thornton[5]研究接合板接合的应力分布以决定在接合板的平均设计应力，并建立有效宽度之设计方法。斜撑面内挫屈之研究有连育群[13]与区玮衡[14]，区玮衡之斜撑与接合板间增加一刀形板，于此刀形板产生挫屈达到斜撑面内挫屈。经实验后发现，接合板除了能接受斜撑受拉降伏外，亦提供斜撑挫屈所需要之凹折区域。本研究旨在设计斜撑使斜撑于受压时之挫屈为面内挫屈，为将斜撑端部翼板切削，使其产生塑铰，达成斜撑面内挫屈。以实尺寸试体反复载重试验，探讨其强度及迟滞循环行为。

2. 斜撑构件反复载重试验

2.1 试体规划

试体仿真斜撑构材子结构，并包括部分梁、柱构材。表1所示为四组试体的设计参数。试体所采用之断面都为组合断面 BH150×150×12×12，由钢板进行切割，再进行组装焊接。图1所示为试体 RS1 设计细节。试体之接合板与梁、柱构材之接合焊道采开槽焊接，于斜撑与接合板接合之焊接则采填角焊接。斜撑构材与接合板于工厂内制作完成，再于国家地震工程研究中心实验场地进行组装及试验。斜撑角度为 53.3°，斜撑与梁、柱中心线交点间距离为 4902 mm，斜撑细长比皆小于 AISC 耐震设计规范[1]之规定值。所有钢材皆采用 A572 Gr.50，钢材拉伸试验所得强度列于表2。

试 体 表   表1

试体	自接合端点至切削起点距离		切削长度		切削深度		接合板厚 (mm)	接合板形式
	"a"	(mm)	"b"	(mm)	"c"	(mm)		
RS1	0	0	$2d$	300	$0.10 b_f$	15	22	单接合板
RS2	b_f	150	d	150	$0.10 b_f$	15	22	单接合板
RS3	$0.5 b_f$	75	d	150	$0.10 b_f$	15	12	双接合板
RS4	$0.5 b_f$	75	d	150	$0.06 b_f$	9	12	双接合板

注：d 为斜撑断面深度，b_f 为斜撑翼板宽度。

钢 材 强 度   表2

Member	Thickness (mm)	Yield strength F_y (MPa)	Tensile strength F_u (MPa)
Brace flange	12	392	525
Brace web	12	394	524
Gusset plate	15	381	515

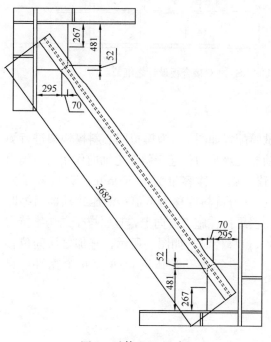

图 1 试体 RS1 尺寸

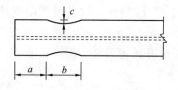

图 2 斜撑端部翼板的切削参数

斜撑端部翼板切削之目的为于斜撑受压时形成铰接，以达到斜撑于断面弱轴方向于平面内挫屈。为了探讨不同的切削细节于斜撑构材行为的影响，试体采用不同的切削尺寸，如表 1 所示。切削参数参考 FEMA-350[15] 切削式弯矩接头之设计，图 2 所示为斜撑端部翼板的切削参数之定义。参数包括自接合端点至切削起点距离"a"、切削的长度"b"与切削的深度"c"。FEMA-350 建议值分别为如下：

$$a \fallingdotseq (0.5 \sim 0.75) b_f \tag{1}$$

$$b \fallingdotseq (0.65 \sim 0.85) d \tag{2}$$

$$c \leqslant 0.25 b_f \tag{3}$$

先前之有限元素分析[16]发现影响比较大的为切削深度"c"与切削长度"b"。若切削深度"c"采取建议设计假设值 $0.2 b_f$，则斜撑切削处之受拉强度将会太低，该处将可能先于斜撑受压挫屈前产生受拉断裂，故试体之设计将切削深度"c"规划为 $0.06 b_f$ 与 $0.10 b_f$。若将切削长度"b"采取式（2）的建议值，塑铰的产生将不明显产生在切削处，有鉴于此，切削长度"b"规划为 d 与 $2d$。

接合板尺寸采 Uniform Force Method[5]进行设计检核。有限元素分析结果显示[16]，采单接合板接斜撑腹板，因斜撑腹板产生局部变形而不易于切削处产生塑铰，须于斜撑腹板以加劲板加劲之；而采双接合板之斜撑皆于切削处产生预期的塑铰。试体 RS1 与 RS2 采用单接合板，接合板尺寸为 365×533(mm)，试体 RS3 与 RS4 为双接合板，接合板尺寸为 241×366(mm)。图 3 所示为试体 RS2 之单接合板与试体 RS3 之双接合板设计之细节。

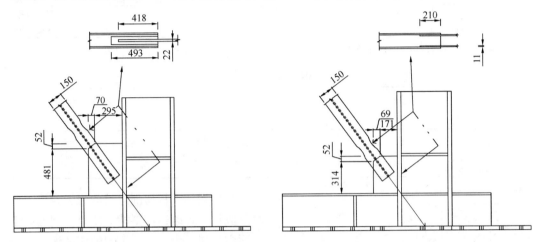

图 3　试体 RS2 之单接合板与试体 RS3 之双接合板设计之细节

2.2　试验设置与程序

本试验于国家地震工程研究中心进行，试验配置如图 4。为模拟同心斜撑构架进行测试，施力端之钢柱底边为铰支承，以油压致动器之水平力传递于斜撑之轴向作用力。施力系统由两支油压致动器施加作用力，油压致动器位移容量为 ±500mm，载重容量共 1922kN。试验时油压致动器伸长时，斜撑受压，力量与位移为负值；油压致动器后缩时斜撑受拉，力量及位移为正值。为使油压致动器维持在施力方向上，且斜撑子结构维持在平面上，于 H 型柱上端安装侧向支撑。试验采位移控制，如图 5 所示，施加之层间位移角（Interstory drift angle）从 0.25％弧度开始，至 0.75％弧度皆执行 6 个循环，自 1.0％弧度开始皆采 2 个循环，直至试体破坏。

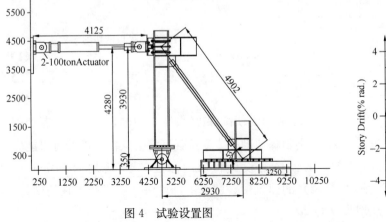

图 4　试验设置图

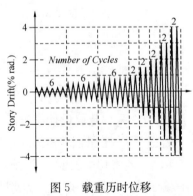

图 5　载重历时位移

3. 试验结果与讨论

3.1 试体行为

试体进行反复载重之试验结果，以试体之载重迟滞行为、破坏模式、面内变形量、挫屈强度等讨论之。

试体 RS1 于 0.25%弧度及 0.5%弧度之层间变位角下皆保持线弹性行为，试体表面未发现明显变化。进入 0.75%之第一个循环时于翼板切削处受拉力端有石膏漆脱落，并且朝斜撑构材端部发展，表示该处已进入降伏状态，同时翼板于斜撑中点也有石膏漆脱落，并且斜撑产生整体挫屈，方向向下，此时面内变形量约为 60mm。于 1%第一个循环时，翼板切削处受压端出现石膏漆脱落的现象，而斜撑中点翼板产生局部挫屈，同部位翼板受拉端石膏漆脱落明显扩大。进入 1.5%第一个循环时，斜撑中点翼板的局部挫屈越发明显。于 2.0%第一个循环时，切削处石膏漆脱落朝斜撑端部发展明显增加，斜撑中点翼板皆发生局部挫屈。于 3.0%第一个循环斜撑受拉时，斜撑中点翼板下端始发生局部挫屈。于 4%第一个循环斜撑受压时，斜撑中点翼板下端产生裂缝，由于该处已产生裂缝，如图 6。继续于同一循环斜撑受拉时，于斜撑中点发生断裂，实验终止，斜撑断裂如图 7，而量测到斜撑最大面内变形量为 413mm。

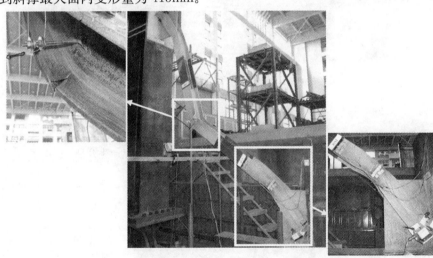

图 6 试体 RS1 于 4.0%弧度层间位移角之情况

试体 RS2 之行为与试体 RS1 约略相同。其破坏为于 4%第一个循环斜撑受压时，斜撑中点翼板下端产生裂缝，继续于同一循环斜撑受拉时，由斜撑中点发生断裂，实验终止，破坏如图 8，而量测到斜撑最大面内变形量为 384mm。

试体 RS3 于 0.25%弧度已有石膏漆脱落。进入 0.5%之第一个循环时，斜撑产生整体挫屈，方向向下，石膏漆脱落明显，此时面内变形量约为 46mm。进入 0.75%之第一个循环时，切削处受拉力端有石膏漆脱落。于 1%第一个循环时，斜撑中点翼板与切削处出现局部挫屈。至 3.0%第一个循环时，斜撑中点与切削处局部挫屈皆明显增加，但无裂缝

图7 试体RS1最终破坏情形

产生。于4%第一个循环斜撑受压时,斜撑中点翼板下端产生开裂,切削处受拉力端也有开裂产生,于同一循环斜撑受拉时,斜撑中点翼板下端始发生局部挫屈,并于翼板上端出现开裂。于5%第一个循环斜撑受压时,斜撑中点翼板下端产生开裂,同一循环斜撑受拉时,于斜撑中点发生断裂,实验终止,破坏情况如图9,而量测到斜撑最大面内变形量为482mm。

试体RS4于0.25%弧度已有石膏漆脱落,进入0.5%之第二个循环时,斜撑产生整体挫屈,方向向下,此时面内变形量约为30mm。进入0.75%之第一个循环时,斜撑中点翼板与切削处出现局部挫屈。于1%第一个循环时,切削处受拉力端有石膏漆脱落。

图8 试体RS2于4.0%弧度层间位移角斜撑中点产生开裂

图9 试体RS3于5.0%弧度层间位移角斜撑之破坏

进入1.5%第一个循环时，斜撑中点与切削处局部挫屈皆明显增加。于2.0%第一个循环受拉时，斜撑中点与切削处翼板受压端发生局部挫屈。于3.0%第一个循环时，斜撑中点与切削处局部挫屈皆明显增加。于4%第一个循环斜撑受压时，斜撑中点翼板下端产生裂缝，于同一循环斜撑受拉时，于翼板上端出现裂缝。进入4.0%第二个循环斜撑受拉时，试体断裂，试验终止，量测到最大面内变形量约为460mm。

3.2 试体迟滞行为与破坏模式

图10为试体之轴力与轴向变形量关系图，为典型斜撑构材之迟滞行为，轴向受拉强度皆大于受压挫屈强度。图中显示试体于每一循环皆有残留变形，导致最终的轴向变形量。试体RS1与RS2之迟滞行为相当类似，发现斜撑最大挫屈载重发生于0.75%弧度层间变位角；此两组试体之轴向变位差异并不明显，显示切削长度"b"并未有效影响其差异性。试体RS3则为唯一达5.0%弧度层间变位角之试体。

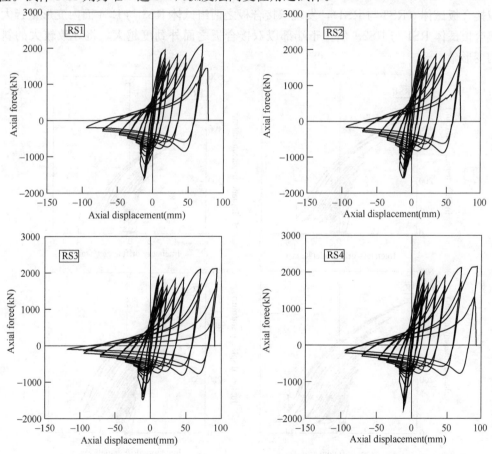

图10 试体之轴力与轴向变位关系图

斜撑之反复载重行为如预期一样，能产生斜撑整体的挫屈，最终并发生断裂。表3所列为试体破坏之描述。试体RS1与RS2为单接合板接合，两者间表现差异不大。试体RS3与RS4为双接合板接合，其中试体RS4的切削量为最小，并没有如同试体RS3能够进入5%弧度层间位移角循环便产生断裂。对于同心斜撑而言，斜撑端部劲度越大，集中于斜撑之变形量就会越大，导致斜撑中点由于变形产生之裂缝较早，因此试体 RS4 早于试体 RS3 断裂而破坏。

试体破坏模式　　　　　　表 3

试　体	破　坏　模　式
RS1	斜撑中点翼板局部挫屈，于斜撑中点发生断裂
RS2	斜撑中点翼板与切削处局部挫屈，于斜撑中点发生断裂
RS3	斜撑中点翼板与切削处局部挫，切削处产生裂缝，于斜撑中点发生断裂
RS4	斜撑中点翼板与切削处局部挫屈，于斜撑中点发生断裂

3.3　试体之面内变形行为

图 11 为各试体之面内变形量与层间位移角之关系。比较单接合板试体（RS1 与 RS2）与双接合板试体（RS3 与 RS4），发现双接合板之两组试体 RS3 与 RS4 面内变形量皆大于单接合板试体 RS1 与 RS2。斜撑于端部以双接合板之面外劲度越大，将导致较大的斜撑面内变形量。

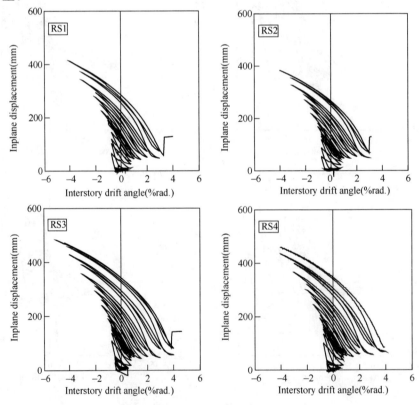

图 11　试体之斜撑中点面内变形量与层间位移角关系图

3.4　斜撑之轴向强度

四组试体斜撑之实验之最大拉力强度与受压强度与计算标称强度[17]如表 4 所示。由表可发现四组试体的实验最大拉力强度十分接近，且与全断面降伏标称拉力强度亦相当接近。于实验最大受压强度，单接合板接合之两组试体RS1与RS2非常接近，显示切削的长度"b"对于斜撑整体挫屈强度并无显著影响。采双接合板接合之两组试体RS3与RS4之实验最大受压强度则随切削的深度"c"减少而增加，因切削部位之切削量减少，斜撑挫屈时于端部相对有较大的塑性弯矩强度，进而反应在试体挫屈强度上。表4内之标称受压强度为假设斜撑两端为铰接，然而斜撑达到实验最大受压强度挫屈时，斜撑两端因产生塑铰而有所束制，因此实验最大受压强度远高于标称受压强度。

斜撑轴向强度　　　　　　　　　　　　表4

试体	实验最大拉力强度 (kN)	实验最大受压强度 (kN)	全断面降伏标称拉力强度 (kN)	净断面断裂标称拉力强度 (kN)	标称受压强度 (kN)
RS1	2100	1594	2009	2304	1336
RS2	2115	1583	2009	2304	1400
RS3	2119	1490	2009	2304	1157
RS4	2151	1754	2009	2455	1157

3.5　实验模拟与分析

为观察试体于往覆载重试验过程中各部位之降伏情形与应力集中现象，借由涂布于试体表面之石膏漆在受力时所造成的表面纹路可得知；而在有限元素仿真分析中，则可借由分析模拟试体的降伏情形与塑性状况以了解试体受力之量化数据。图12为试体RS3实验与分析之层间剪力与层间位移角关系图，显示采有限元素所建立之分析模型于试体最大挫屈载重与其发生挫屈时机皆可有效模拟，试体于分析的循环与试验曲线吻合程度高，而于受拉力加载后之行为则与试验有所偏差，推测为分析模型所采

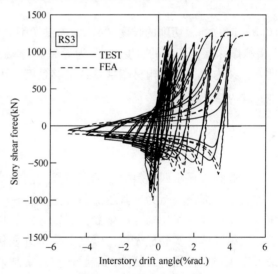

图12　试体RS3实验与分析之层间剪力与层间位移角关系图

之材料应变硬化曲线较实际情形简化。图13为试体斜撑行为与分析结果之等值应力分布对照图，由图显示分析结果成功仿真其塑性铰发生位置以及斜撑中点的局部挫屈，且该处皆有应力集中现象，实验于该处发生断裂破坏。

4. 设计建议

根据有限元素分析数值模型与试体载重试验之结果，以下为切削细节之建议。自接合端点至切削起点距离"a"为确保端部的劲度使其塑铰准确产生于切削处，建议可采用

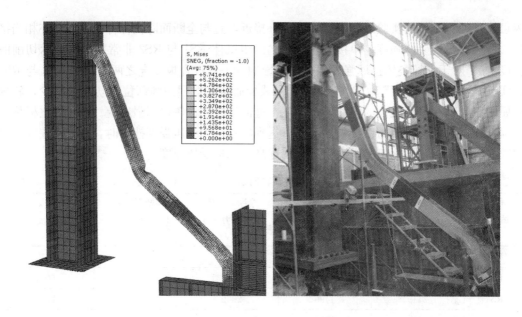

图 13　试体 RS3 试验行为与分析结果比较（3.0%弧度层间位移角）

$(0.5\sim1.0) b_f$。切削的长度"b"为提供切削区域足够空间形成凹折区域产生塑铰，可取 1 倍斜撑深度。切削的深度"c"为导致斜撑端点塑铰的形成，建议 0.1 倍的斜撑翼板宽度 b_f。又为使得斜撑构材端部具备足够劲度，可采用双接合板接合，但因其施工上较为不易，亦可以单接合板接合。

5. 结论

试验结果显示四组斜撑于端点翼板切削的试体，于受拉降伏而受压时挫屈，有典型斜撑构材之迟滞循环行为，其迟滞循环达 4%～5% 弧度之层间位移角，且斜撑于平面内产生挫屈。试体皆于切削区产生断面弱轴方向的塑铰，最终的破坏模式皆于斜撑中点发生断裂。四组试体的实验最大拉力强度十分接近全断面降伏标称拉力强度，然而斜撑达到实验最大受压强度挫屈时，斜撑两端因塑铰之束制，因此实验最大受压强度远高于假设斜撑两端为铰接之标称受压强度。切削之设计参数中，以切削的深度影响行为最为显著，其值增加则受压挫屈强度明显降低；其次为切削的长度，而自接合端点至切削起点距离则不明显。试体采双接合板接合的行为也稍微优于单接合板之试体，双接合板试体可达较大之面内变形量，显示较能将变形集中于斜撑构材上。斜撑端部翼板的切削为可达成斜撑面内挫屈之方式。

致谢

本计划承蒙国科会经费补助，谨此致谢；并感谢国家地震工程研究中心提供试验场地、设备与协助。

参考文献

[1] AISC. Seismic provisions for structural steel buildings. American Institute of Steel Construction, Inc., Chicago, IL. 2010

[2] Popov EP, Takanashi K, Roeder CW. Structural steel bracing systems: behavior under cyclic loading. Report no. UCB/EERC-76/17. Earthquake Engineering Research Centre, 1976

[3] Ghanaat Y. Study of X-braced steel frame structures under earthquake simulation. Report no. UCB/EERC-80/08. Earthquake Engineering Research Centre; 1980

[4] Black RG, Wenger WA, Popov EP. Inelastic buckling of steel struts under cyclic load and reversal. Rep. No. UCB/EERC-80/40, Earthquake Engineering Research Center, Univ. of California, Berkeley, Calif. 1980

[5] Thornton WA. Bracing connections for heavy construction. Engineering Journal, AISC, 1984; 21(3): 139-148

[6] Astaneh-Asl A, Goel SC. Cyclic in-plane buckling of double-angle bracing. Journal of Structural Engineering, 1984; 109: 2036-2055

[7] Astaneh-Asl A, Goel SC, Hanson RD. Cyclic out-of-plane buckling of double-angle bracing. Journal of Structural Engineering, 1985; 111: 1135-1153

[8] Tremblay R, Timler P, Bruneau M, Filiatrault A. Performance of steel structures during the 1994 Northridge earthquake. Canadian Journal of Civil Engineering, 1995; 22(2): 338-360

[9] Tremblay R, Bruneau M, Nakashima M, Prion HGL, Filiatrault A, DeVall R. Seismic design of steel buildings: Lessons from the 1995 Hyogo-ken Nanbu earthquake. Canadian Journal of Civil Engineering, 1996; 23: 727-756

[10] Richards PW, Uang CM. Effect of flange width-thickness ratio on eccentrically braced frames link cyclic rotation capacity. Journal of Structural Engineering, 2005; 131(10): 1546-1552

[11] Tremblay R. Seismic behavior and design of concentrically braced steel frames. Engineering Journal, 2001; third quarter: 148-166

[12] Whitmore R. E. Experimental Investigation of Stresses in Gusset Plates. University of Tennessee Engineering Experiment Station Bulletin No. 16 1952.

[13] 连育群. 斜撑面内挫屈之特殊同心斜撑构架耐震行为研究. 台湾大学土木工程学系硕士论文, 指导教授蔡克铨, 2009

[14] 区玮衡. 斜撑面内挫屈之特殊同心斜撑构件与构架耐震行为研究. 台湾大学土木工程学系硕士论文, 指导教授蔡克铨, 2010

[15] FEMA. Recommended Seismic Design Criteria for New Steel Moment-Frame Building. Report No. FEMA-350, Federal Emergency Management Agency, 2000

[16] 陈诚直、林南交、汤伟干、许需琳、朱致洁. 特殊同心斜撑构架之斜撑构材耐震性能研究(II). 国科会研究成果报告, 2010

[17] AISC. Specification for structural steel buildings. American Institute of Steel Construction, Inc., Chicago, IL. 2010

钢造双核心预力自复位斜撑发展与验证：耐震实验与有限元素分析

周中哲[1]，陈映全[2]

(1. 台湾大学土木工程系教授，台湾大学工学院地震工程研究中心主任，台北 10617；
2. 台湾大学土木工程系硕士，台北 10617)

摘　要：本研究发展双核心预力自复位斜撑（Self-Centering Brace，简称 SCB），是利用斜撑中之预拉力构件束制斜撑中之受压构件，在斜撑受拉与受压下提供自复位能力（在大变形下有回到零残余变形的能力），并利用摩擦消释地震能量；借由两组核心受压构件与两组拉力构件，使斜撑之变形量在拉力构件相同应变下大幅增加（或在相同斜撑变形量下，斜撑之拉力构件应变减少一半）。此文章首先说明双核心自复位斜撑力学及理论，并设计及测试三组 5350mm 长之双核心自复位斜撑，斜撑使用之拉力构件分别为直径 22mm、29mm 玻璃纤维棒与 13 mm 碳纤维棒。实验结果显示使用预力之双核心自复位斜撑有稳定的能量消释及自复位能力外，耐震性能也能符合 AISC（2010）规范针对挫屈束制斜撑要求之最小层间位移角 2%前不破坏，并可继续完成在 1.5%层间侧位移角下 15 圈反复载重下无破坏发生；其中两组试体更可再进一步地完成 2.5%层间位移角试验而不破坏，最大斜撑轴力约 1400 kN。本研究并使用 ABAQUS 有限元素软件分析试体之行为，分析结果与预测及试验结果符合，并对于预力、摩擦力及不同材质之拉力构件进行参数研究，进一步地了解双核心 SCB 在不同预力及摩擦力下的抗震消能行为。

关键词：双核心自复位斜撑；复合纤维棒；反复载重试验；有限元素分析
中图分类号：TP391

DEVELOPMENT OF STEEL DUAL-CORE POST-TENSIONED SELF-CENTERING BRACES: SEISMIC TESTS AND FINITE ELEMENT ANALYSES

C. C. Chou[1], Y. C. Chen[2]

(1. Professor, National Taiwan University, Taipei, 10617, cechou@ntu.edu.tw；
2. Master, National Taiwan University, Taipei, 10617)

Abstract: This work presents mechanics, tests, and finite element analyses of a proposed novel steel dual-core self-centering brace (SCB) with a flag-shaped energy dissipative response under cyclic loads. The axial deformation capacity of the SCB is doubled by serial deformations of two sets of tensioning elements arranged in parallel. The mechanics and cyclic behavior of the brace are first explained; three 5350mm long dual-core SCBs are tested and modeled to evaluate their cyclic performances. All SCBs exhibit excellent performance up to a target drift of 2% with a maximum tensioning element strain of 1%. All SCBs then successfully experience fifteen low-cycle fatigue tests at a drift of 1.5%. All SCBs are cyclically loaded again, and two SCBs achieve a drift of 2.5% with a maximum axial load of 1400 kN. Test results show that the application of dual cores in SCBs reduces significant strain demands on tensioning elements and enables self-centering responses to large deformation. Finite element analysis is then conducted on the specimen to further verify the mechanics and hysteretic responses observed in the test. Finite element analyses are also performed on 16 dual-core SCBs to perform a parametric study to evaluate how tensioning element types, initial PT force, and friction force affect the cyclic performance of the brace.

Keywords: dual-core self-centering brace; composite fiber; cyclic test; finite element analysis

1. 引言

钢造建筑物中常见的制震系统诸如抗弯矩构架、斜撑构架或剪力墙构架，均利用结构构件非线性韧性行为消释结构物承受之地震能量，但反复的非线性行为易产生破坏与残余变形，使结构物修复的困难度大幅增加，因此如何确保结构物在大地震后的抗震能力以承受未来地震是一项重要议题。传统钢造弯矩构架系统的梁柱接头可借由在钢梁翼板内侧与柱间使用加劲板，大幅增加梁柱接头韧性，使钢造梁柱接头能历经二次 AISC 2010[1]耐震规范定义之反复载重历时至近 5% 弧度层间侧位移角才有明显梁挫屈发生[2-4]，或是采用能避免梁挫屈的韧性梁柱接头[5]。作者等人曾发展可更换核心板式的挫屈束制消能斜撑[6]，此种挫屈束制斜撑将核心单元与两组独立分离之围束单元利用螺栓栓接组合而成（图 1a），由于在核心单元与围束单元间预留间隙，所以无须使用脱层材料，斜撑耐震行为（图 1b）远高于 AISC 2010[1]要求，地震过后亦可将围束单元与核心单元相接之螺栓直接松开分离，检查消能斜撑之核心单元是否发生破坏，加速震后检测及更换机制；由过去斜撑及构架实验可发现虽然核心单元破坏，但此种斜撑之围束单元仍可重复再使用，达到节省成本及环保目的。虽然挫屈束制斜撑之良好迟滞行为能加强斜撑构架之耐震性能[7-11]，但结构物在地震作用下会因斜撑消释能量而造成残余变形[12-14]。

为减少结构物因地震灾害造成之残余变形，作者等人曾使用以梁或柱加载预力的方式使结构物具抗震自复位的能力[15-18]，学者曾使用记忆合金制成之自复位斜撑，但记忆合金斜撑过于昂贵且难以根据结构物受地震力进行设计。学者 Christopoulos 等人[19]提出预力式自复位斜撑，由钢受压构件、消能组件及施加于斜撑构件预力之拉力构件所组成。此自复位斜撑虽可承受轴向变形，并提供一稳定之能量消散与自复位之能力，但由于该自复

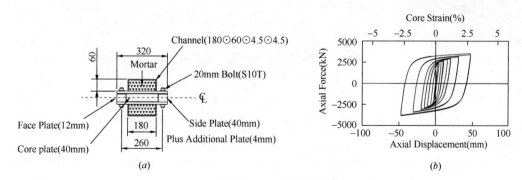

图 1 夹合式 BRB[6, 25]
(a) 夹合式 BRB；(b) 反复载重试验行为

位斜撑之变形量受制于所使用之拉力构件变形能力，因此对拉力构件之线弹性变形需求极高，无法使用于大变形需求下之建筑斜撑构件。

本研究[20-23]首次提出一新式双核心自复位斜撑（Dual-Core Self-Centering Brace，简称 SCB），利用增加一组钢受压构件与一组拉力构件以改变斜撑中之传力机制，使双核心自复位斜撑之变形能力为传统自复位斜撑之两倍，或是在相同变形量下大幅降低对拉力构件线弹性变形量需求。本研究首先介绍双核心自复位斜撑之力学行为，接着设计及测试三支 5350mm 长之双核心自复位斜撑，测试方式采由小至大变形的 AISC 2010[1]标准反复载重方式及 15 圈疲劳载重测试，已检核斜撑的力学机制，变形能力及耐用性能，最后并以有线元素分析模型进行双核心自复位消能斜撑参数研究。

2. 双核心自复位斜撑

2.1 力学行为

图 2 所示为本研究提出之双核心自复位斜撑，其中第一核心构件（1st Core）是由 H 型钢所构成，第二核心构件（2nd Core）是由两方形钢管所组成，放置于由长方形钢管所组成之外围构件（Outer Box）内，各构件之长度皆相同，并于构件两端盖上内与外层端板。在端部由第一核心伸出之钢板与外围构件之角钢间利用螺栓固定，借由界面之相对位移产生摩擦消能。双核心自复位斜撑力学行为如图 3 所示，当拉力未超过消能构件与预力所提供之力量时，斜撑不会有明显的变形量。当力量超过消能构件与预力所提供之力量 F_{dt}：

$$F_{dt} = P_{dt} + P_f = \frac{n}{2}(T_{in} + \delta_{in} \times K_f) + P_f \cong \frac{n}{2}T_{in} + P_f \tag{1}$$

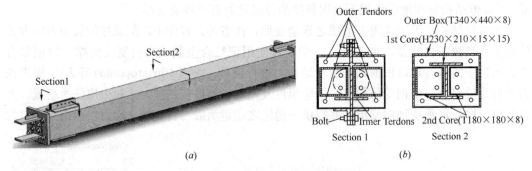

图 2 本研究发展之双核心自复位斜撑
(a) 试体整体图；(b) 试体断面图

式中，K_f 为一支拉力构件之轴向劲度；而 $n/2$ 为一组拉力构件之数量（内外层拉力构件数量皆为 $n/2$）；P_f 为摩擦力；T_{in} 为单根拉力构件之初始预力；δ_{in} 为钢构件初始压缩量。外围构件会顶着右侧外端板向右方移动（图 3a），借由外层拉力构件将力量从右侧外端板传至左侧外端板，并由第二核心构件将力量传至右侧内端板，再由内层拉力构件将力量传至左侧内端板，此时斜撑之伸长量 2δ 为内层与外层拉力构件之伸长量（δ）之合，显示即使拉力构件有 δ 的伸长量，整体斜撑会有 2δ 变形量。拉力构件之伸长量造成的额外拉力

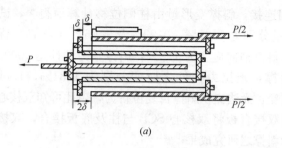 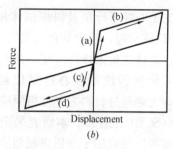

图 3 双核心自复位斜撑行为与轴力与轴位移关系
(a) 斜撑受力变形；(b) 轴力与轴位移关系

便可在斜撑卸除时将两受压构件拉回至初始位置，拥有自复位能力（图 3b）。受拉分离力 P_dt 下之变形量 δ_dt 为：

$$\delta_\mathrm{dt} = 2\delta_\mathrm{in} + \delta_{2c} = 2\delta_\mathrm{in} + \frac{P_\mathrm{dt} - P_{2c,\mathrm{in}}}{K_{2c}} \tag{2}$$

式中，δ_{2c} 为第二核心构件力量从初始预力 $P_{2c,\mathrm{in}}$ 上升至 P_dt 产生的压缩量；K_{2c} 为第二核心构件之轴向劲度，由受拉分离力 F_dt 除以分离位移 δ_dt 可得初始轴向劲度，当斜撑受压力时，会有类似力学行为。

3. 双核心自复位斜撑试体规划

3.1 双核心 SCB 试体试验

本研究设计及测试三支双核心 SCB 试体，试体所使用的拉力构件分别为直径 D22 mm 玻璃纤维棒、直径 D29mm 玻璃纤维棒以及直径 D13mm 的 T-700 碳纤维棒，所有试体之钢构件及端板尺寸均一样，第一核心构件为 H230×210×15×15mm 之钢构件，第二核心构件为两支 T180×180×8mm 钢管，外围构件为 T340×440×8mm 钢管。试体之主要设计参数列于表 1，初始预力（8 支复合材料棒合力）及摩擦力分别为 260kN、250kN。图 4 为 SCB 试体测试构架，此测试构架与先前测试挫屈束制斜撑试体相同，由一支箱型

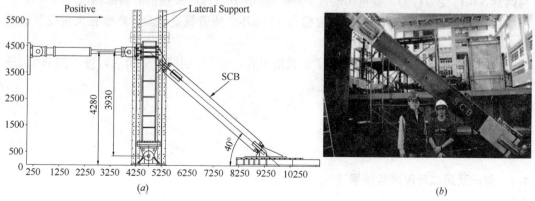

图 4 试验构架及试体变形图（单位：mm）
(a) 试验构架；(b) 2.5% drift

柱与 2 支 100 吨千斤顶及斜撑试体相连接，斜撑变形量由柱侧位移计算而得[6]，试体于 2％层间位移角时，拉力构件之应变量分别为 1.07％、0.93％及 1.06％（表1），皆小于拉力构件试体之极限弹性应变，此时斜撑轴向应变为 1.2％，斜撑轴力预计达 1000kN 左右，但若使用传统单核心自复位斜撑，则拉力构件之应变分别为 1.84％、1.71％及 1.83％，大幅超过材料之弹性极限应变，无法提供斜撑自复位行为，由此可知双核心自复位斜撑（SCB）之优点。本研究采用双接合板将双核心 SCB 与柱及底板接合，双接合板设计则参考以往应用于挫屈束制消能斜撑之研究成果[10, 24, 25]。

试体实验资料表　　　　　　　　　　表1

Specimen	Tendon Type	Tendon Diameter (mm)	Initial PT Force (kN)	2% Drift			
				Axial Force		Brace Strain (%)	Tendon Strain (%)	
				Tension (kN)	Compression (kN)		Double Core	Single Core
1	E-Glass Fibers	22.2	260 (260)	1025 (1060)	1000 (1133)	1.2 (1.16)	1.07 (1.02)	1.84
2	E-Glass Fibers	28.7	260 (290)	1225 (1415)	1180 (1359)	1.2 (1.09)	0.93 (0.86)	1.71
3	Carbon Fibers	12.7	260 (285)	1010 (1035)	990 (1193)	1.2 (1.2)	1.06 (1.07)	1.83

注：（ ）内为试验值。

3.2　四阶段试验程序

为了检验双核心 SCB 之抗震及耐用性，本研究针对每组 SCB 试体进行四阶段载重试验，各阶段试验分述如下：

（1）进行 SCB 试体未加摩擦消能器之反复加载试验至 0.36％层间位移角，目的为求得 SCB 试体初始预力，并与拉预力时所量测的预力值比较。

（2）进行 SCB 试体标准反复载重试验至 2％层间位移角，观察双核心 SCB 试体反应与内部各构件受力行为，载重历时如 AISC（2010）对挫屈束制消能斜撑测试历时一样。

（3）进行 SCB 试体 1.5％层间位移角 15 循环之疲劳载重试验，观察在大量反复载重下双核心 SCB 之耐用性。

（4）再一次进行 SCB 试体标准反复载重试验，过 2％层间位移角后，每增加 0.5％进行两循环测试直到斜撑试体破坏为止。

4. 双核心 SCB 试体试验结果

4.1　第一及第二阶段试验结果

各组试体于第一阶段实验下求得施加初始预力，分别列于表1之括号内，发现由实验求得 SCB 试体初始预力与设计值相差不大。图5为3组双核心 SCB 试体于第二阶段标准反复载重下之侧力与侧位移角关系，SCB 轴力可由千斤顶侧力及斜撑与水平之夹角关系转换而得，轴向变形由 SCB 端部之位移计求得，SCB 试体于 2% 层间位移角时之轴力亦列于表 1 中。试体 1 试验时之最大轴力为 1060kN 与预测值 1025kN 接近，无预力损失现象，且有良好的自复位行为（图 5a），试体 2 试验时之最大轴力 1415kN 较预测值 1225kN 大，为摩擦力较预估大所导致，试体 3 试验时之最大轴力 1035 kN 与预测值 1010 kN 接近，但其中一内层拉力构件于斜撑受拉至 2% 层间位移角时破坏，造成预力些许损失，残余变形于最后一循环（2% 层间位移角）明显增加（图 5c），原因为螺杆未完全锁进续接器中，使得螺杆与续接器传力面积不足导致螺杆抽出，并非碳纤维棒破坏。

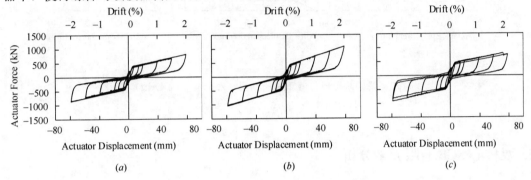

图 5　第二阶段标准试验千斤顶力量位移图
(a) 试体 1；(b) 试体 2；(c) 试体 3

4.2　第三及第四阶段试验结果

进行完双核心 SCB 试体第二阶段试验后，所有试体均进行在 1.5% 层间位移角的 15 圈试验（第三阶段），所有试体迟滞反应均很稳定（图 6），并无强度递减也无破坏发生，因此在完成 15 圈试验后均再次进行由小至大的反复载重测试（第四阶段）。试体 1 重复标准反复载重历时至 2.5% 层间位移角后无破坏发生，迟滞循环亦保持良好自复位行为（图 7a），此时斜撑轴向变形为 1.4%，D22mm 玻璃纤维棒拉应变约为 1.25%，小于极限应变 1.47%，所以无破坏发生。试体 2 于 2.5% 层间位移角斜撑受压时无破坏，但受拉时四支外层拉力构件同时断裂，造成力量下降（图 7b），D29mm 玻璃纤维棒断裂时之拉应变为 1.02%，与 D29mm 玻璃纤维棒极限应变 1.04% 接近，所以拉力构件破坏，但所有钢构件

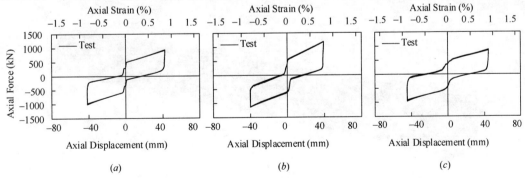

图 6　第三阶段 15 圈疲劳载重试验下斜撑轴力与轴位移图（1.5% 层间位移角）
(a) 试体 1；(b) 试体 2；(c) 试体 3

均无破坏发生。试体 3 亦可完成 2.5％层间侧位移角载重历时 2 圈而无破坏发生，但当进行 3％层间位移角试验时有两支内层拉力构件破坏，造成力量下降（图 7c）。

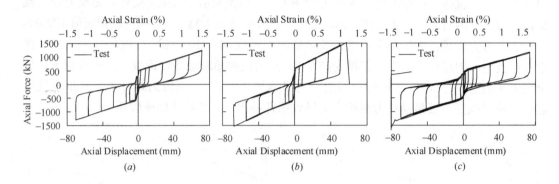

图 7 第四阶段反复载重试验结果（2.5％～3％ Drift）
(a) 试体 1；(b) 试体 2；(c) 试体 3

5. 双核心 SCB 有限元素分析

5.1 试体分析模型

为了解双核心 SCB 之整体力学机制与各构件实际受力情形，本研究使用有限元素分析软件 ABAQUS[26] 进行 3 组双核心自复位消能斜撑分析。模型中之钢材使用双线性曲线，复合材料纤维棒则使用线弹性定义，其中弹性模数皆由试片试验结果所得。受压构件中之第一核心、第二核心、外围构件以及端板皆采用 8 个结点之砖元素，而内外层拉力构件则采用两个结点的桁架元素。

图 8 为 SCB 模型的组装图，束制斜撑其中一端，利用位移控制反复位移量模拟斜撑反复载重试验。双核心 SCB 模型 1 与 2 之轴力与轴位移分析结果与试体试验结果比较如图 9 (a) 与 (b) 所示，结果显示预测循环与 ABAQUS 分析结果几乎重合，证明双核心 SCB 之传力机制与预测相符。借由上述可知有限元素分析可充分仿真出 SCB 受力情形，因此便可使用此模型进行参数分析。

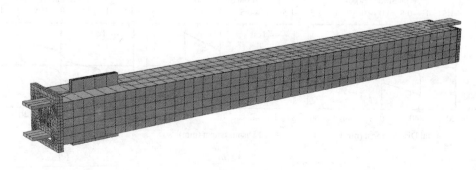

图 8 双核心 SCB 有限元素分析模型

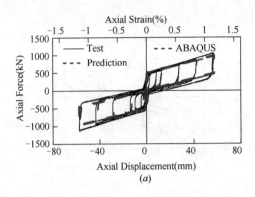
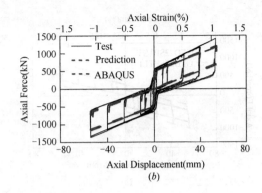

图 9 双核心 SCB 实验与分析结果比较
(a) 试体 1；(b) 试体 2

5.2 参数研究

参数分析主要是为了了解双核心 SCB 在不同预力、摩擦力以及拉力构件之影响。模型 1 至模型 16 分别以大小预力、大小摩擦力以及不同材料之拉力构件为参数作设计，主要参数列于表 2。图 10a 为其中 2 组双核心 SCB 模型的力量与位移图，可知双核心 SCB 模型具大摩擦力的迟滞消能循环较具小摩擦力的迟滞消能循环大，但由于摩擦力大于预力，因此无法自复位，但具小预力及小摩擦力之 SCB 试体模型即可自复位，在增加初始预力之后，SCB 模型仍可达自复位能力（图 10b）。

ABAQUS 有限元素模型参数　　　　表 2

No.	Model ID	Tendon Material	Diameter (mm)	Initial PT Force (kN)	Friction Force (kN)
1	SSD16SPTSF	Steel Strand	15.2	130	128
2	SSD16SPTLF				256
3	SSD16LPTSF			260	128
4	SSD16LPTLF				256
5	GFD22SPTSF	E-Glass Fiber	22.2	160	158
6	GFD22SPTLF				316
7	GFD22LPTSF			320	158
8	GFD22LPTLF				316
9	GFD29SPTSF		28.7	160	158
10	GFD29SPTLF				316
11	GFD29LPTSF			320	158
12	GFD29LPTLF				316
13	CFD13SPTSF	T700-Carbon Fiber	12.7	160	158
14	CFD13SPTLF				316
15	CFD13LPTSF			320	158
16	CFD13LPTLF				316

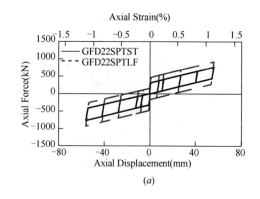 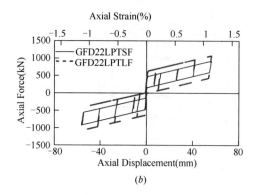

图 10 双核心 SCB 模型参数分析结果比较
(a) 具小预力之双核心 SCB 模型；(b) 具大预力之双核心 SCB 模型

6. 结论

本研究首次成功地研发双核心自复位斜撑（SCB），结合预力之自复位功能及摩擦消能能力发展之双核心 SCB，可运用于结构物中降低地震下最大侧位移及残余变形，经由 3 支实尺寸 SCB 试体（5350mm 长）试验成功地证明其传力机制及消能行为，并如同理论预测大幅降低拉力构件之应变需求量。自复位斜撑构架于 2％层间位移角下，斜撑应变量达 1.2％，3 支试体中拉力构件应变量最大为 1.09％，约为传统单核心自复位斜撑拉力构件之半，在 4 次试验中斜撑之最大轴力也可达约 1400kN。

除了验证双核心 SCB 机制外，试验结果显示出双核心 SCB 稳定的迟滞消能能力及自复位的能力，三组双核心 SCB 耐震性能均能符合 AISC（2010）规范针对挫屈束制斜撑要求之最小层间位移角 2％前不破坏原则，并也能在 1.5％层间侧位移角下历经 15 圈反复载重无破坏发生；试体 1 使用直径 D22mm 玻璃纤维棒及试体 3 使用直径 D13mm 碳纤维棒之试体更可进一步地完成 2.5％层间侧位移角载重试验而不破坏。一般而言，自复位斜撑构架在地震下之最大层间侧位移角小于挫屈束制斜撑构架，且残余变形几乎为零，远低于挫屈束制斜撑构架在地震后之残余变形量[13, 14]。

有限元素分析软件 ABAQUS 分析结果与理论预测及试验结果符合，代表有限元素软件可用来有效的模拟双核心 SCB 之行为，且证明双核心之传力机制与理论一致，确实降低拉力构件之弹性应变需求量，并由参数分析之 16 组模型中，得知拉力构件之材料及尺寸仅影响循环之后劲度，对于消能能力并无影响。

参考文献

[1] AISC（American Institute of Steel Construction），Seismic provisions for structural steel buildings，Chicago，IL，2010

[2] 周中哲，饶智凯．钢造建筑梁柱梁翼内侧加劲板补强接头耐震试验及有限元素分析．建筑钢结构发展，2010，12(1)，18-26

[3] Chou C-C，Jao C-K. Seismic rehabilitation of welded steel beam-to-box column connections utilizing internal flange stiffeners. Earthquake Spectra，2010，26(4)，927-950

[4] Chou C-C, Tsai K-C, Wang Y-Y, Jao C-K. Seismic rehabilitation performance of steel side plate moment connections. Earthquake Engineering and Structural Dynamics, 2010, 39, 23-44

[5] 周中哲, 吴家庆. 削切盖板钢骨梁柱接头设计与耐震性能. 建筑钢结构发展, 2008, 10(2), 11-18

[6] Chou C-C, Chen S-Y. Subassemblage tests and finite element analyses of sandwiched buckling-restrained braces. Engineering Structures, 2010, 32, 2108-2121

[7] Watanabe A, Hitomi Y, Yaeki E, Wada A, and Fujimoto M. Properties of braces encased in buckling-restraining concrete and steel tube. 9th World Conference on Earthquake Engineering. Tokyo-Kyoto, Japan, 1988, 719-724

[8] Usami T, Ge HB, Kasai A. Overall buckling prevention condition of buckling-restrained braces as a structural control damper. 14th World Conference on Earthquake Engineering, Beijing, China. 2008

[9] Tsai K-C, Hsiao B-C, Wang K-J, Weng Y-T, Lin M-L, Lin K-C, Chen C-H, Lai J-W, Lin S-L. Pseudo-dynamic tests of a full scale CFT/BRB frame-Part I: Specimen design, experiment and analysis. Earthquake Engineering and Structural Dynamics, 2008, 37: 1081-1098

[10] Chou C-C, Liu J-H, Pham D-H. Steel buckling restrained braced frames with single and dual corner gusset connections: seismic tests and analyses. Earthquake Engineering and Structural Dynamics, 2012, 7(41): 1137-1156

[11] Chou C-C, Liu J-H. Frame and brace action forces on steel corner gusset plate connections in buckling-restrained braced frames. Earthquake Spectra, 2012, 28(2), 531-551

[12] Uang C-M, Kiggins S. Reducing residual drift of buckling-restrained braced frames. Int. Workshop on Steel and Concrete Composite Construction, Report No. NCREE-03-026, National Taiwan University, Taiwan. 2003

[13] Tremblay R, Lacerte M, Christopoulos C. Seismic Response of Multistory Buildings with Self-Centering Energy Dissipative Steel Braces J. Structural Engineering, ASCE, 2008, 134, 108-120

[14] Chou C-C, Chen Y-C, Pham D-H, Truong V-M. Experimental and analytical validation of steel dual-core self-centering braces for seismic-resisting structures. 9th International Conference on Urban Earthquake Engineering/4th Asia Conference on Earthquake Engineering, Tokyo, Japan. 2012

[15] Chou C-C, Chen Y-C. Cyclic tests of post-tensioned precast CFT segmental bridge columns with unbonded strands. Earthquake Engineering and Structural Dynamics, 2006, 35, 159-175

[16] Chou C-C, Chang^H-J, Hewes J. Two-plastic-hinge and two dimensional finite element models for post-tensioned precast concrete segmental bridge columns. Engineering Structures, 2012 (accepted for publication 2012/7, in press)

[17] Chou C-C, Chen J-H. Seismic design and shake table tests of a steel post-tensioned self-centering moment frame with a slab accommodating frame expansion. Earthquake Engineering and Structural Dynamics, 2011, 40 (11), 1241-1261

[18] Chou C-C, Chen J-H. Analytical model validation and influence of column bases for seismic responses of steel post-tensioned self-centering MRF systems. Engineering Structures, 2011, 33(9), 2628-2643

[19] Christopoulos C, Tremblay R, Kim H-J, Lacerte M. Self-centering energy dissipative bracing system for the seismic resistance of structures: development and validation. J. Struct. Engrg. , ASCE, 2008, 134(1), 96-107

[20] Chou C-C, Chen Y-C. Development of steel dual-core self-centering braces with E-glass FRP composite tendons: cyclic tests and finite element analyses. The International Workshop on Advances in Seismic Experiments and Computations, Nagoya, Japan. 2012

[21] Chou C-C, Chen Y-C. Development and seismic performance of steel dual-core self-centering braces. 15th World Conference on Earthquake Engineering, Lisbon, Portugal. (Paper No. 1648), 2012

[22] 周中哲, 陈映全. 双核心自复位斜撑发展与耐震实验. 结构工程期刊(100-033, 接受刊登), 2012

[23] 周中哲, 陈映全. 钢造双核心自复位斜撑发展与耐震实验：应用复合纤维材料棒为预力构件. 土木工程学报, 45(2), 202-206, 2012

[24] Chou C-C, Liou G-S, Yu J-C. Compressive behavior of dual-gusset-plate connections for buckling-restrained braced frames. J. Constructional Steel Research, 2012, 76, 54-67

[25] 周中哲, 刘佳豪. 含消能斜撑构架效应之接合板耐震设计与试验分析. 建筑钢结构进展. 2011. 13(5), 44-49

[26] ABAQUS, standard user's manual version 6.3, Hibbitt, Karlsson & Sorensen, Inc., Pawtucket, RI. 2003

波纹腹板 H 型钢的研究 *

李国强[1]，张 哲[2]

(1. 同济大学 土木工程防灾国家重点实验室，上海 200092；
2. 郑州大学 土木工程学院，河南 郑州 450001)

摘 要：为研究波纹腹板 H 型钢作为横向受力构件的性能，在理论分析的基础上，借助试验和有限元方法分别对此类型钢的抗剪、抗弯、疲劳性能和局部承压强度进行了研究，并对楔形波纹腹板 H 型钢梁的弹性和弹塑性稳定承载力进行了专项研究。通过研究，分别明确了各类构件的受力特点，并给出了其设计表达式。设计表达式能够对试验和有限元分析结果进行较好的预测，且形式简单，适用于实践设计过程。

关键词：波纹腹板 H 型钢；剪切屈曲；整体稳定；局部稳定；疲劳强度；局部承压强度

中图分类号：TU392.1

INVESTIGATION OF H-BEAMS WITH CORRUGATED WEBS

G. Q. Li[1], Z. Zhang[2]

(1. State Key Laboratory for Disaster Reduction in Civil Engineering,
Tongji University, Shanghai 200092 China;
2. School of Civil Engineering, Zhengzhou University, Zhengzhou 450001, China.)

Abstract: To investigate the performance of the H-beams with corrugated webs as lateral stress components, the test and finite element methods were adopted to study the shearing, bending, fatigue property and the strength under partial compressive edge loading, based upon the theoretical analysis. Then, the stability of tapered H-beams with corrugated webs were researched particularly. From the works above, the mechanical models of various members are established, and the design expressions are put forward respectively. These expressions can forecast the results of finite element imitation and tests effectively, and they are simple in form to be accepted by the designers.

Keywords: H-beam with corrugated webs; shearing buckling; overall stability; local stability; fatigue strength; strength under partial compressive edge loading

---

* 基金项目：国家科技支撑计划资助项目（2006BAJ01B02）；国家自然科学基金资助项目（51008284）.
第一作者：李国强（1963—），男，博士，教授，博导，主要从事结构工程方面的研究，E-mail：gqli@tongji.edu.cn
通讯作者：张哲（1977—），男，博士，讲师，主要从事结构工程方面的研究，E-mail：zzhe@zzu.edu.cn

1. 引言

对波纹腹板 H 型钢力学性能的研究，可追溯至 20 世纪 50、60 年代对波纹金属板的受力分析。波纹金属板的使用最初是在航天器制造中，随后应用到了工业民用建筑和桥梁结构领域。20 世纪 80 年代，日本住友公司首次采用焊接的方法，生产出中间部分波纹腹板 H 型钢[1]。中国东北重型机械学院于 1985 年成功地轧制出了世界上第一根全波纹腹板 H 型钢[2]。近年来，随着自动焊接技术的发展，焊接的全波纹腹板 H 型钢在欧美国家发展较快，较多地应用于桥梁、大跨度房屋和工业厂房等结构中。在欧洲规范中已有关于此类型钢的专项规定，如瑞典轻钢规范[3]、Eurocode 3[4] 等。

波纹腹板 H 型钢的技术改进主要在于将平腹板改为波纹腹板（图 1），从而能够以较薄的腹板厚度，获得较大的平面外刚度及较高的抗剪切屈曲承载能力，因此该类型钢具有较高的承载能力及经济优势。据统计，波纹腹板 H 型钢梁的腹板用钢量一般仅占总量的 25% 左右，最低可至 4%，腹板高厚比可以达到 600，较之现有规格的热轧工字钢、热轧 H 型钢和焊接 H 型钢梁，经济优势非常明显。由于截面更为开展，受弯承载力更高，同时局部承压能力和疲劳强度也有所提高，所以波纹腹板 H 型钢梁非常适合作为横向受力构件。

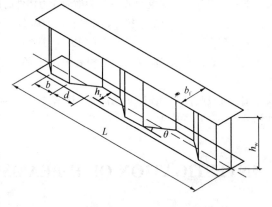

图 1 波纹腹板钢梁几何尺寸示意图

波纹腹板 H 型钢梁的理论受力模型与平腹板钢梁有显著区别（如图 2 所示），可以总结为下面 4 个方面：①由于"折叠效应"，波纹腹板上基本无弯曲正应力分布；②竖向剪力 V_y 完全由腹板承担，且剪应力均匀分布；③由于腹板的波纹形状，腹板与翼缘之间的剪力流将在翼缘中形成附加横向弯矩，因此翼缘存在附加应力；④在面外弯矩 M_y 作用下，腹板不产生正应力和剪应力，弯矩完全由翼缘承担。

为明确波纹腹板 H 型钢作为横向受力构件的力学性能，本文拟从其剪切性能、弯曲

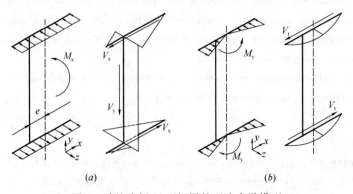

图 2 波纹腹板 H 型钢梁的理论力学模型
(a) 面内弯矩作用下应力分布示意图；(b) 面外弯矩作用下应力分布示意图

性能和疲劳性能等几个角度对该类型钢梁进行理论分析和试验研究,并辅助以有限元分析手段。在此基础上,拟提出设计方法和构造要求等。

2. 波纹腹板 H 型钢的剪切性能

2.1 弹性理论剪切强度

波纹腹板 H 型钢主要的技术创新在于通过腹板形式的改变,从而提高了腹板的面外刚度和剪切屈曲强度。通过对试验发现,波纹金属板的剪切破坏模式分别包括局部剪切屈曲、整体剪切屈曲及材料的屈服。其中局部剪切屈曲发生在某个板带宽度范围内,可以按照板均匀受剪的弹性稳定理论进行分析;以图1所示的波纹尺寸为例,梯形波纹钢板弹性局部屈曲极限应力[5]可以表示为:

$$\tau_{cr,l} = k_s \pi^2 E / [12(1-\mu^2)(w/t_w)^2] \tag{1}$$

式中,k_s 为屈曲系数,与边界条件有关,若长边为简支,短边固结,则:

$$k_s = 5.34 + 2.31(w/h_w) - 3.44(w/h_w)^2 + 8.39(w/h_w)^3 \tag{2}$$

对于四边固结:

$$k_s = 8.98 + 5.6(w/h_w)^2 \tag{3}$$

式中,$w = \max\{b, d/\cos\theta\}$,$\mu$ 为材料泊松比;E 为材料弹性模量;h_w 为腹板的高度。

波纹板整体屈曲发生在整个板的高度范围内,屈曲波纹可能贯穿若干个波长,可以按照各向异性板的弹性稳定理论进行整体分析。弹性整体屈曲极限应力可以表示为:

$$\tau_{cr,g} = k_s D_x^{0.25} D_y^{0.75} / (t_w h_w^2) \tag{4}$$

式中,$D_x = qEt_w^3/12s$,$D_y = EI_y/q$,$I_y = 2bt_w(h_r/2)^2 + t_w h_r^3/(6\sin\theta)$,$t_w$ 为腹板厚度,k_s 为边界条件屈曲系数,Easley[6] 取为 36,Galambos[5] 则定义:简支边界条件,$k_s = 31.6$,固结边界条件,$k_s = 59.2$。q 为腹板波形的一个波长,s 为波长的展开长度。

2.2 弹塑性极限承载力

为得到波纹腹板 H 型钢梁的极限剪切承载力,各国学者进行了大量的研究。1992年,Smith[7] 进行了 4 根梁的试验,包括两种腹板厚度 0.455mm 及 0.75mm。腹板破坏模式除了屈曲外,还包括腹板由于较薄,被焊穿后造成的过早失效,及间断焊缝的连接破坏,试验证明间断焊缝的连接形式是不可取的。Hamilton[8] 进行了 42 根梁的试验,这些梁包括 4 种不同的波纹尺寸和 2 种板厚:0.633mm 和 0.775mm。Elgaaly[9] 总结了 Smith 和 Hamilton 的试验结果,并进行了有限元模拟,发现将初始缺陷加入有限元模型后,将其计算结果与理论分析和试验结果三者进行分析比对,认为结果是令人满意的。

R. Luo (1996)[10-12] 进行了波纹腹板 H 型钢的数值分析,考虑下列因素对屈曲强度的影响:腹板的长度、高度;腹板厚度;腹板波纹高度;波纹的角度;腹板各板带的宽度等。并将分析结果与试验结果和理论算式进行了分析。

Abbas[13,14] 进行了 2 根足尺波纹腹板 H 型钢梁的试验及有限元分析,并将 Hamilton 的 42 根梁试验、Lindner 的 25 根梁试验、Peil 的 20 根梁试验结果进行了总结。作者认为在弹性阶段理论算式偏于安全,而当非弹性屈曲或者是材料屈服起控制作用时,理论算式则过高估计了其承载力。因此提出同时考虑整体屈曲和局部屈曲的相互作用,并给出了承载力计算方法。

对于波纹腹板的弹塑性屈曲,各国研究者提出了不同的设计表达式,为方便对比,本文将极限剪应力统一表达为腹板通用宽厚比的函数。经过形式转换,Elgaaly 提出的算式可以表示为:

$$\frac{\tau_{cr}}{\tau_y} = \frac{0.894}{\lambda_s} \leqslant 1.0 \tag{5}$$

式中,$\lambda_s = \sqrt{\tau_y/\tau_{cr,g}}$ 或 $\sqrt{\tau_y/\tau_{cr,l}}$,算式(5)可以分别计算整体屈曲和局部屈曲极限承载力。

EuroCode 3 提供的计算整体屈曲极限剪应力和局部屈曲极限剪应力分别为:

$$\frac{\tau_{cr}}{\tau_y} = \frac{1.5}{0.5 + \lambda_s^2} \leqslant 1.0 \tag{6}$$

$$\frac{\tau_{cr}}{\tau_y} = \frac{1.15}{0.9 + \lambda_s} \leqslant 1.0 \tag{7}$$

而 Abbas 提出的用于计算局部屈曲极限承载力算式可以转化为:

$$\frac{\tau_{cr}}{\tau_y} = \begin{cases} \dfrac{1}{\sqrt{2}} & \lambda_s < 0.89 \\ \sqrt{\dfrac{1}{1+1.25\lambda_s^2}} & 0.89 \leqslant \lambda_s < 1.12 \\ \sqrt{\dfrac{1}{1+0.98\lambda_s^4}} & \lambda_s \geqslant 1.12 \end{cases} \tag{8}$$

按照式(8)计算最高抗剪承载力为剪切屈服强度控制抗剪承载力的 0.707 倍,材料强度未能充分利用。

Yi J[15,16] 提出的基于相关屈曲算式可以表示为:

$$\frac{\tau_{cr}}{\tau_y} = \begin{cases} 1 & \lambda_s < 0.6 \\ 1 - 0.614(\lambda_s - 0.6) & 0.6 \leqslant \lambda_s < \sqrt{2} \\ 1/\lambda_s^2 & \lambda_s \geqslant \sqrt{2} \end{cases} \tag{9}$$

而瑞典规范相关公式也是基于局部屈曲作为控制条件,可以表示为:

$$\frac{\tau_{cr}}{\tau_y} = \begin{cases} 1.16 & \lambda_s < 0.73 \\ 0.84/\lambda_s & 0.73 \leqslant \lambda_s < 1.38 \\ 1.16/\lambda_s^2 & \lambda_s \geqslant 1.38 \end{cases} \tag{10}$$

在国内,李艳文,张文志等则探讨了此类型钢的轧制工艺[17]及产品的优化设计[18]。常福清,李艳文等利用能量方法[19-22]和数值模拟方法[23]研究了波纹腹板 H 型钢分别作为受弯构件及弹性地基梁情况下腹板的屈曲临界应力。

宋建永[24,25]通过非线性有限元方法分析波纹腹板 H 型钢剪切屈曲极限荷载和屈曲模态,在此基础上研究了波纹形状、腹板整体外形尺寸和腹板厚度等因素对波纹腹板剪切屈曲极限荷载和屈曲模态的影响。李时,郭彦林[26]基于 ANSYS 有限元软件,研究了波折腹板梁的剪应力作用下的破坏机理及基本性能。在计算时,以腹板厚度相同的平腹板工字钢作为比较对象。通过采用一致缺陷模态法模拟波形尺寸缺陷,发现波折腹板梁的抗剪承载力明显高于普通工字钢梁。

综合上述研究成果可以发现,对于波纹腹板 H 型钢腹板的整体屈曲极限承载力各个计算算式差别不大。Abbas 算式较其他算式更为保守,而瑞典规范比其他算式高估了构件的承载力,其他算式对于极限应力达到材料屈服强度所对应的通用宽厚比取值不同,相比之下,EUROCODE 3 提供的算式最为保守。同时,将各个计算式的计算结果与国际范围内试验结果进行对比后可以发现,局部屈曲极限承载力的计算方法较为有效,整体屈曲承载力的计算式未获得理想的结果[27],而这主要是由于整体稳定屈曲极限承载力的控制因素较多,不同的波形及腹板高厚比会造成不同的破坏模式:整体屈曲、局部屈曲、相关屈曲及屈服。基于此,本文设计若干波纹腹板 H 型钢梁抗剪试验,希望能够通过试验对其受力性能和破坏模式进一步加深理解,并用于推导设计表达式。

2.3 试验概况

在设计试验构件时共考虑了 4 种波形进行对比,波形具体参数见表1。

波形尺寸表　　　　　　　　表1

波形	b (mm)	d (mm)	h_r (mm)	q (mm)	s/q
1	64	23.5	38	175	1.25
2	70	80	50	300	1.10
3	70	50	50	240	1.17
4	40	25	30	130	1.22

按照这 4 种波形,共设计了 11 抗剪试件,具体尺寸见表2。

试验构件基本几何参数　　　　　　　　表2

	$b_f \times t_f$ (mm)	波形	t_w (mm)	h_w (mm)	质量 (kg/m)	惯性矩 (cm^4)	a (m)	L (m)
GJ1	200×10	1	1.5	500	36.8	26010	0.5	1.0
GJ2	200×10	1	1.5	500	36.8	26010	1.0	2.0
GJ3	200×10	1	1.5	500	36.8	26010	3.0	7.0
GJ4	280×14	2	2.0	1000	78.8	201526	1.0	2.0
GJ5	280×14	2	2.0	1000	78.8	201526	2.0	4.0
GJ6	280×14	2	2.0	1000	78.8	201526	4.0	10.0
GJ9	150×10	1	2.0	1000	39.3	76507	1.0	2.2
GJ10	150×10	3	2.0	500	32.7	19507	0.5	1.3
GJ11	150×10	3	2.0	500	32.7	19507	1.0	2.2
GJ12	150×10	3	2.0	1000	39.3	76507	1.0	2.2
GJ13	150×10	4	2.0	500	33.1	19507	1.0	2.2

表中 a 代表剪跨段长度;L 代表梁的跨度。所有试件均在工厂内制作,钢材设计采用 Q235 钢,波纹腹板的弯折采用折弯机人工操作,弯折后还需进行矫形,并进行检测,确定误差小于规定值后进行焊接。腹板与翼缘之间采用单面角焊缝,CO_2 气体保护焊,焊

丝为0.8mm。在试验支座位置和加载位置设置加劲肋，加劲肋与腹板及翼缘之间均为单面角焊缝。

所有试验均设计为简支梁试验方案，试验采用液压千斤顶和反力加载架加载。在距离构件边缘0.2m位置（有加劲肋处）布置钢管，保证两端铰接约束。在试验构件跨中侧面用约束构件施加侧向支撑，构件与夹肢之间通过螺栓杆固定，并在夹肢与构件之间填塞PTFE板，PTFE板上涂抹黄油，保证构件在竖直方向自由移动。所有构件加载时采用两步加载：第一步预加载至设计屈服荷载的10%，然后停止卸载至零；随后进行第二步正式加载至破坏。分级缓慢加载，每级10kN，至构件屈服后采用位移控制。

构件上、下翼缘贴单向应变片，测量弯曲正应变，腹板的不同高度贴应变化花，腹板应变片均单侧放置。在布置应变片位置时，兼顾腹板波纹的平面和斜面。位移计主要布置在构件的加载点位置的下翼缘，支座位置的竖向及水平方向，以测量构件的加载处的位移和支座的水平、竖向位移。试验前首先对所采用的材料进行了材性试验，测量内容包括：板材厚度、屈服强度、抗拉强度和伸长率等，其中板材厚度和屈服强度实测值见表3。

2.4 试验结果

由于试件较多，选择其中具有代表性的试件描述其试验现象和过程。同时，取构件的剪力V与理论屈服剪力V_y的比值为纵坐标，加载点位移为横坐标，绘制成荷载-位移曲线，这种曲线能够直观而且明确反映试件的力学性能。将有限元程序ANSYS的分析结果在图中标示为虚线作为参照。

2.4.1 GJ1试验现象

GJ1荷载-位移曲线见图3a，可以观察到，当荷载达到约0.8V_y时，曲线出现非线性发展特征，曲线达到极值前有一定的塑性发展过程。试验过程中，荷载达到极值点之后，在构件一侧腹板突然发生整体剪切屈曲，屈曲波纹倾斜角度约为45°，从加载点处下翼缘指向支座处上翼缘，为典型的剪切屈曲形态（图3b所示），由于变形过大，腹板局部出现撕裂。极限剪力大于屈服剪力，证明GJ1的腹板满足屈服强不发生屈曲的基本要求。极值过后，承载力逐渐下降，但保持有一定的屈曲后承载力，其破坏模式属于极值点失稳[28]。为与试验结果进行对比，进行了有限元分析，模拟中采用壳单元SHELL181，材料选用双折线弹塑性模型，弹性模量E取为$206\times10^3\text{N/mm}^2$，切线模量取为$0.01E$，按照第一特征值屈曲模态施加缺陷，局部缺陷（面外变形）的最大值分别取1mm和5mm。

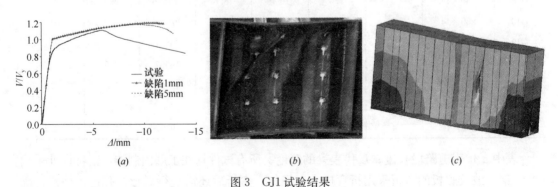

图3 GJ1试验结果

(a) GJ1荷载位移曲线；(b) 破坏形态（试验）；(c) 破坏形态（有限元）

经反复试算，腹板的水平和倾斜板带沿高度划分为 20 个单元，沿跨度方向划分为 3 个单元。翼缘与腹板相交的区域内，单元划分与腹板相互对应。在分析中同时考虑材料非线性和几何非线性。采用自动荷载步，并控制最小荷载步保证足够的精度。

将有限元结果绘入图 3a 中，可以发现，有限元方法得到的试验曲线和破坏形态与试验非常接近，说明有限元分析能够较准确地模拟构件受力过程，对初始刚度和极限承载力的模拟也较为合理，所以有限元方法可以作为试验手段的有效补充。但试验曲线中构件更早进入非线性阶段，而塑性发展过程也较短，说明有限元方法对材料模型仍然有一定误差。同时，从图中可以看到不同的缺陷水平模拟得到的曲线非常接近，证明局部缺陷的影响并不十分显著。而有限元得到的结构最终破坏形态与试验非常接近。

2.4.2 GJ4 试验现象

GJ4 的荷载位移曲线见图 4a，可以看到该构件在达到剪切屈服强度前就发生了破坏，因此属于弹性段内的脆性破坏。试件在破坏前几乎没有塑性发展阶段，破坏形式见图 4b。GJ4 的承载力到达极值点后承载力迅速下降至极限荷载的 50% 左右，并在这个水平的基础上维持。同时，从荷载-位移曲线也可以看到典型的"snap-back（弹性回跳）"现象，破坏模式属于不稳定分岔屈曲。

图 4 中可以看到有限元模拟能够准确反映出这种试验现象，但极限值与试验结果存在较大差异。不同的初始缺陷水平对模拟得到的极限承载力有显著影响，这同样是不稳定分岔屈曲的特征，缺陷值越大承载力越接近试验值，但对屈曲后承载力影响不大。

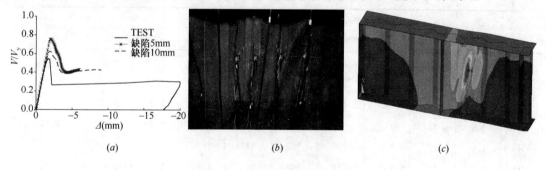

图 4　GJ4 试验结果
(a) 荷载位移曲线；(b) 试验结果；(c) 有限分析结果

2.4.3 GJ13 试验现象

GJ13 采用波形 4，是一种优化的波形，该波形波长较小，波纹较为稠密。从试验曲线图 5a 来看，承载力较高，其承载力达到了腹板的剪切屈服强度，且具有较好的塑性发展能力。GJ13 的破坏形态见图 5b 和图 5c。

综合来看，首先所有试件均为整体屈曲破坏，其次，除采用波形 2 的 GJ4~6 外，其余采用波形 1、3、4 的试件极限剪切应力都达到或超过了材料的屈服强度，受力发展过程和最终的破坏现象也较为类似，都表现为极值点失稳；而 GJ4~6 在弹性段发生了失稳，现象属于弹性回跳。达到极值后，承载力下降较快。因此，可以得到结论，腹板波形对构件受力性能和破坏形态有非常重要的影响，此外，有限元方法可以作为试验的有效补充工具。

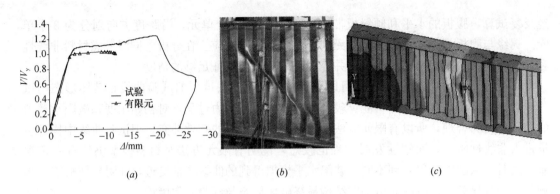

图 5 GJ13 试验结果
(a) 荷载位移曲线；(b) 试验结果；(c) 有限分析结果

2.5 试验结果讨论

为得到波纹腹板 H 型钢抗剪承载力与构件参数的关系，将各构件的试验结果和理论分析结果列在表 3 中进行分析。表中 τ_t 为试验得到的腹板极限剪应力，τ_y 为腹板材料的实测剪切屈服强度，τ_E 为式（5）计算得到的弹塑性极限承载力，τ_{EC3} 为式（6）和式（7）计算得到的较小值，τ_{FEM} 为有限元方法得到的结果。

承载力理论值与试验结果比较   表 3

	t_w (mm)	τ_t (MPa)	τ_y (MPa)	τ_{FEM} (MPa)	τ_E (MPa)	τ_{EC3} (MPa)	$\dfrac{\tau_y}{\tau_t}$	$\dfrac{\tau_{FEM}}{\tau_t}$	$\dfrac{\tau_E}{\tau_t}$	$\dfrac{\tau_{EC3}}{\tau_t}$
GJ1	1.7	126	115	137	115	102	0.91	1.09	0.91	0.81
GJ2	1.7	116	115	120	115	102	0.99	1.03	0.99	0.88
GJ3	1.7	113	115	114	115	102	1.02	1.01	1.02	0.90
GJ4	1.9	81	152	94	152	115	1.88	1.16	1.88	1.42
GJ5	1.9	95	152	109	152	115	1.60	1.15	1.60	1.21
GJ6	1.9	93	152	124	152	115	1.63	1.33	1.63	1.24
GJ9	2.0	153	153	152	153	136	1.00	0.99	1.00	0.89
GJ10	3.0	164	153	167	153	147	0.93	1.02	0.93	0.90
GJ11	2.0	173	153	150	153	131	0.88	0.87	0.88	0.76
GJ12	2.0	157	153	150	153	131	0.97	0.96	0.97	0.83
GJ13	2.0	189	153	158	153	153	0.81	0.84	0.81	0.81

表 3 中，t_w 为腹板实测厚度，从表中可以看出不同构件表现出的不同承载力特性。采用波形 2 的 GJ4～GJ6，试验得到的抗剪承载力显著低于材料的剪切屈服强度，名义剪切屈曲强度仅是材料强度的 60% 左右，而有限元结果得到的结果则介于理论值和试验值之间。分析原因，GJ4～6 波形较为稀疏，各板带之间的相互支撑作用相对较弱，弹性极限强度也相对较低，同时导致破坏更加突然，未达到屈曲荷载前，直接从弹性未屈曲平衡位形转到非邻近的屈曲平衡位形，这与受到轴压作用的圆柱壳非常相似。而且这种屈曲受初始缺陷的影响严重，现有算式无法得到较为接近的解，有限元方法也不能准确模拟实际

构件受力状况，其承载力难以预测。

与此对应的则是，其余构件采用了不同的波形，试验的结果腹板的极限强度均达到或者超过了材料屈服强度，满足了屈服前不发生屈曲这一基本原则。特别留意，与 GJ4～GJ6 一样，GJ9 和 GJ12 的腹板高厚比同为 500，其结果也都达到了材料屈服强度，证明腹板高厚比不是导致 GJ4～6 承载力较低的主要原因。

从上述试验的分析结果，可以发现现有算式对整体稳定极限承载力的计算方法存在一定误差。尤其是对于 GJ4～GJ6，所提出的算式计算结果均远高于试验值。因此本文提出新的计算式（11）作为波纹腹板 H 型钢腹板整体屈曲时抗剪承载力的设计算式：

$$\frac{\tau_{cr}}{\tau_y} = \frac{0.68}{\lambda_s^{0.65}} \leqslant 1.0 \tag{11}$$

将上式与其他算式绘制在一起与试验数据进行对比，图中试验数据除来自本文试验数据外，还包括了现有研究资料中，破坏形式均为腹板整体屈曲的试验结果。

由图 6 可见，本文提出的算式对于大多数试验结果具有较好的预测性和安全性，且形式简单便于采用。同时，图中少数未能落入算式预测范围的试验数据点，主要是本文试验中的 GJ4～6。对于这些因设计不合理而导致腹板在弹性段发生屈曲的波形，可以通过一些构造要求避免此类波形的出现。如，主要的波形尺寸应当控制在：$h_r = 30 \sim 100$mm，$b/(d/\cos\theta) \approx 1$，$\theta = 45° \sim 60°$，$s/q \geqslant 1.15$，$h_w/t_w \leqslant 600$。

从用钢量角度来看，波形 1 的 $s/q=1.25$，波形 2 的 $s/q=1.10$，波形 3 的 $s/q=1.17$，波形 4 的 $s/q=1.22$。由于波纹腹板的用钢量在整个构件的比重已经相当小，所以通过略微提高褶皱率来提高剪切刚度是依然经济的。

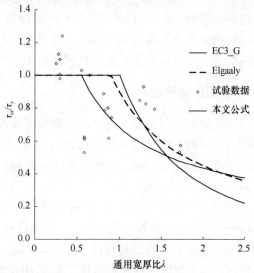

图 6 计算式与试验结果的对比

综合上述分析内容，本文认为波纹腹板 H 型钢的抗剪承载力应按照下列原则进行设计：

（1）若有充分的试验能够证明波纹腹板能够满足屈服前不发生屈曲，则抗剪承载力可以用下式计算：

$$V \leqslant f_v h_w t_w \tag{12}$$

（2）若无试验证明，则抗剪承载力可以用下式计算：

$$V \leqslant \chi_c f_v h_w t_w \tag{13}$$

式中，χ_c 为考虑剪切屈曲影响的承载力折减系数，取 $\chi_{c,l}$ 和 $\chi_{c,g}$ 的较小值：

$$\chi_{c,l} = 1.15/(0.9 + \bar{\lambda}_{c,l}) \leqslant 1.0 \tag{14}$$

$$\chi_{c,g} = 0.68/\bar{\lambda}_{c,g}^{0.65} \leqslant 1.0 \tag{15}$$

式中，$\bar{\lambda}_{c,l} = \sqrt{(f_y/\sqrt{3})/\tau_{cr,l}}$，$\bar{\lambda}_{c,g} = \sqrt{(f_y/\sqrt{3})/\tau_{cr,g}}$。需要注意的是，对于上式的试用要配

合一定波形构造要求才能做到安全经济。

3. 波纹腹板 H 型钢梁的抗弯承载力

在波纹腹板 H 型钢梁的受弯性能研究方面，Elgaaly 等[29]曾进行了 6 根梁的试验和大量有限元分析，认为腹板对受弯承载力的贡献较少，可以忽略。Chan 等[30]用有限元方法研究了腹板波纹的几何尺寸对梁屈曲性能的影响。Johnson 等[31]研究了梁的整体弯曲性能及受压翼缘局部屈曲。Abbas 等[32-34]采用试验及有限元方法，研究了波纹腹板 H 型钢梁的弯曲性能，认为梁的弯曲不能单独按照传统的理论进行求解，还需要对翼缘的横向弯曲进行分析，并提出了相应的求解方法。Jiho 等[35]给出了波纹腹板 H 型钢截面剪切中心和翘曲常数的求解方法，对所提出的弯扭屈曲承载力的计算方法进行了有限元验证。欧洲规范 EuroCode 3 在计算波纹腹板 H 型钢梁受弯承载力时，分别考虑了翼缘的横向弯曲和侧向扭转的影响。郭彦林等[36]提出了波折腹板工形构件翼缘稳定性简化模型，探讨了翼缘弹性屈曲应力与翼缘名义宽厚比、腹板波形的关系。认为基于翼缘屈曲时可能出现的 2 种屈曲模态，分别解释了其发生机理并给出了相应的临界荷载计算算式。

3.1 截面抗弯承载力

上述研究成果中仅文献［31］进行了梁的面内弯曲破坏试验研究，但试件仅在纯弯段采用波纹腹板，剪跨段采用了加劲肋加固的平腹板，与实际使用情况尚存在一定差异。因此有必要进行波纹腹板 H 型钢梁的受弯强度试验研究，从而对理论分析进行进一步的验证。

设计 2 个波纹腹板 H 型钢梁试件，试件腹板尺寸：500mm×3mm，翼缘尺寸：150mm×10mm。腹板采用波形 1。试验已证明该波形腹板剪切屈曲强度可以达到材料的剪切屈服强度。试件钢材设计采用 Q235 钢，实测腹板屈服强度 $f_{wy}=260$MPa，翼缘屈服强度 $f_{fy}=265$MPa。腹板与翼缘之间采用单面角焊缝，焊接方法为 CO_2 气体保护焊，焊丝为 0.8mm。试件的基本参数见表 4。

受弯试件基本参数　　　　　　　　　表 4

编号	$b_f \times t_f$ (mm)	$h_w \times t_w$ (mm)	l_1 (m)	l_2 (m)	f_{fy} (MPa)	I_x (cm^4)
GJ7	150×10	500×3	1.5	1.0	265	19507
GJ8	150×10	500×3	2.0	1.0	265	19507

表 4 中，l_1 和 l_2 分别为剪跨段和纯弯段长度，I_x 为截面惯性矩。试验设计为简支梁两点对称加载。为了防止发生梁的水平侧向扭转变形，在梁跨中设置侧向支撑。在剪跨段，试件上、下翼缘贴单向应变片，腹板的不同高度贴三向应变花。在纯弯段跨中截面，试件上、下翼缘及腹板的不同高度贴单向应变片，腹板应变片均单侧放置。7 个位移计分别测量试件的跨中位移，支座水平位移、竖向位移。

试验过程中首先预加载至 $0.1P_u$（P_u 为试件的预估极限荷载），卸载后正式加载，每 10kN 一级，分级加载到 $0.4P_u$，然后连续加载，每分钟加载 10kN，每秒采样一次。试验在同济大学建筑工程系试验室完成，所采用加载设备最大压力可达到1000kN。位移的测量采用YHD2100型位移传感器，其最大量程为10cm，力和位移的采集均由DH3815型静态应变测试系统完成。通过数据采集和控制系统对试件的荷载-挠度曲线进行监测。

为了与试验结果进行对比，并支持进一步的参数分析，本文采用有限元软件ANSYS进行了波纹腹板H型钢梁受弯性能的有限元分析，模型参数取值同前。板件的几何尺寸和材料的强度按照材性试验的实际结果取值。

GJ7在加载的初始阶段呈线性特征，当翼缘弯曲应力达到材料屈服强度的1.2倍左右时，曲线开始呈现较明显非线性特征，可以明显观察到梁端转角增速加快。当接近破坏荷载时，在纯弯段内，可以观察到梁受压翼缘出现了局部的鼓曲现象，随着荷载逐渐加大，这种现象更加明显。同时，可以观察到侧向支撑的两个夹肢有逐渐分离的趋势，说明夹肢受到了梁面外扭转变形的压力。当荷载达到一定值后，受压翼缘发生波浪形的鼓曲，并造成了腹板的局部变形。随即承载力迅速下降，试件破坏。破坏现象如图7（a）所示，观察破坏现象可知，GJ7的破坏发生在纯弯段，破坏现象为受压翼缘板屈曲后产生的上下交错的波浪变形。

 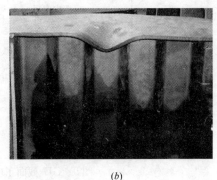

(a) (b)

图7 GJ7和GJ8受弯试验破坏形态
(a) GJ7；(b) GJ8

GJ8的破坏是纯弯段受压翼缘的局部受压向下凹曲，呈现局部屈曲的特征如图7（b）。由于试件变形过大，导致试验设施无法继续工作，所以破坏未能充分发展。但是可以观察到破坏发生在翼缘自由外伸宽度较大的一侧，说明波形腹板对这个部位的翼缘约束较弱。

图8为试件跨中截面翼缘应力与梁端转角（$\sigma_z/f_{fy}-\varphi$）关系曲线，横坐标为试件的梁端转角φ，取为跨中挠度和梁跨度一半的比值，纵坐标为翼缘的弯曲应力σ_z和翼缘屈服强度f_{fy}的比值。为对比方便，将有限元结果同时绘制在图中。

由图8可以看到，有限元方法得到的翼缘极限应力与试验结果较为接近，翼缘的极限应力均超过了材料屈服强度的20%以上，而且初始刚度与试验值也几乎一致。但有限元方法得到的$\sigma_z/f_{fy}-\varphi$曲线显示试件具有较好的塑性性能，这主要是由于有限元方法对约束的施加更为理想，而试验过程中侧向支撑对钢梁的约束有限，钢梁的破坏一定程度受到失稳因素的影响。

为了验证波纹腹板H型钢梁的理论受力模型，以GJ7为例，跨中截面的正应力的实测数据绘制于图9中。可以看出，随着荷载的发展，无论是试验还是有限元方法的数据，都证明了在竖向荷载作用下波纹腹板几乎不产生任何正应力作用的推断。

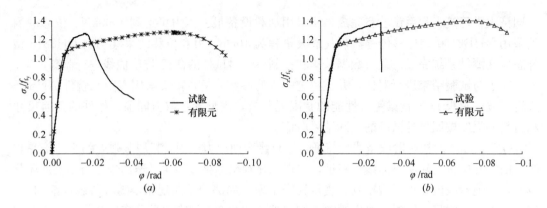

图 8 试件应力比-梁端转角试验曲线
(a) GJ7；(b) GJ8

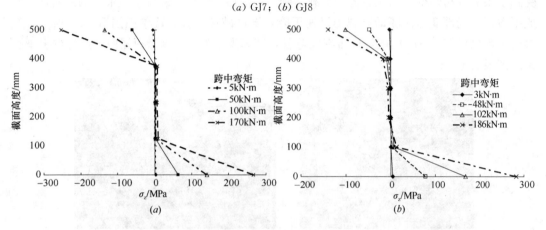

图 9 GJ7 截面正应力发展分布
(a) 试验结果；(b) 有限元分析

将截面的塑性弯矩 M_{cr}、试验结果 M_t 和有限元结果 M_{FEM} 列入表5。可以看到，试验和有限元得到的极限弯矩都大于理论极限塑性弯矩，且幅度都达到20%以上。因此，以截面塑性弯矩作为波纹腹板H型钢梁的受弯承载力设计值是有安全保证的。

理论与试验结果对比 表 5

试件编号	M_{cr} (kN·m)	M_t (kN·m)	M_{FEM} (kN·m)	M_t/M_{cr}
GJ7	203	257	260	1.27
GJ8	203	280	284	1.38

为更深入了解影响波纹腹板H型钢梁受弯承载力的因素，本文用有限元方法进行了参数分析。主要考虑的参数包括：材料屈服后切线模量、腹板波形、翼缘宽厚比、翼缘的强度和腹板高厚比等，具体参数及分析得到的极限弯矩见表6。分析以GJ7、GJ8为基本模型，改变部分参数得到GJ7-1、GJ8-1……GJ7-9、GJ7-10等有限元模型。其中，GJ7-1、GJ8-1的翼缘切线模量 $E_t=0$，GJ7-2 的翼缘切线模量 $E_t=0.005E$；GJ7-3 的腹板采用波形 2 ($h_r=50mm$, $b=70mm$, $d=50mm$)；GJ7-4 采用波形 3 ($h_r=30mm$, $b=40mm$, $d=25mm$)；GJ7-5、GJ7-6 的翼缘厚度分别为12mm、15mm；GJ7-7、GJ7-8 翼缘屈服强

度分别为 235MPa、300MPa；GJ7-9、GJ7-10 腹板高度分别为 750mm，1000mm。

波纹腹板 H 型钢梁参数分析结果 表6

计算模型	波形	h_w (mm)	t_w (mm)	b_f (mm)	t_f (mm)	f_{fy} (MPa)	E_t/E	M_{cr} (kN·m)	M_{FEM} (kN·m)	$\dfrac{M_{FEM}}{M_{cr}}$
GJ7	1	500	3	150	10	265	0.01	203	260	1.28
GJ8	1	500	3	150	10	265	0.01	203	284	1.40
GJ7-1	1	500	3	150	10	265	0	203	211	1.04
GJ7-2	1	500	3	150	10	265	0.005	203	228	1.12
GJ7-3	2	500	3	150	10	265	0.01	203	242	1.19
GJ7-4	3	500	3	150	10	265	0.01	203	291	1.43
GJ7-5	1	500	3	150	12	265	0.01	244	338	1.39
GJ7-6	1	500	3	150	15	265	0.01	307	438	1.43
GJ7-7	1	500	3	150	10	235	0.01	180	250	1.39
GJ7-8	1	500	3	150	10	300	0.01	230	280	1.22
GJ7-9	1	750	3	150	10	265	0.01	302	380	1.27
GJ7-10	1	1000	3	150	10	265	0.01	401	528	1.31
GJ8-1	1	500	3	150	10	265	0	203	227	1.12

通过分析表 6 数据可以观察到：

（1）从材料模型的角度考察，切线模量越小，有限元极限弯矩 M_{FEM} 越接近塑性弯矩 M_{cr}，若采用理想弹塑性材料模型，则两者结果最为接近，这也能够验证文中提出的力学模型的合理性。

（2）腹板波形对极限弯矩的影响具有以下规律：波形越稠密，则 M_{FEM}/M_{cr} 越大。这种现象的原因是由于试件最终的破坏形态为受压翼缘屈服后的屈曲，而稠密的波形能够为翼缘提供更强的约束和支撑作用。

（3）翼缘宽厚比越小，M_{FEM}/M_{cr} 越大。这一点仍然与试件的破坏形态有关，较小的翼缘宽厚比具有更高的局部屈曲强度。

（4）翼缘的屈服强度越小，M_{FEM}/M_{cr} 越大。原因仍然在于最终控制破坏的因素是翼缘的屈曲，所以低屈服点钢材能够更充分的利用材料的强度。

（5）腹板高厚比对极限弯矩无显著影响。

由于波形越稠密，梁的极限弯矩越大，而且从荷载-位移曲线中可以观察到其塑性发展过程更长，所以在选择腹板波形时，在兼顾经济性的前提下，应尽可能的选择较稠密的波形。翼缘宽厚比的减小虽然能够提高极限弯矩，但从经济角度考虑，仍建议满足承载力和局部稳定的构造要求即可。通过试验和有限元分析可知，合理的参数能够有效地提高极限弯矩，但是实际设计工程中，仍建议保守地取截面塑性弯矩作为波纹腹板 H 型钢梁的设计弯矩：

$$\frac{M_x}{W_{nx}} \leqslant f \tag{16}$$

式中，$W_{nx}=b_f t_f h$ 为净截面塑性抵抗矩。

3.2 整体稳定承载力

Lindner[37]研究了波纹腹板 H 型钢的侧向扭转性能,研究认为截面的扭转常数与平腹板钢梁相同,但截面的翘曲常数是不同的:

$$I'_w = I_w + c_w L^2/(E\pi^2) \tag{17}$$

式中,I'_w 为波纹腹板 H 型钢截面的翘曲常数,$I_w = t_f b_f^3 h_w^2/24$,为平腹板 H 型钢截面的翘曲常数,$c_w = (h_r^2 h_w^2)/(8\beta(b+d))$,$\beta = h_w/(2Gbt_w) + h_w^2(b+d)^3/(25b^2 E b_f t_f^3)$,$L$ 为梁的跨度。

对于波纹腹板钢梁的整体稳定承载力,Lindner 提出用下式来计算:

$$M_u = \sqrt[n]{\frac{1}{1+\overline{\lambda}_M^{2n}}} M_p \tag{18}$$

式中,M_p 为全截面塑性弯矩,n 为计算参数,对热轧型钢梁 $n=2.5$,焊接梁 $n=2.0$,$\overline{\lambda}_M = (M_p/M_k)^{0.5}$,为通用长细比,$M_k$ 为弹性侧向弯扭屈曲弯矩。

欧洲规范 EUROCODE 3[38]采用下列算式波纹腹板钢梁的抗弯承载力:

$$M_{Rd} = \min \begin{cases} \dfrac{b_2 t_2 f_{wy,r}}{\gamma_{M0}} \left(h_w + \dfrac{t_1+t_2}{2}\right) \\ \dfrac{b_1 t_1 f_{wy,r}}{\gamma_{M0}} \left(h_w + \dfrac{t_1+t_2}{2}\right) \\ \dfrac{b_1 t_1 \chi f_{wy}}{\gamma_{M1}} \left(h_w + \dfrac{t_1+t_2}{2}\right) \end{cases} \tag{19}$$

式中,γ_{M0} 和 γ_{M1} 均为抗力分项系数,b_1、b_2、t_1、t_2 分别为上、下翼缘的宽度和厚度。$f_{wy,r}$ 为翼缘屈服强度,此强度考虑了翼缘的横向弯矩引起的强度的降低:

$$f_{wy,r} = \left(1 - 0.4\sqrt{\frac{\sigma_x(M_z)}{f_{yT}/\gamma_{M0}}}\right) f_{wy} \tag{20}$$

式中,$\sigma_x(M_z)$ 为横向弯矩在翼缘内引起的应力,χ 是由于侧向扭转屈曲引起的强度折减系数:

$$\chi = \frac{1}{\Phi_{LT} + \sqrt{\Phi_{LT}^2 - \lambda_{LT}^2}} \leqslant 1 \tag{21}$$

$$\Phi_{LT} = 0.5[1 + \alpha_{LT}(\lambda_{LT} - 0.2) + \lambda_{LT}^2] \tag{22}$$

式中,α_{LT} 为弯扭屈曲初始缺陷系数,对于波纹腹板 H 型钢,可以按照 c 类屈曲曲线取 $\alpha_{LT} = 0.49$,λ_{LT} 为通用长细比。

Jiho Moon 认为截面翘曲常数是沿梁长度方向周期性变化的,所以为了简化计算,提出用平均波纹高度来计算截面的翘曲常数,其中平均波高用下式计算:

$$e_{avg} = \frac{(2b+d)h_r}{4(b+d)} \tag{23}$$

Zeman 公司的产品技术手册[39]认为,计算波纹腹板钢梁的整体稳定,可以忽略腹板对受压翼缘的约束作用,将受压翼缘作为"独立"的轴压杆件进行分析,借鉴德国规范的相关算式:

$$N_u = \frac{0.5\pi}{\sqrt{12}} \sqrt{E \cdot f_{yk}} \frac{b_f^2 \cdot t_f}{k_c \cdot c} \quad (24)$$

式中，N_u 为受压翼缘的稳定承载力；f_{yk} 为翼缘的强度标准值；k_c 为规范中规定的压力系数；c 为受压翼缘侧向支撑之间的间距。从上式也可以反推出不需要考虑整体稳定的侧向支撑间距。这种方法概念简单，但可能低估钢梁稳定承载力。

对波纹腹板钢梁整体稳定性的研究，首先需要确定弹性屈曲稳定承载力的计算方法，进而确定梁弹塑性稳定承载力设计算式。

3.2.1 弹性阶段整体稳定承载力

目前对波纹腹板钢梁弹性稳定的主要研究成果是对截面翘曲常数的求解。对于截面翘曲常数的算法有若干种，其中 Lindner 提出的算式（17）是目前最受认可的方法，但该式较为复杂，且将翘曲常数定为梁跨度的函数，不同长度、不同截面的梁的翘曲常数是不同的，所以不方便使用。Jiho Moon 提出了取平均波高的方法，简化了计算。但本文认为将翘曲常数沿长度方向进行平均更为合理。下面以图 10 所示的单轴对称截面为例：

若假设图中 S 点为剪力中心，由于腹板上剪应力均匀分布，所以，当腹板偏离上下翼缘中心连线距离 e 时，则 S 点到腹板的距离也等于 e。若以剪力中心 S 为扇性主极点（原点），腹板中点 M_0 点为扇性零点。则截面的扇性坐标 $\overline{\omega}_d$ 如图 11 所示。

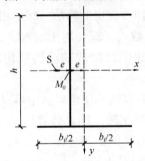
图 10 截面参数图

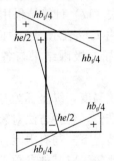
图 11 截面扇性坐标

截面翘曲常数可以通过下式计算得到：

$$I'_w = \frac{t_f h^2 b_f^3}{24} + \frac{t_w h^3 e^2}{12} = I_w + \frac{t_w h^3 e^2}{12} \quad (25)$$

由式（25）可见，腹板的偏移可以增大截面的翘曲常数，波纹波高越大，截面翘曲常数越大。若将 Jiho Moon 提出的平均波纹高度式（23）代入式（25）可得：

$$I'_w = I_w + \frac{t_w h^3 h_r^2}{12} \left(\frac{2b+d}{2q}\right)^2 \quad (26)$$

当腹板位于翼缘中心连线时（$e=0$），其翘曲常数等于双轴对称工字型截面，当腹板位于波纹的波峰或波谷位置时：

$$I'_w = I_w + \frac{t_w h^3 h_r^2}{48} \quad (27)$$

由于波纹腹板钢梁截面沿长度方向呈周期性的变化，采用一个周期内翘曲常数的加权平均值作为平均截面翘曲常数：

$$I^*_w = \left(\int_0^q \frac{t_f h^2 b_f^3}{24} + \frac{t_w h^3 e^2}{12} dx\right)/q = I_w + \frac{t_w h^3 h_r^2}{12} \frac{(b+d/3)}{2q} \quad (28)$$

算式（27）比算式（28）形式较为简单，其基本思想与算式（23）较为类似。

为了比较上述截面翘曲常数的计算方法，本文以受纯弯曲作用的波纹腹板简支梁为基本模型，根据弹性稳定的相关理论，其弹性屈曲极限弯矩为：

$$M_{cr} = \frac{\pi}{l}\sqrt{EI_y\left(GI_t + EI_w\frac{\pi^2}{l^2}\right)} \tag{29}$$

通过有限元方法求解出弹性稳定承载力 M_{cr}，可以反推出截面翘曲常数：

$$I_w = l^2\left(\frac{M_{cr}^2}{EI_y}\frac{l^2}{\pi^2} - GI_t\right)/(\pi^2 E) \tag{30}$$

若以上式得到的结果作为基准，可以对比各种算法的准确程度。在有限元分析过程中，首先考察不同波纹高度情况下，截面翘曲常数的对比情况。按照 Jiho 文中提供的构件参数建立模型：$b=330\text{mm}$，$d=270\text{mm}$，$t_w=12\text{mm}$，$h_w=2000\text{mm}$，$b_f=500\text{mm}$，$t_f=40\text{mm}$，$L=15600\text{mm}$，波高 h_r 包括 50、100、200、250、350 共 5 种情况。

有限元模型中，仅在梁端上、下翼缘分别施加水平方向拉压荷载，同时在两端支座截面设置约束，使其满足"夹支"边界条件，使支座截面可以自由翘曲，但不能绕转动，且不能侧向移动。在此条件下，可以得到波纹腹板钢梁的弹性弯扭屈曲第一阶特征值模态为典型的弯扭屈曲。将有限元得到 M_{cr} 带入式（30）计算得到的翘曲常数用 I_{w-F} 表示，而将按照 Lindner 提出的式（17）计算结果用 I_{w-L} 表示，Jiho 提出的算法（26）用 I_{w-M} 表示，本文提出的式（28）用 I_{w-Z} 表示，将截面最大翘曲常数式（27）用 $I_{w-Z'}$ 表示。随后将有限元方法获得的弹性屈曲极限结果与各算法结果的比值绘制成随波高变化的曲线（图12）。

图中横坐标为波纹高度，纵坐标为有限元结果与各算法的比值。从中可以看出，若以有限元计算结果为标准值，各种方法对翘曲常数的计算比较准确，各算法精度较为接近，尤其是在波高较小的情况下。与有限元分析结果的差值比例都在 $\pm 10\%$ 之内。

为了进一步进行比较，取其他参数不变，取较大波高的情况：$h_r=350\text{mm}$，分析当梁的长度从 12m 变化至 17.6m 时，有限元结果和各算法的比值，并绘制成图13。从可以看到 Lindner 提出的式（17）算法最为准确。算式（28）理论上更完备，当实际情况却是算式（27）更准确，考虑该算式形式简单，准确度可以接受，因此本文认为可以将算式（27）作为波纹腹板 H 型钢截面翘曲常数的计算方法。

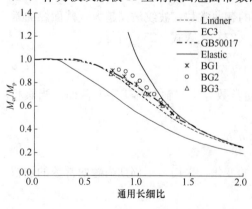

图 12 翘曲常数计算方法比较 1

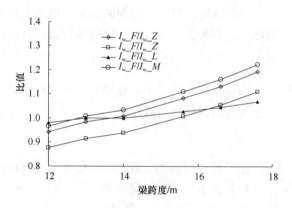

图 13 翘曲常数计算方法比较 2

3.2.2 弹塑性极限稳定承载力

本文拟通过一系列有限元分析，分析波纹腹板 H 型钢梁弹塑性稳定承载力。有限元模型采用 3 类构件，分别用 BG1、BG2 和 BG3 表示，其基本参数见表 7。

整体稳定有限元分析模型参数　　　　　　　表 7

	b (mm)	d (mm)	h_r (mm)	h_w (mm)	t_w (mm)	b_f (mm)	t_f (mm)	f_y (MPa)	L (m)
BG1	180	140	100	1500	8	300	25	250	5～11
BG2	330	270	200	2000	12	500	40	250	11～18
BG3	64	23.5	38	500	1.73	200	9.63	235	4～8

每一类构件都通过改变梁的跨度 L 来改变弹性承载力，进而改变通用长细比，BG1、BG2 和 BG3 分别包含 6、8、5 个试件。需要注意的是，梁的跨度不能过短，否则第一阶屈曲模态可能为受压翼缘的局部屈曲，与分析目标不符。在有限元分析中，模型为承受纯弯曲的简支梁，单元类型为 SHELL181，梁端两个截面均设置理想夹支约束，材料采用切线模量为 $0.01E$ 的双折线模型，经模拟得到各个试件的极限弯矩用 M_{cr} 表示，并将其与塑性弯矩的比值作为纵坐标绘制在图 14 中。

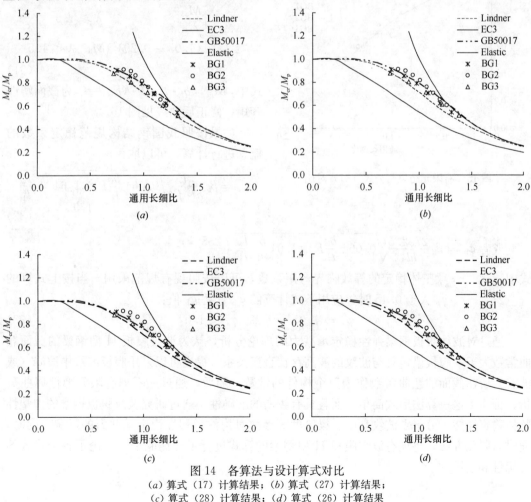

图 14　各算法与设计算式对比
(a) 算式（17）计算结果；(b) 算式（27）计算结果；
(c) 算式（28）计算结果；(d) 算式（26）计算结果

图中横坐标为梁的通用长细比 $\lambda = (M_p/M_{ecr})^{0.5}$，其中梁的弹性稳定承载力 M_{ecr} 是分别通过不同方法的截面翘曲常数 I_w 计算得到的，分别对应图 14 (a)、(b)、(c)、(d)。每个小图中 4 条曲线分别为 Lindner、EUROCODE 3、GB50017 的设计算式，及弹性屈曲曲线，散点为有限元分析得到试验点。

从上图的对比中可以发现，有限元计算结果都低于弹性理论弯矩，而 EUROCODE 3 给出的算式是基于 c 类梁屈曲曲线的，显得过于保守。从图形中看我国规范现有设计算式若直接用于波纹腹板梁略偏不安全，需要修正。虽然截面翘曲常数算法的准确性各有差异，但通过上图的对比可以发现，若用于弹塑性稳定设计，各种算法其实差别很小。因此，在波纹腹板 H 型钢梁的设计中，本文依然建议采用算式（27）来计算梁的翘曲常数。

此外，若将上述所有试件用式（24）计算，得到的结果均远小于有限元分析结果，所以在设计使用中不建议将翼缘作为孤立构件来衡量梁的整体稳定性。

Lindner 的设计算式较为合理，能对大多数试验点进行有效预测，但也存在计算结果高于有限元结果的情况，且在实践过程中需要配合使用其翘曲常数计算公式，所以本文建议，波纹腹板钢梁的极限稳定承载力采用下列修正后的式子进行计算：

$$M \leqslant \begin{cases} \dfrac{M_p}{\lambda^2} & \lambda \geqslant 1.5 \\ (1.05 - 0.29\lambda^2)M_p & \lambda < 1.5 \end{cases} \quad (31)$$

式中，$M_p = b_f \cdot t_f \cdot f \cdot (h_w + t_f)$ 为截面塑性弯矩；修正后曲线见图 15。

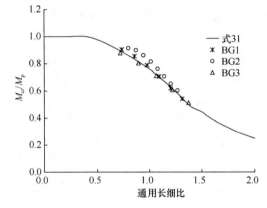

图 15 有限元结果与公式（31）对比

如果按照我国钢结构规范稳定系数的概念进行计算，可以取：

$$\varphi_b = \beta_b \frac{\pi E}{h l_1 f_y} \sqrt{0.064 \frac{b_f}{t_f} I_t + 1.64 \frac{b_f}{t_f} \frac{I_w}{l_1^2}} \quad (32)$$

或者 $\varphi_b = \beta_b \dfrac{2752.5}{h l_1} \sqrt{0.064 \dfrac{b_f}{t_f} I_t + 1.64 \dfrac{b_f}{t_f} \dfrac{I_w}{l_1^2}} \cdot \dfrac{235}{f_y}$

式中 β_b——梁整体稳定的等效临界弯矩系数，可以按照现有规范采用。当按上式算的 φ_b 大于 0.45 时，应用下式计算的 φ_b' 代替 φ_b 值：

$$\varphi_b' = 1.05 - 0.29/\varphi_b \leqslant 1.0 \quad (33)$$

通过对波纹腹板钢梁弹性稳定承载力的理论分析，认为波纹腹板 H 型钢梁的截面翘曲常数大于平腹板梁，且与波纹的波高存在直接关系，提出可以采用腹板波纹中波峰（或波谷）所在截面的翘曲常数作为整个构件的常数。随后，通过一系列有限元弹性屈曲分析，证实了这一算法形式简单，而且具有较高的准确性。通过研究波纹腹板钢梁的弹塑性整体稳定性能，分析了波纹高度、梁跨度等参数对构件弹塑性稳定承载力的影响。同时，通过有限元方法，将现有波纹腹板 H 型钢梁的算式进行了对比验证，讨论了各个算式的准确性和适用性。

4. 楔形波纹腹板 H 型钢梁的整体稳定

楔形波纹腹板 H 型钢是对这种型钢的进一步优化，其在单层轻型工业厂房中应用日渐广泛。楔形波纹腹板 H 型钢梁的弯扭失稳理论非常复杂，有必要进行深入研究。

对于最常用的楔形构件，当其翼缘尺寸不变，仅腹板高度随着轴线线性变化，因其几何参数随着轴线变化，用能量法很难求解，一般采用数值方法计算[40]。现有数值方法是基于平腹板的情况推导得到的，对楔形波纹腹板 H 型钢梁尚不能确定是否适用[41-43]。

本文拟通过求解楔形梁在不等端弯矩作用下的平衡微分方程，用数值方法求解，最终拟合得到翼缘尺寸不变，仅腹板高度随轴线线性变化的楔形波纹腹板 H 型钢梁的弹性临界弯矩计算算式。

4.1 弹性稳定承载力

4.1.1 平衡方程

波纹腹板 H 型钢梁截面翘曲常数建议采用（27）的计算表达式，其扭转常数与平腹板相同。根据薄壁结构理论，H 型钢截面扭转常数表达为轴线坐标 z 的函数（图 16）：

$$I_k(z) = I_{k0}(1 + d_k\, z/l) \tag{34}$$

$$d_k = (h_0 + t_f)t_w^3 \gamma/(3 I_{k0}) \tag{35}$$

$$\gamma = H_l/H_0 - 1 = (h_l + t_f)/(h_0 + t_f) - 1 \tag{36}$$

其中，I_{k0} 为小端截面的惯性矩；h_0，h_l，分别为两个截面上、下翼缘形心距离；H_0 和 H_l 分别为两个截面的全高；γ 为楔形梁的楔率。对于 H 型钢梁，其绕弱轴的惯性矩 I_y 主要由翼缘提供。为了简化计算，可偏于安全地近似取作：

$$I_y = b_f^3 t_f/6 \tag{37}$$

如图 16 所示，双轴对称楔形波纹腹板 H 型钢梁，在不等端弯矩作用下，其在刚度大的 yz 平面内承受沿轴线变化的弯矩。现采用固定的右手坐标系 x，y，z 和移动坐标系 ξ，η，ζ，截面的形心 O 和剪心 S 都在对称轴 y 上。当构件在弯矩作用的平面外有微小的侧扭变形时，任意截面的变形和受力如图 16b 和 c。

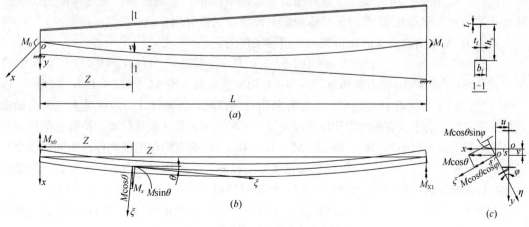

图 16　简支楔形梁内力与弯扭变形

根据经典弹性稳定理论，对于图 16 所示楔形构件，做如下基本假定：（1）构件为弹性体；（2）侧扭变形时，构件截面形状不变；（3）构件侧扭变形微小；（4）忽略构件在弯矩作用平面内的变形；（5）不考虑残余应力。

根据以上假定，S Kitipornchai[44]推导了渐变工字钢构件（腹板高度、翼缘宽度均发生变化）的弹性稳定微分方程：

$$EI_y u'' + M_x \phi = 0 \tag{38}$$

$$M_x u' - Q_y u = \left[EI_k + \frac{3}{2}\lambda^2 EI_w - \lambda \frac{d}{dz}(EI_w)\right]\frac{d\phi}{dz} - \frac{d}{dz}(EI_w)\frac{d^2\phi}{dz^2} - EI_w \frac{d^3\phi}{dz^3} \tag{39}$$

$$\lambda = \frac{2}{h}\frac{dh}{dz} \tag{40}$$

对双轴对称楔形波纹腹板 H 型钢梁，截面弯矩可以表示为：

$$M_x = M_0 + (M_l - M_0)/z = M_0 + Q_y z \tag{41}$$

翼缘形心距变化率：

$$\gamma' = h_l/h_0 - 1 = \gamma(h_0 + t_f)/h_0 \tag{42}$$

翼缘形心距：

$$h = h_0 + \frac{h_l - h_0}{l}z = h_0\left(1 + \gamma' \frac{z}{l}\right) \tag{43}$$

所以：

$$\frac{dh}{dz} = \frac{\gamma' h_0}{l} \tag{44}$$

对式（38）微分二次，对式（39）微分一次，并结合式（27）、式（34）、式（40）～式（44）整理可得适合于任意边界条件的弯扭平衡方程：

$$EI_y u^{IV} + M_x \phi'' + 2Q_y \phi' = 0 \tag{45}$$

$$E(I_w + I_{ww})\phi^{IV} + \left(\frac{4h_0 \gamma' EI_w}{lh} + \frac{5h_0 \gamma' EI_{ww}}{lh}\right)\phi''' + \left(\frac{2h_0^2 \gamma'^2 EI_{ww}}{l^2 h^2} - GI_k\right)\phi'' - \left(\frac{2h_w^3 \gamma'^3 EI_{ww}}{l^3 h^3} + \frac{Gh_w \gamma' t_w^3}{3l}\right)\phi' + M_x u'' = 0 \tag{46}$$

式（39）微分过程中出现的 $Q_y u'$ 为平行于剪心轴的力，不产生扭矩，故在式（46）中不出现，I_{ww} 相关的项为波纹腹板对翘曲常数的贡献量产生的影响，也是楔形波纹腹板 H 型钢梁与楔形平腹板 H 型钢梁整体稳定平衡方程的差异所在。

对图 16 所示两端简支的楔形梁，其边界条件为：

$$u(0) = u''(0) = 0; \phi(0) = \phi''(0) = 0; u(l) = u''(l) = 0; \phi(l) = \phi''(l) = 0 \tag{47}$$

式（38），式（39）没有解析解，需要用数值方法求解。设 $M_0 = kM_l = kM$，则 $Q_y = (1-k)M/l$。对于变截面 H 型钢梁，为充分利用材料强度，截面设计时应使最大应力沿梁轴线接近于常数。对于仅有端弯矩作用的简支梁，一种可能的最优设计是要求构件达侧扭屈曲时大端的最大应力与小端相同，即 $M_0/M_l = W_{x0}/W_{xl}$（两端弯矩作用使构件产生同向曲率）。因此，取 $-W_{x0}/W_{xl} \leqslant k \leqslant W_{x0}/W_{xl}$ 可以囊括实际工程的大多数情况。由于在计算其绕强轴的截面抵抗矩时，腹板作用可以忽略，所以，实际求解时，取 $-h_0/h_l \leqslant k \leqslant h_0/h_l$ 即可。

利用式（45）～式（47）编写有限积分程序可以求解楔形波纹腹板 H 型钢梁的弹性整体稳定承载力。

4.1.2 算式拟合及验证

通过壳体有限元理论或能量法得到楔形工字钢梁的稳定临界荷载表达式：

$$M_{crl} = \beta_1 \frac{\pi^2 EI_y}{l^2} \sqrt{\frac{I_{w0}}{I_y}\left[A_\gamma + \frac{GI_{k0}(1+0.5d_k)l^2}{E\pi^2 I_{w0}}\right]} \tag{48}$$

$$A_\gamma = 1 + 1.6\gamma + 0.4\gamma^2 \tag{49}$$

$$\beta_1 = 1.84 - 0.84\sin[(W_{xl}/W_{x0})^{0.15} \times 0.5k\pi] \tag{50}$$

式（48）形式上与 GB 50017[45]中等截面 H 型钢梁在不等端弯矩作用下的临界弯矩表达式相同，概念明确，且只需要令 $\gamma=0$，$d_k=0$，$W_{xl}/W_{x0}=1$ 算式即可自动退化为等截面梁在不等端弯矩作用下的临界荷载算式，应用非常方便。正是基于以上优点，本文在式（48）～式（50）的基础上，通过有限元分析计算，进行回归分析，调整相关系数，得到适用于常用规格楔形波纹腹板 H 型钢梁的临界弯矩计算算式。数值拟合主要包括以下两个方面：

（1）弯矩系数 β_1 的调整：对常用规格等截面波纹腹板 H 型钢梁在不等端弯矩作用下的临界弯矩进行了大量的数值计算，利用式（51）的形式进行回归分析。弯矩系数的拟合过程如图 17 所示。

对于变截面的情形，仍采用式（51）的形式，在算式中考虑 W_{xl}/W_{x0} 项，只是结合波纹腹板 H 型钢的特点，对算式进一步简化如下：

$$\beta_1 = 1.88 - 0.88\sin[(h_l/h_0)^{0.15} \times 0.5k\pi] \tag{51}$$

（2）楔率修正项 A_γ 调整：对常用规格变截面波纹腹板 H 型钢在楔率小于 2 的范围内进行大量的回归分析。分析后认为常用规格变截面波纹腹板 H 型钢在楔率小于 2 的范围内其楔率修正项可调整如下（图 18）：

$$A_\gamma = 1 + 2\gamma - 0.1\gamma^2 + 0.15\gamma^3 \tag{52}$$

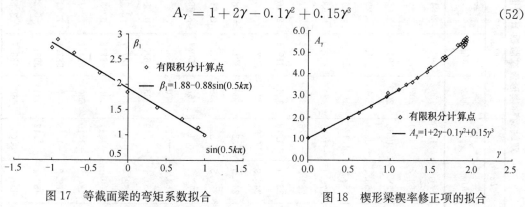

图 17　等截面梁的弯矩系数拟合　　　　图 18　楔形梁楔率修正项的拟合

综上所述，本文建议常用规格楔形波纹腹板 H 型钢在不等端弯矩作用下侧扭失稳的临界弯矩的计算算式可以表示为：

$$M_{crl} = \beta_1 \frac{\pi^2 EI_y}{l^2} \sqrt{\frac{I_{w0}^*}{I_y}\left[A_\gamma + \frac{GI_{k0}(1+0.5d_k)l^2}{E\pi^2 I_{w0}^*}\right]} \tag{53}$$

式中，β_1 按式（51）计算，A_γ 按式（52）计算，且 $0 \leqslant \gamma \leqslant 2$，$I_{w0}^*$ 为小端截面处的翘曲常数。

下面将由算式（53）计算所得的结果 M_{crl} 和采用通用有限元程序 ANSYS 特征屈曲分析得到 M_{FE} 进行对比，其中 k 分别取 h_0/h_l，$0.5h_0/h_l$，0，$-0.5h_0/h_l$，$-h_0/h_l$，对比结果见表 8，弯矩单位为 kN·m。表 8 中所列构件，其腹板波形均采用表 1 中所列波形 3。ANSYS 进行特征屈曲分时，基本参数与前同。

表 8 双轴对称楔形波纹腹板 H 型钢梁临界弯矩对比（$L=12\mathrm{m}$）

截面规格	$k=h_0/h_l$			$k=0.5h_0/h_l$			$k=0$			$k=-0.5h_0/h_l$			$k=-h_0/h_l$		
	M_{crt}	M_{FE}	误差 %	M_{crt}	M_{FE}	误差 %	M_{crt}	M_{FE}	误差 %	M_{crt}	M_{FE}	误差 %	M_{crt}	M_{FE}	误差 %
CWB500~1500-200×10	169	171.0	-1.15	199	197.6	0.84	233	230.3	1.05	266	269.0	-1.05	296	303.9	-2.47
CWB750~1500-200×10	162	163.2	-0.94	202	197.5	2.45	252	243.8	3.56	303	302.2	0.14	343	355.1	-3.32
CWB1000~1500-200×10	157	158.4	-0.68	204	199.0	2.61	271	268.4	1.15	339	337.4	0.40	386	399.5	-3.47
CWB1250~1500-200×10	155	155.1	-0.31	201	201.0	0.20	284	277.3	2.57	367	373.4	-1.59	414	423.0	-2.07
CWB1500~1500-200×10	154	152.5	1.17	194	201.0	-3.43	290	288.0	0.74	386	399.2	-3.27	426	427.2	-0.30
CWB500~1500-220×12	271	275.4	-1.55	320	317.9	0.58	374	370.2	0.91	427	431.8	-1.03	476	497.0	-4.23
CWB750~1500-220×12	259	262.3	-1.26	324	317.0	2.27	405	390.8	3.54	485	483.5	0.33	550	573.9	-4.11
CWB1000~1500-220×12	251	254.4	-1.18	326	319.1	2.26	434	423.6	2.44	542	538.6	0.54	616	636.9	-3.21
CWB1250~1500-220×12	247	248.9	-0.95	321	322.1	-0.31	454	446.8	1.51	586	594.8	-1.49	661	672.8	-1.82
CWB1500~1500-220×12	246	244.9	0.27	309	321.3	-3.87	462	459.9	0.39	614	630.6	-2.56	678	677.8	0.00
CWB500~1500-250×12	389	387.5	0.52	459	447.6	2.61	537	521.6	2.88	614	609.0	0.81	684	701.4	-2.52
CWB750~1500-250×12	373	372.3	0.15	467	450.1	3.69	583	565.0	3.10	698	686.8	1.68	792	814.9	-2.78
CWB1000~1500-250×12	362	363.3	-0.27	470	455.6	3.23	625	605.3	3.32	780	768.5	1.56	888	907.2	-2.07
CWB1250~1500-250×12	355	357.1	-0.49	463	461.7	0.25	654	635.7	2.84	845	851.0	-0.75	952	959.8	-0.80
CWB1500~1500-250×12	354	352.4	0.38	445	457.7	-2.79	665	660.5	0.69	885	916.0	-3.37	976	968.1	0.85
CWB500~1500-250×15	502	509.1	-1.41	592	587.3	0.83	692	683.2	1.32	792	796.0	-0.47	883	915.3	-3.58
CWB750~1500-250×15	479	484.2	-1.10	600	584.6	2.57	749	729.5	2.63	898	888.5	1.05	1019	1052.8	-3.25
CWB1000~1500-250×15	464	469.2	-1.15	602	587.8	2.43	801	780.2	2.64	1000	987.0	1.27	1138	1164.1	-2.26
CWB1250~1500-250×15	454	459.1	-1.12	591	593.0	-0.29	835	811.9	2.87	1079	1087.2	-0.74	1216	1225.1	-0.70
CWB1500~1500-250×15	452	451.5	0.00	568	588.7	-3.53	849	843.4	0.65	1130	1165.7	-3.08	1246	1229.8	1.33

从以上各表临界弯矩对比可知，式（53）具有较高精度，与ANSYS特征屈曲分析结果相比，误差均在5%以内，满足工程要求，且式（53）形式与等截面梁算式一样，便于工程应用。

4.2 弹塑性稳定承载力

对于变截面波纹腹板H型钢梁的弹塑性稳定承载力，本文拟通过有限元软件ANSYS进行分析，验证式（31）是否适用，随后用试验方法进一步验证。

本文弹塑性分析模型的腹板采用波形3，在楔形波纹腹板H型钢梁端部上下翼缘分别施加轴向力偶以模拟端弯矩作用。其他参数设置与前同。

本文弹塑性分析涉及荷载和边界条件包括以下五大类：（1）不等端弯矩作用下的简支梁；（2）跨中集中荷载作用下的简支梁；（3）满跨均布荷载作用下的简支梁；（4）自由端集中荷载作用下的悬臂梁；（5）满跨均布荷载作用下的悬臂梁。

图19给出了ANSYS弹塑性稳定分析计算结果和利用式（31）计算结果的对比情况。其中图19（a）各有限元计算点所用算例截面规格如表9所示，图19（b）～（e）各有限元计算点所用算例截面规格如表10所示。

图19（a）算例截面规格说明　　　　　　　　　　　　　　表9

h_0（mm）	h_l（mm）	t_w（mm）	$b_f \times t_f$（mm×mm）			
500	1500	3	200×10	220×12	220×15	250×15
750	1500	3	200×10	220×12	220×15	250×15
1000	1500	3	200×10	220×12	220×15	250×15
1250	1500	3	200×10	220×12	220×15	250×15
1500	1500	3	200×10	220×12	220×15	250×15

注：1. 对每种规格的截面，分别取8.16m，10.08m和12m三种跨度进行分析。

2. 每种规格截面每种跨度下，弯矩系数分别取$k=h_0/h_l$，$0.5h_0/h_l$，0，$-0.5h_0/h_l$，$-h_0/h_l$五种情况进行分析。

图19（b）～（e）算例截面规格说明　　　　　　　　　　　　表10

h_0（mm）	h_l（mm）	t_w（mm）	$b_f \times t_f$（mm×mm）			
300	900	3	200×10	220×12	220×15	250×15
450	900	3	200×10	220×12	220×15	250×15
600	900	3	200×10	220×12	220×15	250×15
750	900	3	200×10	220×12	220×15	250×15
900	900	3	200×10	220×12	220×15	250×15

注：对每种规格的截面，分别取6.24m，8.16m，10.08m和12m四种跨度进行分析。

从图19可知，楔形波纹腹板H型钢梁的弹塑性稳定极限承载能力的有限元计算结果均不小于式（31）计算的结果，且随着通用长细比λ的减小，有限元分析结果与式（31）计算的结果相比呈增大的趋势，但整体上偏差并不大。因此，用式（31）计算楔形波纹腹板H型钢梁弹塑性整体稳定极限承载能力结果是偏于安全的。

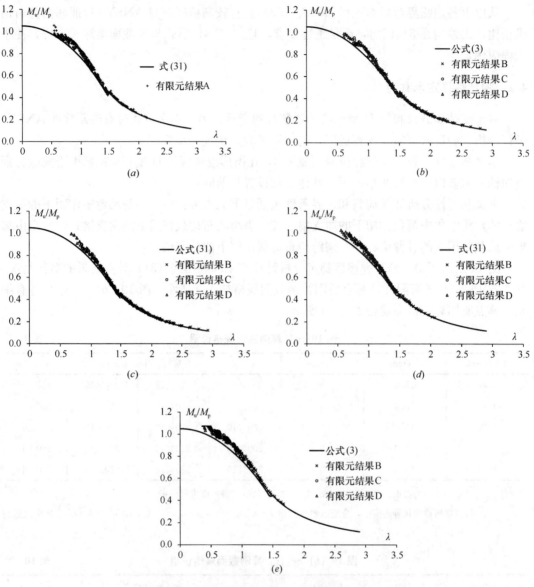

图 19 有限元计算结果与公式（31）的对比
(a) 端弯矩作用下的简支梁；(b) 横向集中荷载作用下的简支梁；
(c) 横向均布荷载作用下的简支梁；(d) 横向集中荷载作用下的悬臂梁；
(e) 横向均布荷载作用下的悬臂梁

4.3 试验验证

受弯构件稳定试验的关键和难点主要有二：其一是边界条件的模拟，其二是荷载条件的确定[46-48]。相对而言，悬臂梁稳定试验边界条件较为明确。对于单点加载的稳定试验，最佳的加载方式是在加载点悬挂重物，但这样受到重物体量限制，无法进行大吨位试验。本文即采用在悬臂梁端部悬挂重物的方式加载。利用试验室现有的条件，自制钢筋笼盛放重物（钢筋笼自重约 330kg），加载时利用吊车进行吊装，在此情况下，最大加载吨位可达 30kN。

4.3.1 试验概况

共5个试件,腹板的波形尺寸为,$h_r=40mm$,$b=63mm$,$d=31mm$。试件均采用大端截面处固定,小端为自由端,长度取3.0m和4.0m。试件具体参数见表11。

试件几何参数   表11

参数 编号	$b_f \times t_f$ (mm×mm)	t_w (mm)	h_w (mm)	楔率 γ	楔角 θ	L /m
TPL1	110×8	2.0	250~400	0.564	2.5%	3.0
TPL2	110×8	2.0	250~400	0.564	2.5%	3.0
TPL3	110×8	2.0	250~400	0.564	2.5%	3.0
TPL4	110×8	2.0	200~400	0.926	2.5%	4.0
TPL5	110×8	2.0	300~400	0.316	1.25%	4.0

采用在梁自由端悬挂重物的方法加载。加载点设置90mm宽加劲板,通过四根绳索悬挂吊篮,用于加载。实际加载时依据试验室中废钢板重量确定每级荷载的大小。第一级荷载为钢筋笼自重(约330kg),此后先加载150kg钢板若干块(对于试件TPL1~TPL3,加11块;TPL4~TPL5加3块);加载完150kg钢板后,继续加载36kg钢板12块;36kg钢板加载完成后,采用23kg重物进行加载,直到试件破坏。试验测量包括两部分内容:应变和位移。应变片共10个测点(S1~S10),其中S3和S8为直角应变花,其余为应变片。位移计共9个测点(D1~D9),分别测量距离固定端$L/4$、$L/2$和$3L/4$处截面的挠度、面外位移和侧向扭转角。位移计和应变片布置如图20所示。

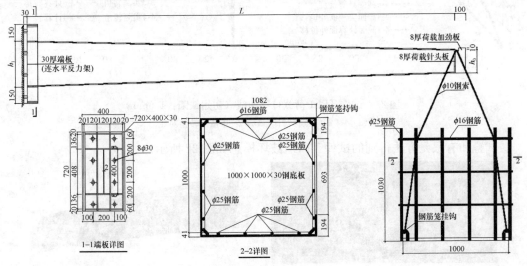

图20 悬臂端加载试验装置

为了对比分析需要,采用ANSYS软件对试验进行数值模拟,材料模型等与前同。屈服强度、腹板和翼缘厚度表12的实测结果取值。按照第一特征值屈曲模态施加初始缺陷,初始缺陷最大值按实测结果,最大(试件TPL2)为8.5mm,最小为1mm。有限元分析还考虑了梁的自重和加载点钢索与加劲板之间的摩擦,经多次尝试,取竖向荷载的1.2%施加侧向摩擦力,重力加速度通过施加惯性力的方法施加,故图中箭头方向与实际重力加速度的方向相反。

试件材性试验结果　　　　　　　　　　　　　　表 12

批次	板材	部件	数量	厚度(mm)	δ_5(%)	f_y(MPa)	f_u(MPa)	强屈比
1	2mm	腹板	3	2.30	43.0	270.4	420.7	1.56
1	8mm	翼缘	3	7.75	38.8	306.5	426.3	1.39
2	2mm	腹板	3	2.56	37.5	362.0	490.6	1.36
2	8mm	翼缘	3	7.76	35.2	315.0	448.2	1.42

4.3.2 试验结果

限于篇幅，下文仅给出 TPL1 的荷载位移曲线对比，如图 21 所示。其中，空心线为实测值曲线，实心线条为有限元计算曲线，面外位移指位移计 D3、D6、D9 所对应的位移；挠度值分别为各截面竖向位移测点位移的平均值；各截面转角分别为各截面竖向位移测点位移之差的绝对值与翼缘宽度之比。

从各试件的荷载-位移曲线可知，五个试件各测量截面的面外位移及截面扭转均较明显，表现出典型的弯扭失稳特征。有限元分析得到的荷载-位移曲线与试验结果吻合良好，有限计算的极限荷载略大于试验结果，但误差较小，在 3% 以内。

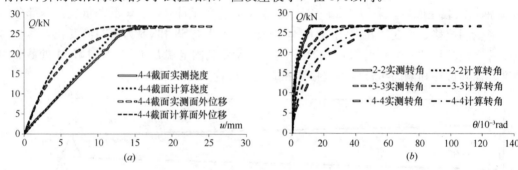

图 21　试件 TPL1 荷载位移曲线（初始缺陷：1.5mm）
(a) 4-4 截面位移对比；(b) 2~4 截面转角对比

试验和有限元方法得到的试件 TPL1 破坏模态如图 22 所示。

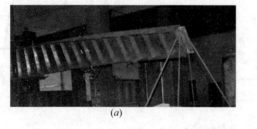 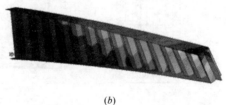

图 22　试件 TPL1 破坏模态
(a) 试验结果；(b) 有限元分析结果

最后统计试验结果可以发现，5 个试件的最终破坏模态均为一阶侧扭屈曲，失稳破坏后构件变形很大，呈现明显的侧弯和扭转效应。ANSYS 有限元弹塑性稳定分析得到的试件最终变形图与试验的变形照片相比非常相似，说明有限元分析能够较准确的模拟构件受力过程，对初始缺陷和极限承载力的模拟也较为合理，所以有限元方法可以作为研究楔形波纹腹板 H 型钢梁整体稳定性能的工具。

4.3.3 算式验证

为验证变截面波纹腹板 H 型钢梁整体稳定极限承载能力算式的可靠性,将各试件的试验结果和理论分析结果列在表 13 中进行分析。Q_t 为试验得到的各试件的极限荷载,Q_{cr} 为弹性临界荷载,Q_p 为悬臂梁塑性荷载,Q_u 为弹塑性稳定极限承载能力。

极限承载力理论值与试验结果比较　　　　　表 13

试件编号	t_f (mm)	t_w (mm)	f_{fy} (MPa)	Q_t (kN)	Q_{cr} (kN)	Q_p (kN)	Q_u (kN)	Q_t/Q_u
TPL1	7.75	2.30	306.5	26.5	30.3	35.5	25.2	1.052
TPL2	7.75	2.30	306.5	25.7	30.3	35.5	25.2	1.020
TPL3	7.76	2.56	315.0	26.3	30.6	36.5	25.7	1.023
TPL4	7.76	2.56	315.0	15.1	15.3	27.4	14.5	1.041
TPL5	7.76	2.56	315.0	13.4	14.2	27.4	13.4	1.000

从表 13 可以知,试验得到的极限荷载略大于理论弹塑性稳定极限荷载,且误差在 5% 以内,可见本文提出的算式(31)计算变截面波纹腹板 H 型钢梁整体稳定极限承载能力具有较高的精度,且结果略偏于安全。

5. 波纹腹板 H 型梁的疲劳

影响钢构件疲劳寿命的因素很多,主要是应力集中[49,50]。梯形波纹腹板焊接 H 形钢在波纹转折处存在一定程度的应力集中,但应力的整体变化趋势受截面弯矩控制[51],因此,波纹腹板 H 形钢受弯时最大正应力出现在最大弯矩附近应力集中区域,其疲劳性能与腹板波形有关。

5.1 试验概况

试验设计了 4 个简支梁试件,试件编号为 GJ1~GJ4。每个试件长 4.4m,跨度 4.0m。支座处设置加劲肋,跨中加载点附近设 1m 长钢轨。试件腹板波形尺寸如下:$h_r=36mm$,$b=64mm$,$d=36mm$,但波纹对称性不同:其中 GJ3 波纹关于梁跨中反对称,其余 3 根试件关于跨中正对称。试件截面同为 H524mm×300mm×200mm×3mm×14mm×10mm,其中上翼缘尺寸为 300mm×14mm、下翼缘为 200mm×10mm,以确保下翼缘出现最大正应力,且上翼缘不发生局部失稳。试件材料为 Q235 级钢,材性试验结果如表 14 所示。

试件材性试验结果　　　　　表 14

试件厚度(mm)	屈服强度 (MPa)	拉伸强度 (MPa)	伸长率 (%)
10	260	410	37.5
14	285	445	32.5
3	295	425	36

4根试件均在工厂加工完成,翼缘与腹板之间采用气体保护手工焊,单面角焊缝。试验装置如图 23 所示。GJ1~GJ3 钢轨为中心布置,GJ4 为偏心布置钢轨,偏心距为 10mm。

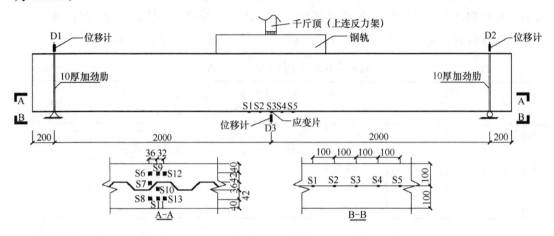

图 23 试验装置图

梯形波纹腹板 H 形钢在弯矩作用下,下翼缘最大弯矩附近正应力最大,故试件疲劳寿命由跨中下翼缘附近的主体金属应力幅控制。为试验测量和设计方便,试验中将跨中下翼缘下表面中心实测正应力幅作为疲劳试验的控制应力幅(见图 22 应变片 S3),并在跨中附近布置直角应变花(如图 22 应变花 S6~S13),以测量应力集中区域应力水平,确保跨中正应力与最大主应力相差不大。根据 GB 50017 条文 6.2.1,在常幅疲劳荷载作用下,钢构件容许应力幅按下式计算:

$$[\Delta\sigma] = (C/n)^{1/\beta} \tag{54}$$

参数 C、β 的取值,可以参照 GB50017 相关规定选取。其中第 2~4 类构件相应的参数值及由此计算出的对应于 $n=2\times10^6$ 次疲劳寿命的允许应力幅如表 15 所示。

2~4 类构件允许应力幅计算参数及应力幅计算值　　　　　表 15

类 别	C	β	允许应力幅(MPa)
2	861e12	4	144
3	3.26e12	4	118
4	2.18e12	3	103

试验设计中,参照 GB 50017 中平腹板焊接 H 形钢(4 类构件)200 万次疲劳寿命对应的允许应力幅确定设计应力幅,并在设计应力幅值下加载超过 200 万次以后采用该应力幅的 1.5 倍进行试验,循环超过 100 万次以后如果仍未破坏则改用初始应力幅的 2.0 倍进行试验,直到破坏。4 根试件疲劳试验实际所采用应力幅值及相应循环次数如表 16 所示。表中 $\Delta\sigma_1$、$\Delta\sigma_2$、$\Delta\sigma_3$ 为各试件在最大弯矩位置下翼缘下表面中心处的应力幅,N_1、N_2、N_3 为各应力幅下的循环次数。对于试件 GJ1,由于千斤顶荷载控制仪器的原因,未能有效控制集中荷载的大小,故表中 GJ1 加载应力幅的实测值大于 103MPa。

GJ1~GJ4 加载应力幅及实际循环次数　　　　表 16

试件编号	$\Delta\sigma_1$ (MPa)	N_1 (万次)	$\Delta\sigma_2$ (MPa)	N_2 (万次)	$\Delta\sigma_3$ (MPa)	N_3 (万次)
GJ1	159.0	112	—	—	—	—
GJ2	103.0	220	154.5	116	206.0	22
GJ3	103.0	232	154.5	82	—	—
GJ4	103.0	247	154.5	52	—	—

5.2　试验结果

试件 GJ1~GJ4 在表 16 所示的加载制度下，实测的应力幅和跨中挠度历程曲线如图 24 所示。

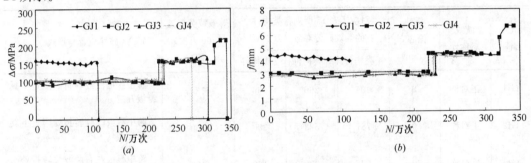

图 24　GJ1~GJ4 实测历程曲线
(a) 跨中下翼缘下表面中心正应力历程；(b) 跨中下翼缘下表面中心挠度历程

结合图 24 挠度和应力历程可知，4 根试件在疲劳破坏前始终处于弹性工作状态，符合疲劳破坏一般要求。各试件破坏位置及形式如图 25 所示。

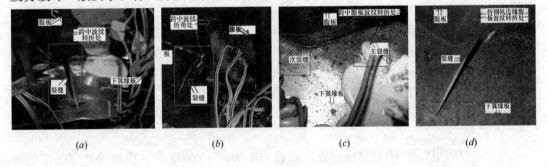

图 25　GJ1~GJ4 破坏形式图
(a) 试件 GJ1；(b) 试件 GJ2；(c) 试件 GJ3；(d) 试件 GJ4

综合 GJ1~GJ4 的试验结果，根据 Palmgren-Miner 线性累积损伤原则，及 GB50017 中疲劳寿命计算方法，试件对应于不同应力幅的等效疲劳寿命按式（54）和下式计算[52]：

$$\Sigma \left(\frac{n_i}{N_i} \right) \leqslant 1 \tag{55}$$

式中，n_i 为循环过程中应力幅水平达到 $[\Delta\sigma]_i$ 的实际循环次数；N_i 为试件在应力幅 $[\Delta\sigma]_i$ 下进行等幅疲劳试验所能达到的循环次数。

根据式（54）、式（55）计算得到试件 GJ1～GJ4 对应于表 15 所列各类构件的等效疲劳寿命和对应于 200 万次疲劳寿命的等效疲劳强度，计算结果如表 17 所示。其中，"对应于表 15 所列各类构件的等效疲劳寿命"指按表 15 所列各类构件的参数进行换算，由表 16 所列 GJ1～GJ4 的实际应力幅及循环次数得到对应于各类构件允许应力幅的疲劳寿命；"对应于 200 万次疲劳寿命的等效疲劳强度"指按表 15 所列各类构件的参数进行换算，由表 16 所列 GJ1～GJ4 实际应力幅及循环次数得到对应于 200 万次疲劳寿命的疲劳强度平均值。

从表 17 可知，4 根试件对应于 200 万次疲劳寿命的疲劳强度均超过 GB 50017 规定的 3 类构件，疲劳强度介于 3 类构件和 2 类构件之间。因此，疲劳验算可按 GB 50017 所述 3 类构件考虑。

试验结果 表 17

试件编号	波纹对称性	钢轨偏心 e (mm)	对应于表 15 所列各类构件的等效疲劳寿命（万次）			等效疲劳强渡 (200 万次) (MPa)	破坏位置与形式
			4 类构件	3 类构件	2 类构件		
GJ1	正对称	0	401	266	166	133	跨中腹板波纹转折处附近下翼缘板开裂
GJ2	正对称	0	787	523	300	160	跨中腹板波纹转折处附近下翼缘板开裂
GJ3	反对称	0	508	338	169	139	跨中腹板左右两波纹转折处附近下翼缘板开裂
GJ4	正对称	10	451	300	133	133	钢轨边缘腹板波纹转折处附近下翼缘板开裂

GJ1～GJ4 裂缝均在下翼缘腹板波纹转折处开始，这是因为在腹板波纹转折处存在应力集中。对于 GJ4，在钢轨存在 10mm 偏心的情况下，破坏仍然发生在下翼缘，且钢轨偏心虽然导致试件疲劳寿命降低，但降低程度不大，这是因为波纹腹板在梁平面外方向有一定的宽度，可大大削弱钢轨偏心在梁上翼缘与腹板连接处产生的弯曲应力，使得该处主拉应力较小，不易发生疲劳破坏。波纹腹板梁作为整体抵抗荷载偏心引起的扭矩，这使得其最大主应力仍出现在下翼缘板且其值较中心加载时增大有限。

5.3 小结

（1）波纹腹板焊接 H 形钢疲劳强度超过 GB 50017—2003 中 3 类构件疲劳强度的标准，疲劳验算可按 3 类构件考虑；

（2）波纹腹板焊接 H 形钢腹板波纹转折处存在一定程度的应力集中，疲劳破坏易发生在最大弯矩附近应力集中处，但应力集中引起的应力增幅不大，故工程设计中可不考虑应力集中的影响；

（3）波纹腹板 H 形钢在规范允许的最大偏心量的偏心荷载作用下，疲劳寿命有所降低，但仍然有相当的安全富余度，满足 GB 50017—2003 中 3 类构件疲劳强度标准；

（4）实际设计中，可按照 CECS 291：2011 第 5.3.2 计算截面弯曲应力幅，且不必考虑安装偏差等施工因素的影响。疲劳验算的容许应力幅 $[\Delta\sigma]$ 按式（54）计算。

6. 波纹腹板 H 型钢梁局部承压强度

根据现有文献资料研究的结论：集中荷载作用在上翼缘时，波纹腹板 H 型钢局部承压承载力较高。所以在设计和使用时，一般可以不使用加劲肋，根据这一优势，在施工过程中可以减少人工焊接工艺，降低工程造价，且可以提高构件的疲劳寿命。在移动的集中荷载（如吊车轮压）作用下，这一优点将更为突出[53]。20 世纪 70 年代起，日本曾将波纹腹板 H 型钢用做吊车梁。

6.1 研究背景

在波纹腹板 H 型钢梁局部承压性能的研究方面，1997 年，Elgaaly[54]进行了 5 根试件的局部承压试验，试验简图如图 26 所示。集中力施加在构件上翼缘，沿梁轴线方向具有一定的分布宽度 c，集中力作用位置无加劲肋。

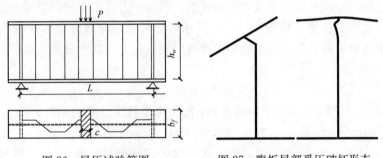

图 26　局压试验简图　　　图 27　腹板局部受压破坏形态

构件共包含 4 种不同的波纹尺寸，及不同的荷载分布宽度和位置。试验中观察到两种不同的破坏模式，如图 27 所示。

（1）腹板的弯折：

当集中力位于波纹中的水平板带，有可能在受压翼缘的某些位置出现塑性铰，从而形成翼缘的塑性铰破坏机制，受压翼缘发生竖向弯曲和扭转，并同时引发的腹板非弹性的局部弯曲。这种破坏模式可以将翼缘看作是支撑于腹板上的弹性地基梁，承载力同时取决于腹板和翼缘，结合有限元参数分析，Elgaaly 提出了极限承载力计算公式：

$$P_u = P_f + P_w \tag{56}$$

式中，$P_w = (Ef_{wy})^{0.5} t_w^2$，代表腹板承载力；$P_f = 4M_{pf}/[a-(c/4)]$，代表翼缘承载力；$M_{pf} = b_f f_{fy} t_f^2/4$，为翼缘的塑性抗弯承载力；$c$ 为荷载分布宽度；a 为负弯矩塑性铰间距或取：

$$a = \left(\frac{f_{fy} b_f t_f^2}{2 f_{wy} t_w}\right)^{0.5} + \frac{c}{4} \geq \frac{c}{2} \tag{57}$$

（2）腹板的屈服：

当集中力位于波纹中的倾斜板带或是位于水平与倾斜板带的交接处时，翼缘可能在两个方向发生竖向弯曲，但不发生扭转，也不形成塑性铰，腹板发生屈服后随即弯曲。其极限承载力可以按照有效宽度内腹板的屈服理论计算：

$$P_u = (b_0 + b_a) t_w f_{wy} \tag{58}$$

式中，当集中力位于倾斜板带时，$b_0 = d/\cos\theta$，当位于波纹的折弯线时，$b_0 = (b+d)/2$，$b_a = \alpha t_f (f_{fy}/f_{wy})^{0.5}$，$\alpha = 14 + 3.5\varphi - 37\varphi^2 \geqslant 5.5$，$\varphi = h_r/b_f$。

随后作者分析讨论了面内弯矩或剪力对局部承压能力的影响，并提出了建议的设计公式：

$$(P/P_u)^{1.25} + (M/M_u)^{1.25} = 1 \tag{59}$$

$$(P/P_u)^{1.25} + (V/V_u)^{1.25} = 1 \tag{60}$$

P 代表作用在构件上翼缘的局部集中力，M 和 V 分别代表相应截面的弯矩和剪力。P_u 为无弯矩和剪力时所对应的极限承载力，M_u 和 V_u 分别为不考虑局压集中力时所对应的极限弯矩和剪力。

R. Luo[55]通过进行非线性有限元分析，研究了下列因素对梁屈曲强度的影响：应变硬化、角部效应、初始几何缺陷、荷载位置、荷载分布宽度等。通过分析发现采用应变强化的 Ramberg-Osgood 模型计算得出的极限承载力比理想弹塑性模型高出 8%~12%，而冷弯所引起的局部效应对极限承载力影响较小。当集中力作用在倾斜板带中点时，其极限承载力最高，当作用在水平板带中点时，其承载力最低。荷载分布形式对承载力也有影响，分布荷载所对应的承载力高于集中力所对应的承载力。基于分析结果提出了承载力经验公式：

$$P_u = \gamma t_f t_w f_{wy} \tag{61}$$

式中，$\gamma = 15.6 \gamma_a \gamma_c$，15.6 为一个经验系数，$\gamma_a$ 为考虑波纹尺寸的系数，当 $t_f/t_w \geqslant 3.82$，$\gamma_a = (1+\cos\theta)/2\cos\theta$，当 $t_f/t_w < 3.82$，$\gamma_a = 1$。$\gamma_c = 1 + c/240$。

Krzysztof R. Kuchta[56]总结了欧洲一些学者的研究成果，如 Broude 曾提出若加载单元刚度较大而且对变形不敏感，则局部承压承载力可以用下列公式计算：

$$P_{Rd} = c_0 t_w f_d \tag{62}$$

式中，$c_0 = \eta \sqrt[3]{I_{xf}/t_w} + c$，为局压荷载在腹板上的有效分布宽度；$\eta$ 为腹板对翼缘的嵌固系数，对焊接梁可取为 3.26；I_{xf} 为受压翼缘的惯性矩；f_d 为腹板的设计强度。上式主要由腹板的塑性破坏机制推导而来，与试验结果较为接近，但主要适用于腹板较为"矮壮"的情况。

Kähönen 提出将受压翼缘作为弹性地基梁，并用图 28 来解释其受力机理，提出的公式相对比较复杂：

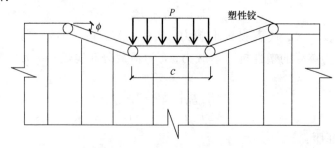

图 28 翼缘塑性铰破坏机制

$$P_{Rd} = (R_{d1} + R_{d2} + R_{d3}) k_0 k_r / \gamma_M \tag{63}$$

式中，R_{d1} 为腹板支座反力，R_{d2} 为翼缘抗弯承载力造成的附加力，R_{d3} 为作用在翼缘上的正应力所引起的压力的增加。由于建立平衡方程是基于结构变形后的状态，所以上式同时考虑了腹板和翼缘的承载力。由于参数过多，上式应用起来较不方便。

Máchacek 调查了波纹腹板 H 型钢在吊车梁中的应用的可能性。在数值分析和试验研究的基础上，提出了有轨吊车梁局压承载力的计算公式：

$$P_{Rd} = (78.9t_w + 3.2t_f - 14.7)\sqrt[3]{\frac{I_f + I_r}{I_f + I_b}}\gamma_M \tag{64}$$

式中，γ_M 为材料安全系数，可以取为 1.15，几何参数按照 mm 取值，I_r 为轨道的截面惯性矩，I_b 为 50×30mm 的块体的惯性矩。上式适用范围为轨道轴线和梁轴线的偏心不超过 ±20mm。上式也可以用来计算没有轨道但加载单元的宽度超过 150mm 时梁的局压承载力。

奥地利的 Zeman 公司在其产品技术手册中规定波纹腹板 H 型钢的局部承压计算方法：

$$P_{Rk} = t_w(c + 5t_f)f_{yk} \tag{65}$$

上式与平腹板梁的局压强度计算方法一致，但是，当荷载分布宽度 c 较小时，如 $c=0$，上式过于保守。

6.2 试验研究

由于波纹腹板 H 型钢梁局压承载力计算方法较多，而且考虑的因素各异，因此有必要通过试验和有限元方法的研究，找到较为理想的设计公式。在抗剪和抗弯试验完成后，作者对各试件未破坏的部分进行局部承压试验研究，局压试验试件参数如表 18 所示。

局压试验构件汇总表　　　　　　　　　　表 18

编号	波形	t_w (mm)	t_f (mm)	b_f (mm)	f_{wy} (MPa)	f_{fy} (MPa)	垫块形式	L (m)	c (mm)	c_0 (mm)
GJ2	1	1.7	10	200	199	317	钢块	1.0	150	200
GJ4	2	1.9	14	280	263	268	钢块	1.0	150	220
GJ5	2	1.9	14	280	263	268	钢块	1.5	150	220
GJ7-1	1	3.0	10	150	260	265	钢轨	1.5	65	335
GJ7-2	1	3.0	10	150	260	265	钢轨	2.0	95	365
GJ8-1	1	3.0	10	150	260	265	钢轨	2.5	95	365
GJ8-2	1	3.0	10	150	260	265	钢轨	2.5	95	365
GJ9	1	3.0	10	150	265	265	钢轨	1.0	95	365
GJ11	3	2.0	10	150	265	265	辊轴	1.0	0	50
GJ13	4	2.0	10	150	265	265	辊轴	1.0	0	50

表中，c 代表加载头宽度，c_0 代表局压荷载在腹板上的有效分布宽度，按照 $c_0 = c + 2h_R + 5t_f$ 计算，h_R 为轨道高度。试验设计为简支梁单调加载，支座处设加劲肋，加载点处不设加劲肋。采用千斤顶在梁中部上翼缘施加荷载，千斤顶包括 50 吨和 100 吨两种规格，其中 50 吨千斤顶加载头直径 65mm，100 吨千斤顶加载头直径 95mm。千斤顶与构件上翼缘之间设垫块，垫块包括 3 种形式：（1）钢块，尺寸为 150mm×150mm，厚度

70mm；（2）钢辊轴，直径为 50mm；（3）钢轨道，长度为 800 或 1000mm，高度 110mm，宽度 85mm。

试验过程中用约束架为梁提供侧向支撑，腹板上侧靠近加载点位置单侧贴应变片，应变片分布长度根据 c_0 计算确定。由于局部承压试验中主要考察的是构件极限承载力，所以在试验中仅在加载点处上翼缘或下翼缘设置一个位移计，所测得数据仅做为绘制荷载位移曲线的一个坐标参数。为了与试验结果进行对比，本文同时采用有限元程序 ANSYS 进行分析。模型参数选择与前同。下面给出几个典型试件的试验过程和现象。

6.2.1 GJ5 局压试验现象

GJ5 采用了刚性垫块，宽度 150mm。将 GJ5 加载点处的上翼缘位移和加载值绘制成荷载位移曲线（图29），图30a 给出了 GJ2 的破坏状态，图30b 为有限元模拟得到的破坏状态。

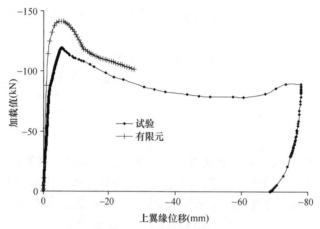

图 29　GJ5 局压承压试验曲线

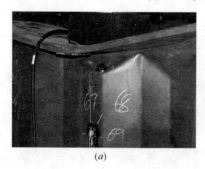 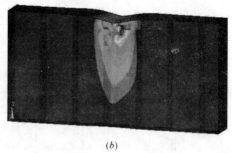

图 30　GJ5 局压试验破坏形态
（a）试验；（b）有限元

GJ5 荷载位移曲线达到极值前，未有塑性发展段，达到极值后承载力下降，并保持一定屈曲后强度。破坏现象为受压部位腹板和翼缘的局部破坏，构件在弹性阶段直接达到极限荷载，局部腹板发生较大的压缩变形，翼缘发生了弯曲，整个过程翼缘没有发生扭转。腹板屈服后，翼缘向下产生较大的位移。GJ5 破坏后构件能保持较高的破坏后强度，屈曲后强度可达到极限强度的 80% 以上。有限元方法得到的极限承载力与试验比较接近，且最终破坏形态也非常逼真，所以有限元方法能够有效预测波纹腹板 H 型钢梁的局部承压受力行为。

6.2.2 GJ7局压试验2现象

GJ7的局压试验分为两部分，分别对构件的左右剪跨段进行局压试验。其中GJ7局压试验2所采用的钢轨长0.8m，采用了100吨的千斤顶施压，加载头直径为95mm。GJ7局压试验2的荷载位移曲线如图31所示，破坏形态如图32所示。

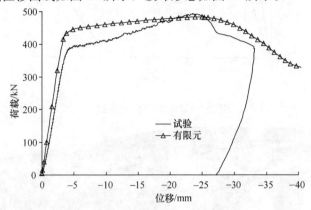

图31 GJ7局压试验2曲线

图32 GJ7局压试验2破坏形态
(a) 试验；(b) 有限元

试验中，波纹腹板靠近上翼缘的部位首先出现了局压鼓曲现象，表现出局部承压破坏的特征。但最终破坏现象却是腹板的剪切屈曲。这是由于所施加荷载已经超过了腹板的剪切屈服强度（约为230kN），所以首先发生了腹板的剪切破坏。同时，该试验结果证明了GJ7的局压承载力超过了460kN。有限元方法得到的极限承载力与试验结果较为吻合，且破坏形态也具有一定的局部剪切破坏特征。

6.2.3 GJ11局压试验现象

GJ11局压试验采用钢辊轴作为加载方式，GJ11的荷载位移曲线见图33，破坏形态见图34。

可以看到采用辊轴加载方式后，由于荷载分布宽度较小，波纹腹板梁的局压承载力显著降低，但破坏形态仍然是腹板的屈服和屈曲，以及翼缘的向下弯曲变形。从受力过程来看，破坏发生于弹性阶段，到达极值后承载力快速下降，属脆性破坏。有限元方法得到的

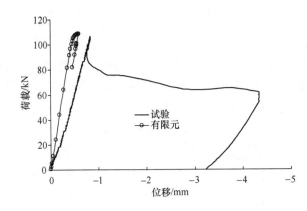

图 33　GJ11 局压试验曲线

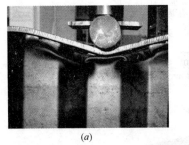
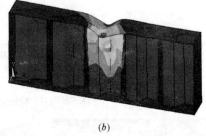

图 34　GJ11 局压试验破坏形态
(a) 试验；(b) 有限元

极限承载力与试验结果较为接近，同时荷载位移曲线也较为准确地反映了试件的受力特征，达到极值后，曲线几乎按照弹性刚度卸载。

6.3　试验结果分析

为了与上文提到的各项计算公式进行对比，将试验、有限元分析结果和公式计算结果一同列入表 19。

局压试验结果表　　　　　　　　　　表 19

	t_w (mm)	f_{wy} (Mpa)	c_0 (mm)	q (mm)	P_1 (kN)	P_2 (kN)	P_3 (kN)	P_4 (kN)	P_5 (kN)	P_t (kN)	P_{FEM} (kN)	P_t/P_5
GJ2	1.7	199	200	175	68	68	122	75	68	108	116	1.60
GJ4	1.9	263	220	300	115	129	196	128	110	135	151	1.23
GJ5	1.9	263	220	300	115	129	196	128	110	119	141	1.08
GJ7-1	3.0	260	335	175	75	84	385	263	261	450	460	1.72
GJ7-2	3.0	260	365	175	111	126	407	287	285	493	483	1.73
GJ8-1	3.0	260	365	175	111	126	407	287	285	500	466	1.76
GJ8-2	3.0	260	365	175	111	126	407	287	285	501	466	1.76
GJ9	2.0	265	365	175	75	84	271	195	194	347	304	1.79
GJ11	2.0	265	50	240	75	85	100	32	27	106	109	4.00
GJ13	2.0	265	50	130	75	84	106	32	27	124	123	4.68

表中 P_1、P_2、P_3、P_4、P_5 分别对应式 56、58、61、62、65 的计算结果，其中 P_5 可以看做是等腹板厚度的平腹板梁局部承压强度。P_t 为试验测到的极限荷载，P_{FEM} 为有限元方法计算得到的极限荷载。

为了获得更具有说服力的结论，本文将能收集到的国内外局压试验结果列入表 20。

国外资料局压试验结果列表 表 20

来源	序号	t_w (mm)	t_f (mm)	f_{wy} (MPa)	f_{fy} (MPa)	c (mm)	c_0 (mm)	q (mm)	P_1 (kN)	P_2 (kN)	P_3 (kN)	P_4 (kN)	P_5 (kN)	P_t (kN)	P_t/P_5
Krzysztof (2007)	GA	2.5	6	267	267	75	105	155	—	—	—	—	70	170	2.43
	GB	2.7	6	257	257	203	233	155	—	—	—	—	162	246	1.52
Elgaaly (1997)	E1	2	10	379	389	146	196	500	95	137	219	153	149	131	0.88
	E2	2	10	379	389	0	50	500	95	137	136	42	38	80	2.11
	E3	2	10	379	389	104	154	500	95	137	195	121	117	102	0.87
	E4	2	10	379	389	0	50	500	95	137	136	42	38	96	2.53
	E5	2	10	379	389	0	50	500	95	137	136	42	38	71	1.87
Aravena (1987)	B1	2.5	12	335	475	0	60	400	159	216	181	66	50	149	2.97
	B2	2.5	12	335	475	0	60	400	159	216	181	66	50	170	3.38
	B3	2.5	12	317	475	0	60	400	155	208	171	63	48	152	3.20
	B4	2.5	12	317	475	50	110	400	155	208	207	102	87	168	1.93
	B5	2.0	10	280	475	0	60	400	128	154	120	48	34	107	3.18
	B6	2.0	10	280	475	50	110	400	128	154	146	76	62	124	2.01

分析试验数据后可以总结出以下一些基本规律：

(1) 波纹腹板 H 型钢梁的局部承压强度都大于等厚度平腹板梁的局压强度（E1、E3 除外），荷载分布宽度 c 越小，这种趋势越明显，例如 $c=0$ 的情况下，$P_t/P_5=1.87\sim 4.68$。这就说明，在荷载分布宽度越小的情况下，波纹的局部效应越明显，紧邻的板带和翼缘对直接受力板带能够提供较强的支撑，相当于平腹板中加劲肋的作用。

(2) 当荷载分布宽度较小时，局部承压的破坏属于脆性破坏；当荷载分布宽度较大时，在荷载作用下，出现了内力的重分布的过程，所以呈塑性破坏特征。因此，当荷载分布宽度较小时，可以取较高的抗力分项系数，而荷载分布宽度较大时，可以取较低的抗力分项系数。

(3) 各研究者提出的计算公式差异较大，且上述公式的提出大多基于荷载分布宽度较小的情况下推出的，在这种情况下，式（56）的计算结果偏于安全。但是，当荷载分布宽度较大时，式（56）估值过低。

(4) 当荷载分布宽度较大时（GJ71-2，GJ81-2，GJ9），极限承载力均达到了平腹板梁局压强度的 1.7 倍以上。各公式的计算结果普遍偏小，而公式（61）计算结果与试验值较为接近。

通过上述分析，首先波纹腹板 H 型钢梁的局部承压能力都显著高于平腹板梁，在集中荷载较小时，可以考虑不设置加劲肋。若用于承受移动的结构，则波纹腹板 H 型钢的局压性能能够得到更好的利用。这种性能在吊车梁中更有实践意义，同时由于不需要设置加劲肋，所以还能提高吊车梁的疲劳强度。

6.4 结论及设计建议

（1）当荷载有效分布宽度 c_0 小于腹板波纹的波长 q 时，按照式（56）计算承载力；

（2）当荷载有效分布宽度 c_0 大于腹板波纹的波长 q 时，按照式（61）计算承载力。

若按照上述设计方法对表 19 和表 20 的试验结果进行验证，可以得到得到如下结论：

当荷载分布宽度 c_0 小于腹板波纹的波长 q 时，共 15 个试验结果数据，P_t/P_1 的均值为 1.08。当荷载分布宽度 c_0 大于腹板波纹的波长 q 时，共 6 个试验结果数据，P_t/P_1 的均值为 1.17。

所以，所提出的设计方法能够较为准确的反映波纹腹板钢梁的局压承载力，且具有一定的安全余量。

7. 结论

（1）波纹腹板 H 型钢梁抗剪承载力可以按照式(12)～式(15)进行计算；

（2）波纹腹板 H 型钢梁抗弯强度可以取截面塑性抵抗弯矩；

（3）波纹腹板 H 型钢梁的截面翘曲常数可以按照腹板波形中波峰（或波谷）位置的单轴对称截面进行计算；

（4）波纹腹板 H 型钢梁整体稳定设计公式可以在现有公式的基础上进行微调，以式（31）进行计算；

（5）楔形波纹腹板 H 型钢梁的弹性稳定承载力可以按式（53）计算，其弹塑性稳定承载力按照（31）计算；

（6）波纹腹板 H 型钢梁的疲劳强度可以按照 3 类构件进行考虑；

（7）波纹腹板 H 型钢梁的局部承压强度，应根据荷载分布宽度与腹板波形波长的关系，分别采用不同的表达式进行计算；

（8）波纹腹板 H 型钢非常适合作为横向受力构件应用在建筑结构领域。

参考文献

[1] Hamada Masaki. Manufacture and manufacturing roll for H-shaped steel possessing corrugated at middle part of web：Japan, 54107778[P]. 1981.04.13

[2] 曹鸿德，才志华，张文志. 波纹腹板 H 型钢梁的热轧工艺：中国，86106315A[P]. 1988.03.30

[3] Swedish Institute of Steel Construction. Swedish Code for Light-Gauge metal Structures[S]. Stockholm, Sweden, 1982

[4] European Committee for Standardisation. prEN 1993-1-5. EUROCODE 3：Design of steel structures；Part 1.5：Plated structural elements. 2004

[5] Galambos, T. V. (ed.). (1988). Guide to stability design criteria for metal structures. John Wiley&Sons, Inc., New York, N.Y

[6] Easley J T. Buckling formulas for corrugated metal shear diaphragms[J]. J. Struct. Div, ASCE,

1975, 101(7), 1403-1417

[7] Smith D. Behavior of corrugated plates subjected to shear[D]. Dept. of Civ. Engrg, Univ. of Maine, Orono, Maine, 1992

[8] Hamilton R. Behavior of welded girders with corrugated webs, PhD thesis, Dept. of Civ. Engrg., Univ. of Maine, Orono, Maine, 1993

[9] Elgaaly M, Hamilton R W and Seshadri A. Shear strength of beams with corrugated webs[J]. J. Struct. Eng., 1996, 122(4), 390-398.

[10] Luo R, Edlund B. Buckling analysis of trapezoidally corrugated panels using spline finite strip method[J]. Thin-Walled Structures, 1994, 18: 209-224

[11] Luo R, Edlund B. Numerical simulation of shear tests on plate girders with trapezoadally corrugated webs. Division of Steel and Timber Structures. Chalmers University of Technology. Sweden. 1995

[12] Luo R, Edlund B. Shear capacity of plate girders with trapezoidally corrugated webs[J]. Journal of Thin-Walled Structures, 1996, 26(1): 19-44

[13] Abbas H H. Analysis and design of corrugated web I-girders for bridges using high performance steel[D]. Lehigh Univ., Bethlehem, Pa, 2003

[14] Driver R G, Abbas H H, Sause R. Shear Behavior of Corrugated Web Bridge Girders [J]. J of structural engineering, 2006, 132(2): 195-203

[15] Yi J, Gil H, Youm K, et al. Interactive shear buckling of trapezoidally corrugated webs[J]. Eng Struct 2008, 30: 1659-1666

[16] Jiho Moon, Jongwon Yi, et al. Shear strength and design of trapezoidally corrugated steel webs[J]. Journal of Constructional Steel Research, 2009, 65: 1198-1205

[17] Zhang W Z, Zhou Q T. Hot rolling technique and profile design of tooth-shape rolls Part 1. Development and research on H-beams with wholly corrugated webs[J]. Journal of Materials Processing Technology: 2000(101): 110-114

[18] Zhang W Z, Li Y W. Optimization of the structure of an H-beam with either a flat or a corrugated web Part 3. Development and research on H-beams with wholly corrugated webs[J]. Journal of Materials Processing Technology: 2000(101): 119-123

[19] 常福清, 李恒伟. 波纹腹板 H 型钢梁腹板的屈曲强度(II)[J]. 东北重型机械学院学报, 1996, 20(2): 150-153

[20] 常福清. 波纹腹板 H 型钢梁腹板的屈曲强度[J]. 机械强度, 1997, 19(1): 42-44

[21] 李艳文. 全波纹腹板 H 型钢屈曲性能分析[J]. 燕山大学学报, 2001, 5(4): 371-374

[22] Li Y W, Zhang W Z. Buckling strength analysis of the web of a WCW H-beam: Part 2. Development and research on H-beams with wholly corrugated webs(WCW)[J]. Journal of Materials Processing Technology: 2000, 101: 115-118

[23] 常福清, 张文志, 吴波. 波纹腹板工字钢强度数值分析[J]. 钢结构, 2005, 20(78): 4-7

[24] 宋建永, 张树仁. 波纹钢腹板剪切屈曲分析中初始缺陷的模拟和影响程度分析[J]. 公路交通科技, 2004, 21(5): 61-64

[25] 宋建永, 任红伟, 聂建国. 波纹钢腹板剪切屈曲影响因素分析[J]. 公路交通科技, 2005, 22(11): 89-92

[26] 李时, 郭彦林. 波折腹板梁抗剪性能研究[J]. 建筑结构学报, 2001, 22(6): 49-54

[27] 张哲. 波纹腹板 H 型钢及组合梁力学性能理论与试验研究[D]. 上海: 同济大学, 2009

[28] 陈绍藩. 钢结构设计原理[M]. 北京: 科学出版社, 2005

[29] Elgaaly M, Seshadri A and Hamilton R W. Bending strength of steel beams with corrugated webs[J]. J. Struct. Eng., 1997, 123(6), 772-782

[30] Chan C L, Khalid Y A. finite element analysis of corrugated web beams under bending[J]. Journal of Constructional Steel Research. 2002, 58: 1391-1406

[31] Johnson R P, Cafolla J. Local flange buckling in plate girders with corrugated webs[J]. Proceedings of the Institution of Civil Engineers, Structures and Buildings, 1997, 22(2)2: 148-156

[32] Abbas H H, Sause R, Driver R G. Behavior of Corrugated Web I-Girders under In-Plane Loads[J]. Journal of Engineering Mechanics, 2006, 132(8): 806-814

[33] Abbas H H, Sause R, Driver R G. Analysis of Flange Transverse Bending of Corrugated Web IGirders under In-Plane Loads[J]. Journal of Structural Engineering, 2007, 133(3): 347-355

[34] Hassan H. Abbas, Richard Sause, et al. Simplied analysis of flange transverse bending of corrugated web I-girders under in-plane moment and shear[J]. EngineeringStructures, 2007

[35] JihoMoon, Jong-WonYi, et al. Lateral-torsional buckling of I-girder with corrugated webs under uniform bending[J]. Thin-Walled Structures, 2009, 47: 21-30

[36] 郭彦林, 张庆林. 波折腹板工形构件翼缘稳定性能研究[J]. 建筑科学与工程学报, 2007, 24(4): 64-69

[37] Lindner J. Lateral torsional buckling of beams with trapezoidally corrugated webs[R]. Proc., Int. Colloquium of Stability of Steel Structures, Budapest, Hungary, 1990: 79-86

[38] European Committee for Standardisation. prEN 1993-1-1. EUROCODE 3: Design of steel structures; Part 1-1: General rules and rules for buildings[S]. 2003

[39] Zeman & Co Gesellschaft mbH. Corrugated web beam(Technical documentation) [OL]. Austria: 2003. http://www.zeman-steel.com

[40] 陈骥. 钢结构稳定理论与设计[M]. 北京: 科学出版社, 2006: 271-335

[41] 朱群红, 童根树. 简支楔形工字钢梁的弹性弯扭屈曲[J]. 建筑结构, 2006, 36(1): 31-34

[42] 周佳. 双轴对称楔形工字钢梁的弹性弯扭屈曲[D]. 杭州: 浙江大学, 2007: 58-62

[43] 童根树. 钢结构的平面外稳定[M]. 北京: 中国建筑工业出版社, 2007: 122-141

[44] Kitipornchai, S and Trahair, N. S. Elastic Stability of Tapered I-beams[J]. Journal of the Structural Division, ASCE, 1972, 98(3): 713-728

[45] GB 50017-2003 钢结构设计规范[S]. 北京: 中国计划出版社: 2003

[46] Fukumoto Y, Itoh Y, Kubo M. Strength variation of laterally unsupported beams[J]. ASCE, Journal of the structural Division, 1980, 106(1): 165-181

[47] XiaJianguo. Inelastic lateral buckling behavior of steel beams[M]. In: A A O Tay, K Y Lam, Computational Methods in Engneering Advance & Application, Singapore: World Scientific Publishing Co Pte Ltd, 1992: 139-144

[48] 夏建国. 工字形钢梁整体稳定性能的研究[J]. 上海铁道大学报, 1999, 20(2): 17-21

[49] Sherif A. Ibrahim. Fatigue analysis and instability problems of plate girders with corrugated webs[D]. Philly: Drexel University; 2001: 50-230

[50] Ichikawa A, Kotaki N, Suganuma H, Miki C. Fatigue performance of the bridge girder with corrugated web[C]. // Proceedings of International Institute of Welding; Copenhagen: International Institute of Welding, 2002. Doc. XIII-1927-02

[51] Abbas H. Analysis and design of corrugated web I-girders for bridges using high performance steel [D]. Bethlehem: Lehigh University; 2003: 239-338

[52] 夏志斌, 姚谏. 钢结构——原理与设计, 中国建筑工业出版社, 2004: 442-446

[53] 张哲,李国强,孙飞飞. 波纹腹板 H 型钢研究综述[J]. 建筑钢结构进展,2008,10(6):41-46
[54] Elgaaly M, Seshadri A. Girders with corrugated webs under partial compressive edge loading[J]. Journal of Structural Engineering ASCE. 1997, 123(6):783-91
[55] Luo R, Edlund B. Ultimate strength of girders with trapezoidally corrugated webs under patch loading[J]. Thin-Walled Structures. 1996, 24:135-156
[56] Krzysztof R. Kuchta. Design of corrugated webs under patch load[J]. Advanced Steel Construction, 2007, 3(4):737-751

BIM 在钢结构制造中的应用

贺明玄，沈　峰

（宝钢钢构有限公司 上海201900）

摘　要：钢结构制造的 BIM 技术已被引入多年，过去通常我们称之为钢结构 3D 模型，且通常停留在深化设计的建模和出图阶段，BIM 产生的信息在后续流程的应用却常常被忽视，本文着重描述钢结构制造 BIM 的创建过程，以及 BIM 信息在后续加工生产中的应用。

关键词：钢结构；BIM；建模；制造

APPLICATON OF BIM IN STEEL STRUCTURE MANUFACTURE

M. X. He，F. Shen

(BaoSteel Construction Co., Ltd Shanghai 201900, China.)

Abstract: BIM technology has been introduced into fabrication of steel structure for many years. It used to be called the steel structure 3D modeling technology, which was at the stage of the development of modeling and automatic detailing. However, the application of information generated from BIM system in the following process has always been neglected. This paper mainly describes the creating process of BIM in steel structure manufacture and the application of BIM information in the following fabrication process.

Keywords: steel structure; BIM; modeling; manufacture

1. 前言

传统结构专业的工作流程（如建筑设计、结构设计、深化设计、加工制作和施工安装之间）常常是不同程度地信息间断的，如图1所示。

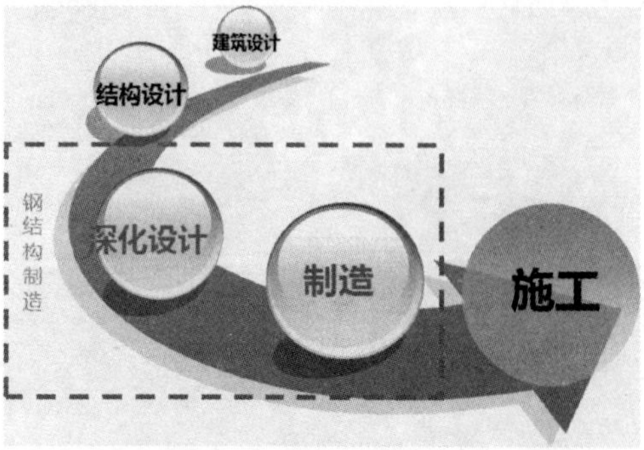

图 1　钢结构制作在结构专业内的工作流程

在工程项目执行的各个阶段，设计、分析、详图和制作管理系统中的模型和数据很多时候会被多次重复输入或重新建模。因此，在项目估算、深化设计、加工生产和施工计划中有时会造成不必要的高成本返工。

BIM 技术的引入应用可以充分共享应用

各阶段模型信息，使流程各阶段紧密结合，信息互通，从而提高效率和降低成本。

　　钢结构加工制造的 BIM 技术被引入至今已经 10 多年，过去，通常我们称之为钢结构 3D 模型，但是 BIM 模型过去通常停留在建模和出图过程中，其产生的信息在后续流程的完整应用正被研究和应用中。本文从钢结构 BIM 模型创建开始，介绍钢结构深化设计和加工制造中的 BIM 应用。

2. 钢结构 BIM 模型及创建

2.1　钢结构 BIM 模型

　　钢结构 BIM 三维实体建模出图进行深化设计的过程，其本质就是进行电脑预拼装、实现"所见即所得"的过程。首先，所有的杆件、节点连接、螺栓、焊缝、混凝土梁柱等信息都通过三维实体建模进入整体模型，该 BIM 三维实体模型与以后实际建造的建筑完全一致；其次，所有施工详图（包括布置图、构件图、零件图等）均是利用三视图原理投影生成，图纸中所有尺寸，包括杆件长度、断面尺寸、杆件相交角度等均是从三维实体模型上直接投影产生的。图 2～图 4 为完全实现电脑预拼装的某工程项目 BIM 三维实体模型和节点模型。

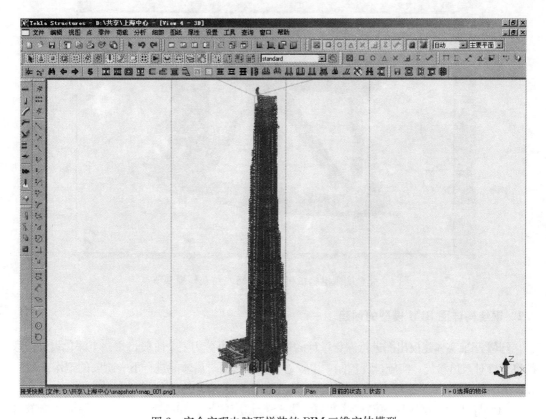

图 2　完全实现电脑预拼装的 BIM 三维实体模型

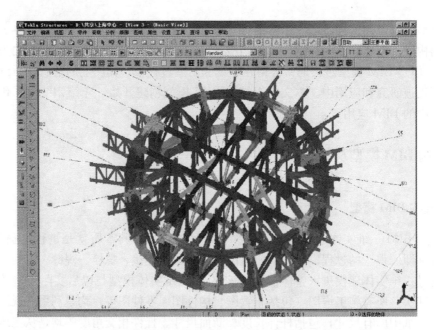

图 3 完全实现电脑预拼装的 BIM 三维实体模型局部

图 4 完全实现电脑预拼装的 BIM 三维实体模型节点

2.2 钢结构详图 BIM 模型的创建

BIM三维实体建模出图进行深化设计的过程，基本可分为三个阶段，每一个深化设计阶段都将有校对人员参与，实施过程控制，由校对人员审核通过后才能进行下一阶段的工作。

2.2.1 第一阶段，根据结构施工图建立轴线布置和搭建杆件实体模型。

（1）导入 AutoCAD 中的单线布置，并进行相应的校合和检查（图5），保证两套软件设计出来的构件数据理论上完全吻合，从而确保了构件定位和拼装的精度。

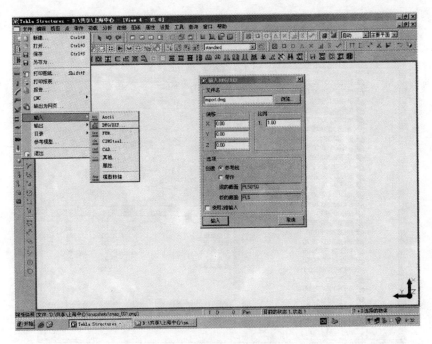

图 5 导入 CAD 对话框

(2) 创建轴线系统及创建、选定工程中所要用到的截面类型、几何参数（图 6、图 7）。

(3) 整体 BIM 三维实体模型的建立与编辑（图 8～图 10）

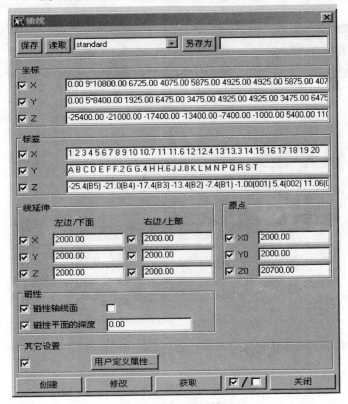

图 6 创建工程的轴网

311

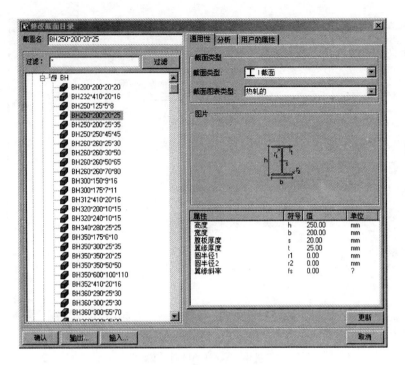

图 7 修改截面对话框

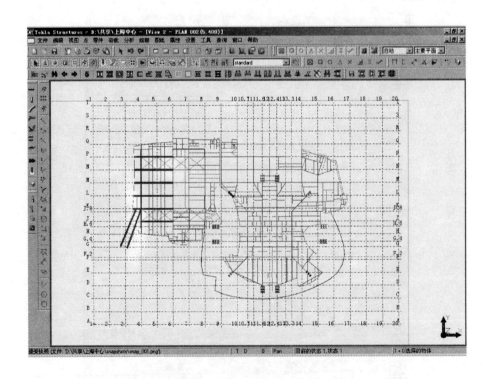

图 8 整体 BIM 三维实体模型平面构件的搭建

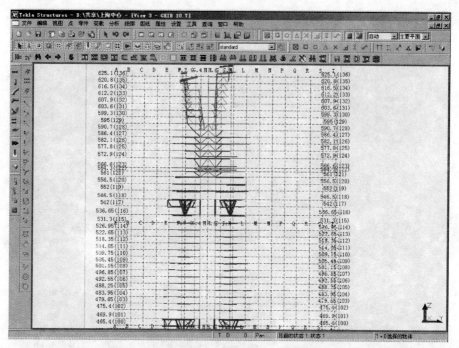

图 9 整体 BIM 三维实体模型立面构件的搭建

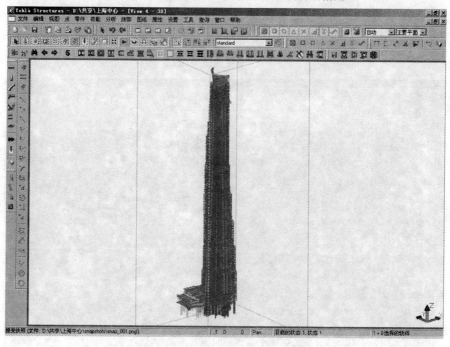

图 10 整体 BIM 三维实体模型的搭建

2.2.2 第二阶段，根据设计院图纸对模型中的杆件连接节点、构造、加工和安装工艺细节进行安装和处理。

（1）在整体模型建立后，需要对每个节点进行装配，结合工厂制作条件、运输条件，考虑现场拼装、安装方案及土建条件（图11～图14）。

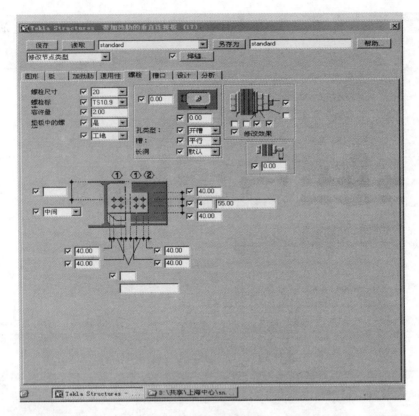

图11 节点参数对话框

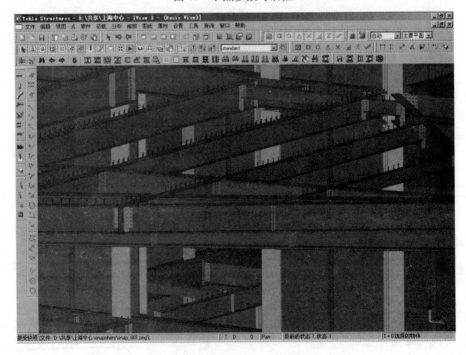

图12 节点装配后的平面梁实体模型

图 13　节点装配后的桁架实体模型

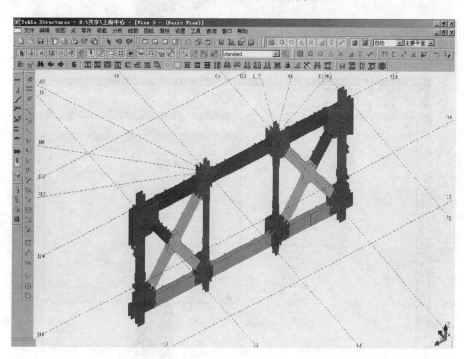

图 14　建好节点并按运输、起重量要求分好段的实体模型

2.2.3 第三阶段，对搭建的模型进行"碰撞校核"，并由审核人员进行整体校核、审查。

所有连接节点装配完成之后，运用"碰撞校核"功能进行所有细微的碰撞校核，以检查出设计人员在建模过程中的误差（图15，图16）。这一功能执行后能自动列出所有结构上存在碰撞的情况，以便设计人员去核实更正，通过多次执行，最终消除一切详图设计误差。

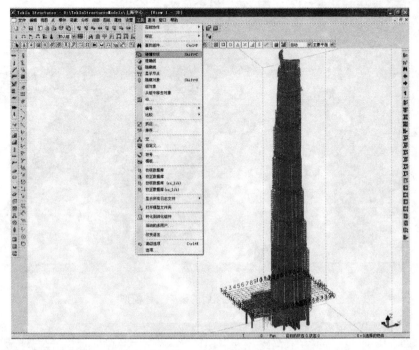

图15　碰撞校核对话框

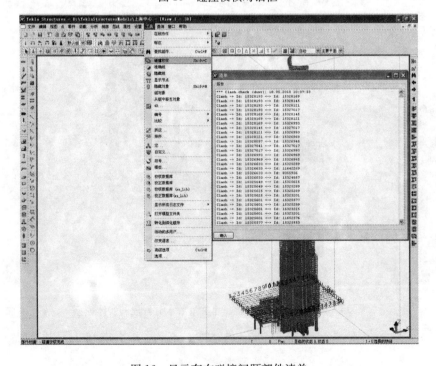

图16　显示存在碰撞问题部件清单

3. 基于 BIM 模型的设计出图

运用建模软件的图纸功能自动产生图纸，并对图纸进行必要的调整，同时产生供加工和安装的辅助数据（如材料清单、构件清单、油漆面积等）。

（1）节点装配完成之后，根据设计准则中编号原则对构件及节点进行编号（图17）。

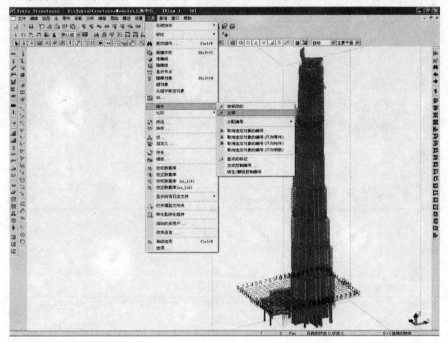

图17　构件编号对话框

（2）编号后就可以产生布置图、构件图、零件图等，并根据设计准则修改图纸类别、图幅大小、出图比例等（图18）。

（3）所有施工详图（包括布置图、构件图、零件图等）均是利用三视图原理投影、剖面生成深化图纸，图纸上的所有尺寸，包括杆件长度、断面尺寸、杆件相交角度均是在杆件模型上直接投影产生的（图19～图21）。因此由此完成的钢结构深化图在理论上是没有误差的，可以保证钢构件精度达到理想状态。

（4）用钢量等资料统计。统计选定构件的用钢量，并按照构件类别、材质、构件长度进行归并和排序，同时还输出构件数量、单重、总重及表面积等统计信息（图22、图23）。

（5）BIM 模型的结构化数据的形成。基于 BIM 理念，可深入挖掘钢结构深化设计软件的功能，充分体现 BIM 的特性，使钢结构深化设计向建筑设计信息化的方向发展。依托 PDM 平台，既可实现深化设计与上游结构设计的集成，又可与下游工艺、制造信息形成传递与集成，最终达到提高生产效率，节省项目成本的目的。将设计信息如 3D 模型、2D 图纸、构件、零件、文档抽象为不同的类的对象实例，所有设计信息存储在统一的数据库中，且始终保持对象之间的关联关系（图24）。

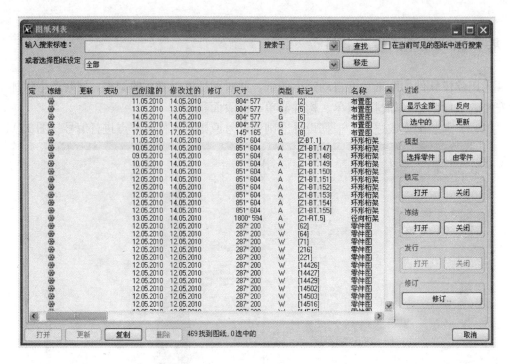

图 18　图纸清单对话框

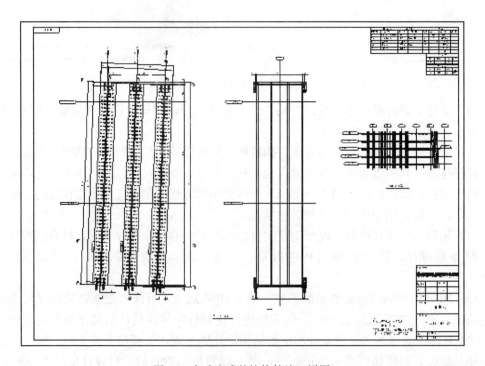

图 19　自动生成的柱构件施工详图

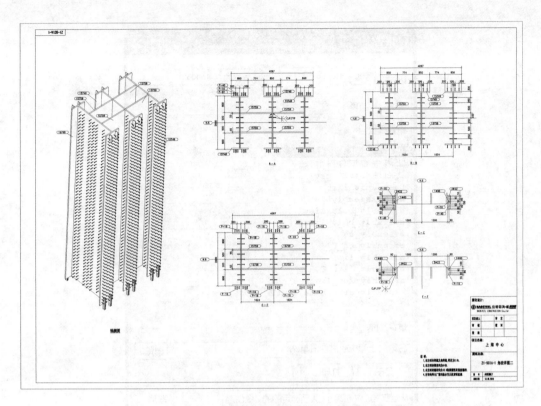

图 20 自动生成的柱构件施工详图

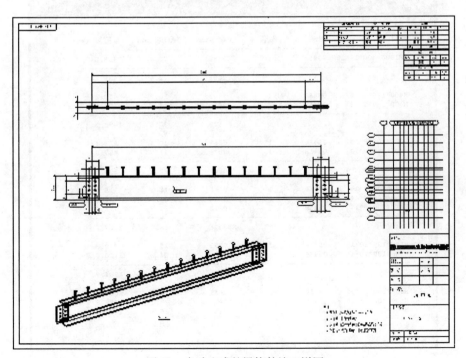

图 21 自动生成的梁构件施工详图

图 22 用钢量及其他统计报表对话框

图 23 材料统计清单

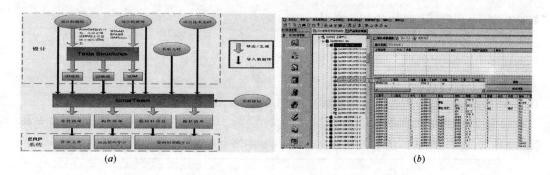

图 24 BIM 模型的结构化数据形成
(a) BIM 模型与结构化数据的形成过程；(b) BIM 模型的结构化数据

4. 钢结构 BIM 模型在生产制造阶段的应用

在信息技术和自动化程度日益发展的今天，手工加工技术和人为管理已日显疲态，逐步被甩在了上个世纪，取而代之的是数字化加工技术和数字化生产管理。

BIM 技术在钢结构深化设计应用中的应用起到了直观、便捷、高效、准确的作用，但是 BIM 模型的应用远不止这些，其在 3D 模型建立过程中所产生的信息，对后续加工中的作用更为显现，通过对这些信息的采集、加工、快速推送和应用，可确保信息流转的高效、有序、精细和可控。

4.1　BIM 模型产生的数据信息

BIM 技术的引入，也使钢结构加工制造流程变得简单，尤其是 BIM 模型产生的各类信息对于工艺路径的设定、排版套料环节及数控自动化的作用显得尤为显著，同时 BIM 技术的应用，使工程项目管理的数字化管理也变成可能。

以下是通过 BIM 模型产生的各类数据格式的文件信息：

（1）CIS/2 格式；
（2）CNC 和 DSTV 格式；
（3）GODATA_ASSY3.RPT 模板；
（4）DXF、DGN 和 DWG 格式。

这些信息将在之后的生产管理和自动化生产中起到极大的作用。

4.2　BIM 数据在生产管理系统中的应用

我们知道，在传统的钢结构加工制作过程中，绝大部分企业通过手工管理图纸、清单、工艺卡片和工作指令来组织部署构件和零件的加工，制定一系列加工计划，但在整个组织管理中，往往对车间内各工位、各设备的具体加工情况很难获取准确的信息，以至于在整个组织管理和计划制订中经常处于被动和不断的计划调整过程中。

钢结构生产管理系统是 BIM 模型数据产生的精细化应用平台，主要依托所有部门、供应商和客户间的钢结构项目合同来掌控从估算投标、采购、生产到施工现场的信息流以及工作流，这里我们切合主题，主要阐述在车间加工生产中的应用过程。

4.2.1 在生产管理系统中的初始数据形成

作为信息管理系统，基础数据的准确性非常重要。如果没有 BIM 模型，我们将不得不通过手工采集的方式，去完成一些图纸数据的录入，准确性和及时性得不到保障。

在这里，我们可以充分利用 BIM 模型产生的各类信息，通过标准化接口方式，完成从 BIM 模型到生产管理系统的快速数据生成。

深化设计 BIM 模型可输出以 Godata_assy3. rpt 为报表模版的 XSR 格式的清单文件、NC 文件（DSTV 格式）的数控数据文件或 dwg 图纸文件等。如图 25 所示。

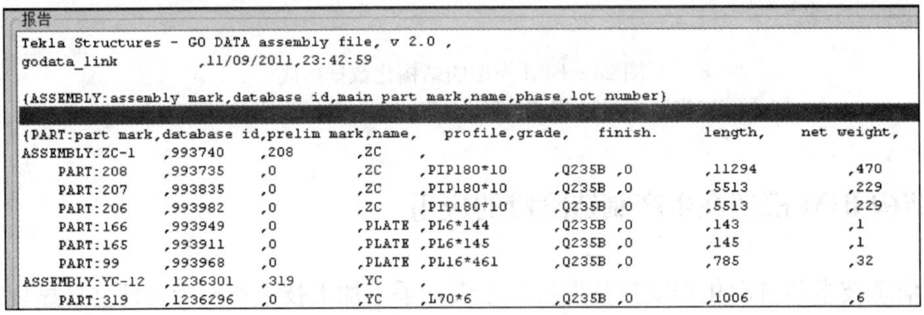

```
报告
Tekla Structures - GO DATA assembly file, v 2.0 ,
godata_link          ,11/09/2011,23:42:59

{ASSEMBLY:assembly mark,database id,main part mark,name,phase,lot number}

{PART:part mark,database id,prelim mark,name,   profile,grade,   finish.     length,    net weight,
ASSEMBLY:ZC-1    ,993740    ,208        ,ZC     ,
    PART:208     ,993735    ,0          ,ZC     ,PIP180*10    ,Q235B ,0      ,11294        ,470
    PART:207     ,993835    ,0          ,ZC     ,PIP180*10    ,Q235B ,0      ,5513         ,229
    PART:206     ,993982    ,0          ,ZC     ,PIP180*10    ,Q235B ,0      ,5513         ,229
    PART:166     ,993949    ,0          ,PLATE  ,PL6*144      ,Q235B ,0      ,143          ,1
    PART:165     ,993911    ,0          ,PLATE  ,PL6*145      ,Q235B ,0      ,145          ,1
    PART:99      ,993968    ,0          ,PLATE  ,PL16*461     ,Q235B ,0      ,785          ,32
ASSEMBLY:YC-12   ,1236301   ,319        ,YC     ,
    PART:319     ,1236296   ,0          ,YC     ,L70*6        ,Q235B ,0      ,1006         ,6
```

图 25　Godata_assy3 文件

生产管理系统提供标准的数据接口，方便地将上述 XSR 文件清单导入系统中，形成初步的系统加工清单，包括图纸清单、构件清单和零件清单等，为后续的工作做好准备，见图 26、图 27。

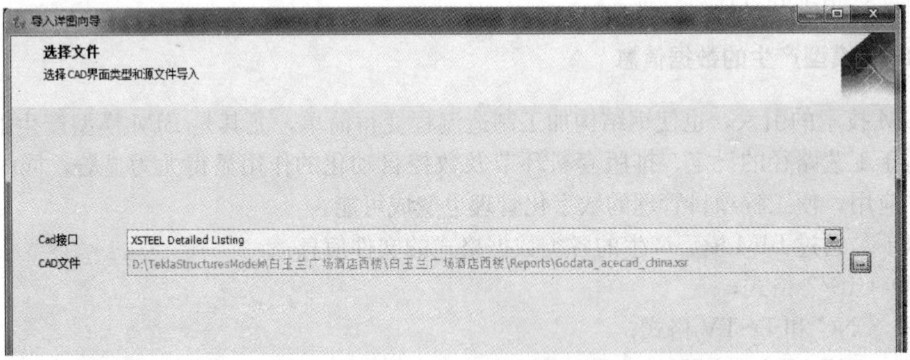

图 26　XSR 格式加工清单导入管理系统

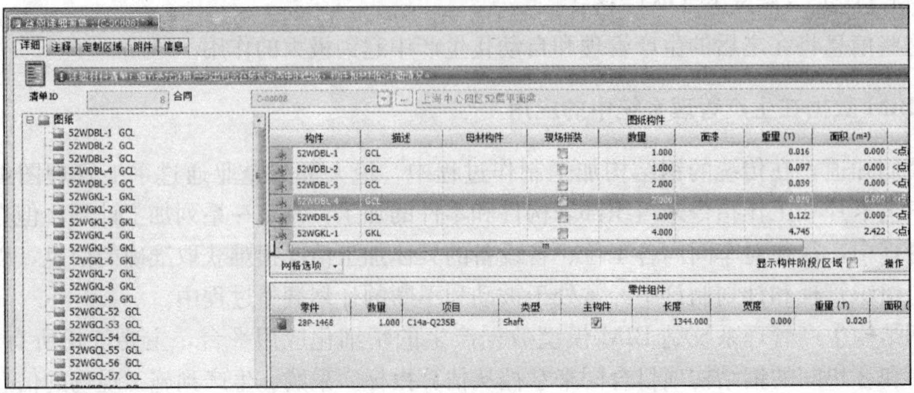

图 27　系统加工清单

4.2.2 分组工作

将导入生产管理系统的构件清单,根据运输、生产批次等各种方式进行分组别管理(图28)。

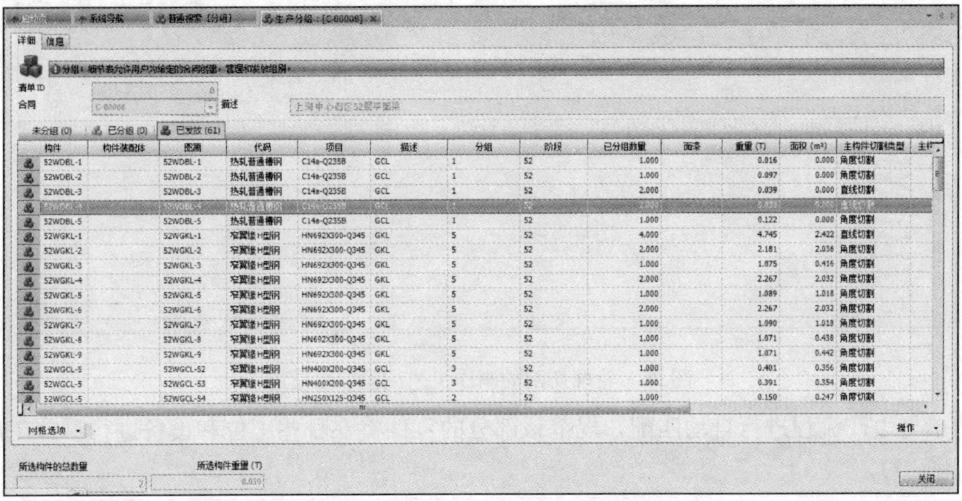

图28 构件分组

4.2.3 工艺流程的设定

工艺流程的设定是组织生产的基础,需要熟悉工艺和设备的人员事先根据实际加工工艺流程和车间设备布置,在生产管理系统中编制和设定好生产流程以及每一个生产流程的参数配置和优先设备指定,如直线切割流程、轮廓切割流程、制孔流程、组立流程和装配流程等,为生产数据工艺路径的自动规划做好准备,如图29、图30所示。

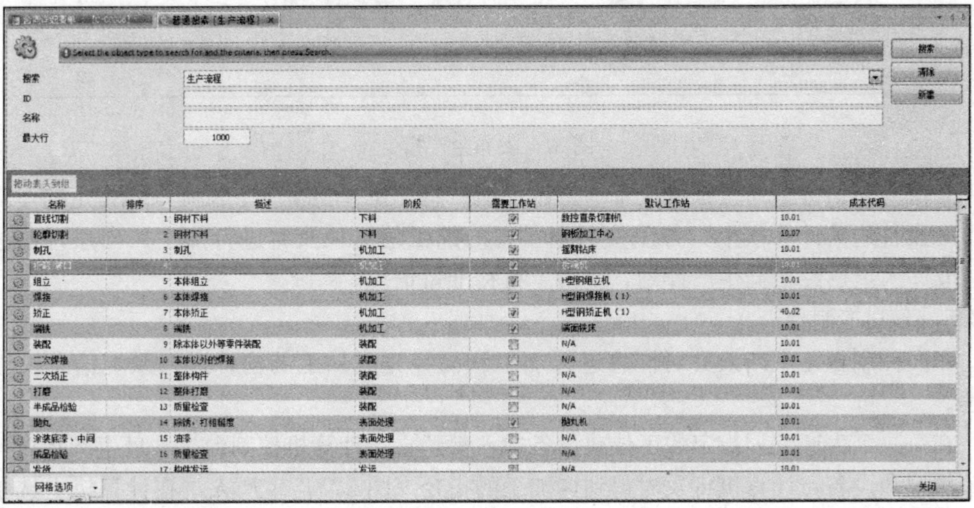

图29 各类制作工艺流程列表

4.2.4 生产指令的发布

当一切生产数据准备完毕,生产指挥人员将在系统内进行生产指令的发布,系统将零

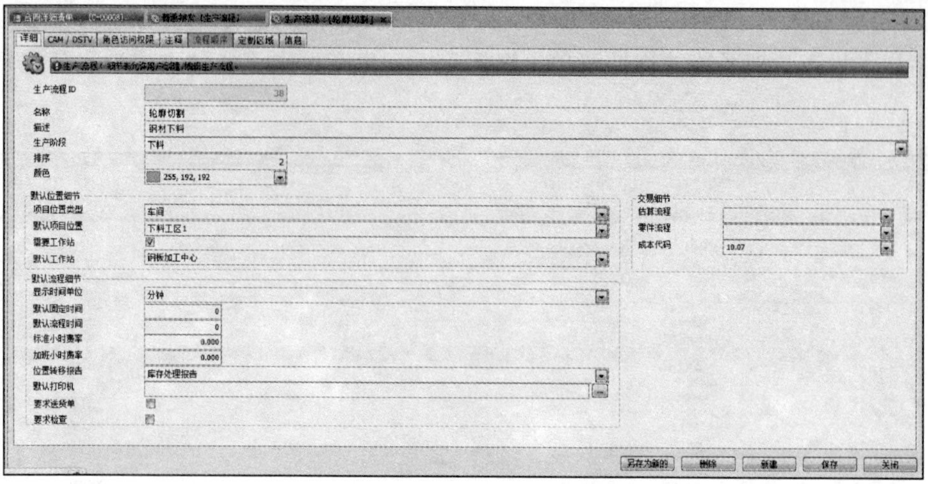

图30 轮廓切割的制作工艺流程参数设定

件清单与生产流程进行自动匹配,规范该部分的零件将在哪些工位和设备进行哪些工序加工（图31）。

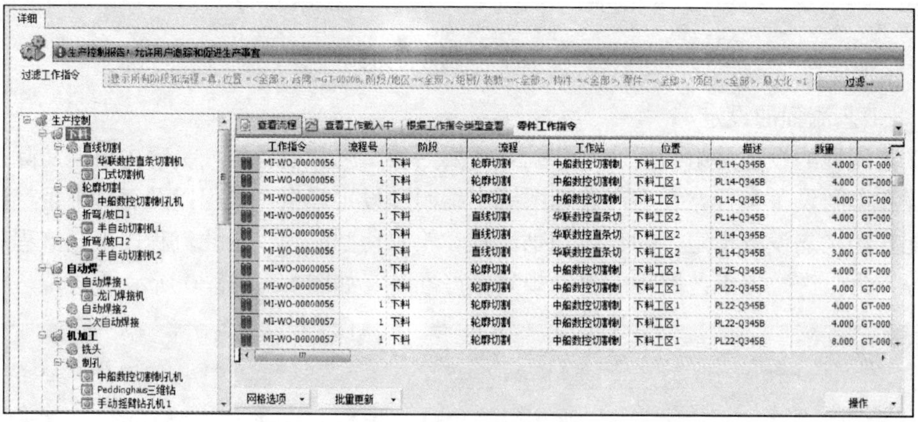

图31 发布生产指令

系统同时结合车间工位或设备的负载反馈信息,快速将指令下发到系统指定的车间各工位和设备边的计算机上,各工位根据获得的加工信息,及时进行加工。当然,在这之前,仓库人员已经根据生产计划,获得了材料准备信息。

4.2.5 车间控制台

车间控制台是生产人员对即将进行加工和已经完成加工的构件进行系统信息反馈的过程。

在每一个生产流程指定的工位或设备边,都建立了计算机控制台,一旦生产指令流转到该工位和设备,计算机将及时得到待加工的指令（图32）,生产人员在计算机上选中该构件（或零件）,按下【开始】状态或【停止】按钮,系统便能及时记录下该构件的过程状态（图33）。这使得生产管理人员能够及时了解整体生产状况,并根据实际情况,及时进行调整。

通过以上一系列对BIM数据的导入、处理和再利用,钢结构生产管理系统完整地完

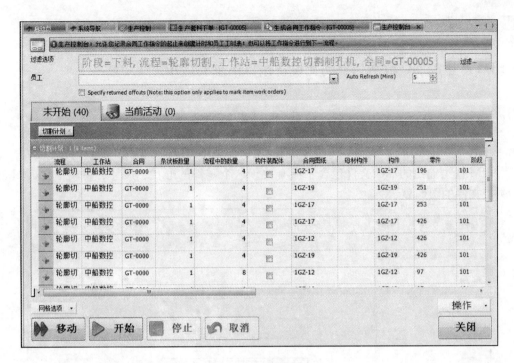

图 32 车间控制台

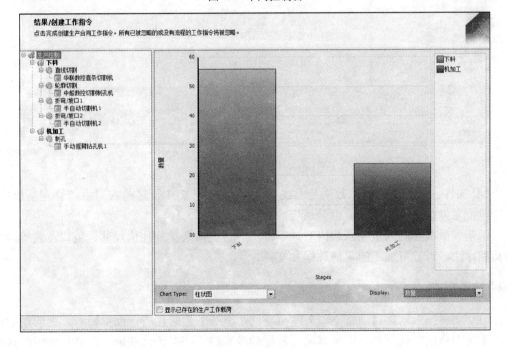

图 33 车间加工负载

成了构件零件加工的自动化管理，确保了生产订单的有序和准时完成。

但是，只有自动化管理还是远远不够的，如果能够充分应用 BIM 模型信息，结合自动化生产设备，降低人工干预程度，才能在钢结构制作过程中，对产品质量和成本降低起到根本性的作用。

以下，我们对钢结构零件板自动套料和自动切割技术进行阐述。

5. BIM 产生的 NC 数据与自动化设备接口信息的应用

5.1 零件工艺信息的形成

在上一节中我们已经提到 BIM 模型能够输出 NC 文件（DSTV 格式）的数控数据文件。BIM 模型输出的 NC 文件包含了所有关于这个零件的形状、尺寸、材质信息，图 34 为数控文件的图形显示。

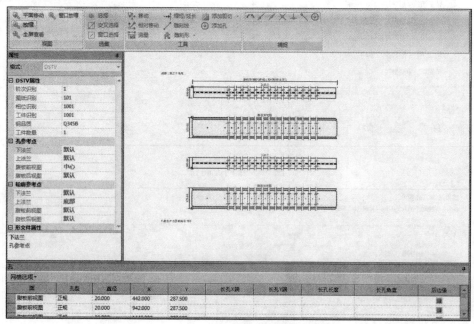

图 34　数控数据预览

NC 文件（DSTV 格式）数据与上述 XSR 格式清单文件一起输入到生产管理系统中，作为单个零件加工的轨迹图，也是系统的初始加工数据。

工艺人员或排版人员在系统内对该零件图进行焊接余量、孔位尺寸、坡口位置等进行编辑和标注，形成可加工的零件数据和文件。

5.2　自动套料

自动套料排版软件自动将 NC 文件和经过处理的工艺信息进行批量转入，为前期数据输入节省大量的时间，并保证所有输入数据的准确性。同时，在获取 NC 文件的零件这些信息后，将输入的所有零件按钢板厚度不同、材质不同的零件自动进行套料分类，完成每组零件的套料任务，大大减少了人为区分钢板厚度和材质进行分组的工作，实现了多种钢板厚度、多种材质的零件同时批量进行套料的功能。

当所有零件作业文件完成之后，排版人员只需简单按下套料系统中"自动套料"按钮，完成对所有零件的套料，并形成可下料切割的排版图和数控数据。见图 35。

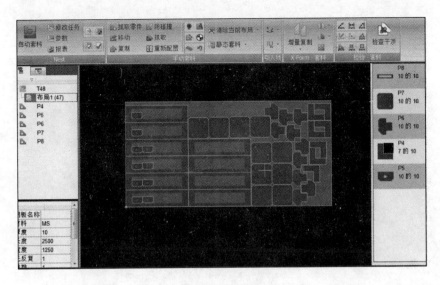

图 35　自动套料

5.3　CAM 数据的形成

CAM 数据不同于 NC 文件，它是数控机床能够识别的加工指令，CAM 数据的形成是通过自动套料系统中内嵌的后置程序将上述自动套料系统所形成的数据文件进行转换。

通过自动套料系统内置的后置程序，可将形成的套料文件快速生成能够驱动数控机床的 CAM 数据，并通过网络下达到数控机床的 PC 中，数控操作人员只需选中该条指令，通过简单操作，即能驱动数控机床完成对零件的切割。

6.　总结

充分应用钢结构 BIM 模型，并贯穿于深化设计到制造流程，不但提高了工作效率，而且改进了制造质量。此外，BIM 模型提供的信息是基于高度精确、协调、一致的数字设计数据，这些数据不仅在钢结构制造行业，并完全值得在相关的建筑活动中共享。

完成施工详图设计和构件制作后，并不代表最后的竣工状况，我们还能通过信息化技术，对制作完成的构件进行虚拟预拼装，确保在安装阶段的构件精确度，BIM 模型信息会被再一次应用。

另外，从跳出钢结构制作 BIM 模型分析看，数字化从设计到制造流程离不开结构工程师、钢结构详图设计人员和钢结构制造商之间的协作。因此，就需要采用不同于以往的项目交付方法来连通设计与制造环节，也就是说由业主、建筑商、工程师和承包商组成跨职能的项目团队，就设计、制造和施工环节中的工作进行协调。原本需要按顺序进行的步骤（设计、详图设计、制造）可以并行展开。设计模型和施工详图可以同时创建。加快施工详图的完成，可以尽早地向钢厂下订单，提前开始生产制造，为现场钢结构安装等各个环节缩短工期创造条件。

总之，BIM 在不久的将来，不仅在钢结构制作中将被广泛应用，也将给整个建筑业引起更深层次的变革。